5th ITB International Geothermal Workshop (IIGW 2016)

IOP Conference Series: Earth and Environmental Science
Volume 42

Bandung, Indonesia
28 March - 2 April 2016

ISBN: 978-1-5108-3229-9
ISSN: 1755-1307

Printed from e-media with permission by:

Curran Associates, Inc.
57 Morehouse Lane
Red Hook, NY 12571

Some format issues inherent in the e-media version may also appear in this print version.

Printed by Curran Associates, Inc. (2016)

For permission requests, please contact the Institute of Physics
at the address below.

Institute of Physics
Dirac House, Temple Back
Bristol BS1 6BE UK

Phone: 44 1 17 929 7481
Fax: 44 1 17 920 0979

techtracking@iop.org

Additional copies of this publication are available from:

Curran Associates, Inc.
57 Morehouse Lane
Red Hook, NY 12571 USA
Phone: 845-758-0400
Fax: 845-758-2633
Email: curran@proceedings.com
Web: www.proceedings.com

Table of contents

Volume 42

5th ITB International Geothermal Workshop (IIGW2016)
28 March to 2 April 2016, Bandung, Indonesia

Accepted papers received: 2 September 2016
Published online: 31 October 2016

Preface

Papers

PREFACE

ITB International Geothermal Workshop (IIGW) is an annual event which was organized by *Prodi Teknik Panas Bumi, Fakultas Teknik Pertambangan dan Perminyakan (FTTM),* ITB. The workshop celebrated its 5th anniversary this year. It was held on March 27th – April ^{2nd} 2016 and has become a special moment in supporting the geothermal development acceleration program in Indonesia.

The objective of the workshop was to create a knowledge sharing forum among the key stakeholders in energy and extractive resource sectors on the status of geothermal development in Indonesia. It focused on the latest condition after the approval of new regulation of geothermal energy law No 2014. The conference featured distinguished speakers, scientists, engineers, financiers, and other key players in geothermal and energy sectors worldwide.

Participants benefit the exchanges of view, knowledges and experiences on latest technologies and researches by bringing together scientists, engineers, academicians, experts, and industrial stakeholders involved in geothermal studies and developments.

As part of this year commitment, outstanding papers are published as an open access volume of *IOP Conference Series: Earth and Environmental Science*. Other papers are published in the conference proceeding in print version. We hope wider geothermal communities will gain the same benefits as our conference attendee.

Sincerely,

Suryantini, Dr. Eng.

Chairman of the ITB International Geothermal Workshop (IIGW) 2016

AUDITORIUM
IPTEKS
The New Era : Initiatives, Strategies,
Opportunities, and Challenges
toward Geothermal Development in Indonesia

IOP Conf. Series: Earth and Environmental Science **42** (2016) 011001 doi:10.1088/1755-1315/42/1/011001

LIST OF SPONSORSHIP 5th ITB INTERNATIONAL GEOTHERMAL WORKSHOP 2016

Sponsored By

Supported By

Organized By

IOP Conf. Series: Earth and Environmental Science **42** (2016) 011001 doi:10.1088/1755-1315/42/1/011001

LIST OF SPONSORSHIP 5th ITB INTERNATIONAL GEOTHERMAL WORKSHOP 2016

1. Chevron
2. PT. Pertamina Geothermal Energy
3. PT. Geodipa Energi
4. PT. Supreme Energy
5. BNI
6. PT. LAPI ITB
7. GEOCAP
8. Ministry of Energy and Mineral Resources of the Republic of Indonesia
9. Indonesia Geothermal Association
10. Indonesian Association of Geologists
11. BAGUS-SATREPS
12. Thermochem Indonesia
13. Institut Teknologi Bandung
14. Geothermal Technology Study Program, Institut Teknologi Bandung

IOP Conf. Series: Earth and Environmental Science **42** (2016) 011001 doi:10.1088/1755-1315/42/1/011001

LIST OF COMMITTEES

Steering Committee	Nenny Miryani Saptadji and Staff of Geothermal Technology Study Program, FTTM – ITB Antonie de Wilde (SATE PTE LTD-Singapore)
Chairman	Suryantini
Vice Chairman	Ali Fahrurrozie, Akhmad Kunio Pratopo
Secretary General	Angga Alfandi Ahmad, Evanda Eko Putra, Fitri Oktaviani Purwaningsih
Treasurer	Rizki Trisna Hutami, Betseba br Sibarani
Creative Team	Hibban Hammka, Ali Fahrurozie, Hendro Wibowo
Sponsorship, Publications & Promotion	Prihadi Soemintadiredja, Fikri Adam Dermawan, Rully Abdul Malik, Fransiscus Xaverius Guwowijoyo, Damar Nandawardhana, Gugun Abdurrahman, Hutra Guswinanda, Andhika Arief Rachman
Pre – Workshop Course	Dimas Taha Maulana, Allen Haryanto Lukman, Fikri Adam Dermawan, Famelia Nurlaela
Technical Session & Planning	Asep Saepuloh, Suryantini, Nurita Putri Hardiani, Juni Yesy Sianipar, Setya Drana Harry Putra, Anggie Rengganis, Agastyo Nugroho, Heru Berian Pratama
Editorial	Asep Saepuloh (ITB), Suryantini (ITB), Hendro Wibowo (ITB), Fiorenza Deon (GFZ), Juni Yesy Sianipar (ITB)
Field Trip & Field Camp	Qodri Syahrur Ramadhan, Silvester Dimas Sanjaya, Mahesa Pradana, Ratih Nurruhliati
Exhibition	Mahesa Pradana, Rully Tri Abdul Malik
Reviewer Team	Suryantini (ITB), Hendro Wibowo (ITB), Sutopo (ITB), Zuher Syihab (ITB), Prihadi Darminto (ITB), Prihadi Soemintadiredja (ITB), Jooened Hendrasakti (ITB), Fiorenza Deon (GFZ), Chris Hecker (Twente University), Asep Saepuloh (ITB), Hendra Grandis (ITB), Rachmat Sule (ITB), Nenny Miryani Saptadji (ITB), Sutopo (ITB), Antonie de Wilde (SATE PTE LTD), Joe Iovenitty (USC), Dasapta Erwin (ITB), Agastyo Nugroho (ITB), Dimas Taha (ITB), Heru Berian Pratama (ITB), Mahesa Pradana (Thermochem Indonesia), Nurita Putri Hardiani (ITB), Setya Drana (ITB), Sutikno Bronto (Geological Agency Indonesia), Akhmad Solikhin (Geological Agency Indonesia), Rian Nurtyawan (PT. Digital Imaging Geospatial)
Logistic	Qodri Syahrur Ramadhan, Welly Todheus, Fhandy Pandey

IOP Conf. Series: Earth and Environmental Science **42** (2016) 011001 doi:10.1088/1755-1315/42/1/011001

EDITORIALCOMMITTEE

Asep Saepuloh, Dr. Eng.

Faculty of Earth Sciences and Technology

Bandung Institute of Technology (ITB)

Jl. Ganesha No. 10, Bandung, West Java, Indonesia. 40132

Suryantini, Dr. Eng.

Faculty of Earth Sciences and Technology

Bandung Institute of Technology (ITB)

Jl. Ganesha No. 10, Bandung, West Java, Indonesia. 40132

Hendro Wibowo, MSc.

Geothermal Technology Master Program, Faculty of Mining and Petroleum Engineering

Bandung Institute of Technology (ITB)

Jl. Ganesha No. 10, Bandung, West Java, Indonesia. 40132

Fiorenza Deon, Ph.D.

Faculty of Civil Engineering and Geosciences

Delf University of Technology

Stevinweg 1 / PO-box 5048, 2628 CN Delf / 2600 GA Delf, Netherlands

Juni Yesy Sianipar, S.T.

Geothermal Technology Master Program, Faculty of Mining and Petroleum Engineering

Bandung Institute of Technology (ITB)

Jl. Ganesha No. 10, Bandung, West Java, Indonesia. 40132

IOP Conf. Series: Earth and Environmental Science **42** (2016) 011002 doi:10.1088/1755-1315/42/1/011002

Peer review statement

All papers published in this volume of *IOP Conference Series: Earth and Environmental Science* have been peer reviewed through processes administered by the proceedings Editors. Reviews were conducted by expert referees to the professional and scientific standards expected of a proceedings journal published by IOP Publishing.

Identification of Surface Manifestation at Geothermal Field Using SAR Dual Orbit Data

Dinul Akbari[1], Asep Saepuloh[2]

[1] Faculty of Mining and Petroleum Engineering, Bandung Institute of Technology (ITB)
[2] Faculty of Earth Sciences and Technology, Bandung Institute of Technology (ITB)
Jl. Ganesha No.10, Bandung, West Java, Indonesia. 40132.

Email: dinul_akbari@yahoo.com

Abstract. The Wayang –Windu Geothermal Field located in West Java, Indonesia is a geothermal field under tropical zone which is identified by high precipitation, dense vegetation, and extensive weathering/alteration. The clouds due to high precipitation and vegetation conditions on the tropical zone inhibit the identification of surface manifestation using optical remote sensing techniques. In this paper, we reduced these inhibiting factors using microwave remote sensing techniques termed as Synthetic Aperture Radar (SAR). The SAR dual orbits were used to observe the targets on the surface by minimizing the effects from the clouds and dense vegetation cover. This study is aimed to identify surface manifestation based on Geomorphologic and Structural Features (GSF) of the SAR in Ascending and Descending orbits. The Linear Features Density of SAR (lifedSAR) method was applied to quantify the linear features of the ground surface and served as basis of surface manifestation identification. Based on the lifedSAR and field observations, the surface manifestations could be detected succesfully at Wayang and Cibolang craters with density about 45%. The soil measurements were used validate the result and to interpret the correlation between LFD and surface manifestations.

1. Introduction

Wayang Windu Geothermal Field termed as WWGF is geographically located in West Java Province, Indonesia. The WWGF is located under tropical zone which has high precipitation and dense vegetation. This condition inhibit identification of surface manifestation using optical remote sensing techniques. Overcoming the problem, we used the Synthetic Aperture Radar (SAR) system to detect the targets by reducing the atmospheric and vegetation conditions [4]. The WWGF is interpreted as transition of geothermal system between water and vapor-dominated system [2]. The reservoir of the WWGF is characterized by the three steam dominated reservoirs with two steam dominated reservoirs in the south associated with Mt. Wayang - Windu and one steam dominated reservoir in the north associated with Mt. Gambung – Puncak Besar. The steam dominated reservoirs are located above the water dominated reservoir with elevation 0 - 400 m above sea mean level in the northern areas [2]. The steam dominated reservoirs are increasingly deep towards to the south associated with Mt. Gambung – Puncak Besar at the north to Mt. Wayang - Windu at the south [2]. The geomorphic of WWGF is composed by structural features of volcanic complex [7]. The Linear Feature Density of SAR (lifedSAR) method could quantify the linear features of the ground surface and served as basis of surface manifestation identification. The measurements data was used to validate the identification surface manifestation and linear features at ground surface, accurately.

2. Data and Methodology

2.1. Data

We used the Phased Array L-band type Synthetic Aperture Radar (PALSAR) onboard Advanced Land Observing Satellite (ALOS) in ascending and descending orbits data. Both data are orthorectified in level 1.5. The ascending orbit indicates that the satellite moves from South to North with looking direction to East. Meanwhile, the descending orbit indicates that the satellite moves from North to South with looking direction to the West. In addition, we also used ground measurement at surface manifestations to validate the SAR analyses. Figure 1 shows the ascending and descending orbits of ALOS PALSAR backscattering intensities images.

2.2. Methodology

Linear Feature density (LFD) is a map which illustrates the density of linear features in particular grid. More linear features detected in a grid cell will produce high density. On the contrary, less linear feature will produce low density. The LFD map was generated following lifedSAR processing [4]. The program required the input from ALOS PALSAR level 1.5 or orthorectified intensity images. The lifedSAR used edge detection process based on the Laplacian of Gaussian (LoG) filter to define the linear feature in two intensities images at different LOS. Density was then calculated based on grid size about 10 x 10 pixels [5]. The lifedSAR result could be confirmed by extracting circular and linear features related to the crater rim and geological structures [4]. In this paper, we focused on the correlation between SAR dual orbits and LFD, obtained from lifedSAR, and the ground measurements of the surface manifestation including magnetic susceptibility, pH and ground temperature. The visual interpretation was taken into account to assess the effectiveness of the SAR dual orbits and LifedSAR in predict of geothermal surface manifestation.

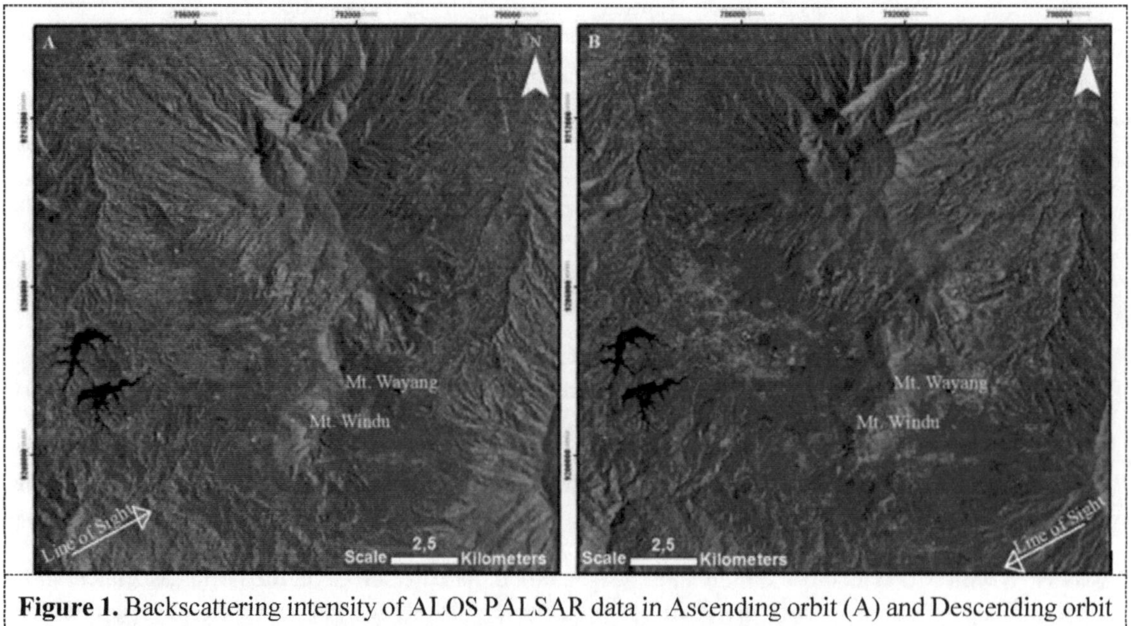

Figure 1. Backscattering intensity of ALOS PALSAR data in Ascending orbit (A) and Descending orbit (B) shows Mt. Wayang dan Windu at West and East Line of Sight (LOS).

3. Result and Discussion

3.1. Linear Features

Linear features are useful to interpret the information related to crater rim, ridge and valley, as well as to provide an overview of the volcanic structures in the study area. Figure 2 shows the detected linear features at study area based on SRTM, ALOS PALSAR dual orbit and geological maps [3]. The visual analyses were showed by red thin line that Mt. Malabar (Northern Area) has density of linear features originated from foothill, ridge, and valley. On the contrary, the linear features at Mt. Wayang - Windu smoother than the linear features at Mt. Malabar because linear features detected at Mt. Wayang – Windu less than Mt. Malabar.

According to the figure 2, the thick red lines are main faults of Mt. Wayang - Windu regions [3]. See figure 3, Bedil fault is located between the Northern of Mt Wayang and Southern of Mt. Bedil. The dominant strikes of the fault are Northeast – Southwest direction. Then, the Cibolang fault located at the Southern of Windu crater, and the dominant strikes of the fault are Northwest – Southeast direction. The dominant strikes of the youngest fault system (Windu fault) are Northwest - Southeast direction and located at North of Cibolang fault [8].

3.2. Ground Measurements at Surface Manifestation

The ground measurements of surface manifestation such as altered ground, mud pool, steaming ground, warm springs and hot springs were measured by thermometer, pH meter, and magnetic susceptibility meter. Wayang crater has only steaming and altered ground surfaces. Figure 3 shows Wayang crater depicted by dotted-lines which is located to the East and South of surface manifestation points. The measured temperature and acidity at the ground surface of Wayang crater are about 60 °C, and pH 3, respectively. It may indicate that the ground surface was affected by hydrothermal activity strongly.

The Cibolang crater presented by dotted-line is located at the North and East of surface manifestation points including steaming ground, altered ground, and hot springs with soil pH about 3.6 and temperature steaming ground at crater about 90°C and hot spring exist far from crater about 40 °C. They indicate hydrothermal activity at the surface of Cibolang crater.

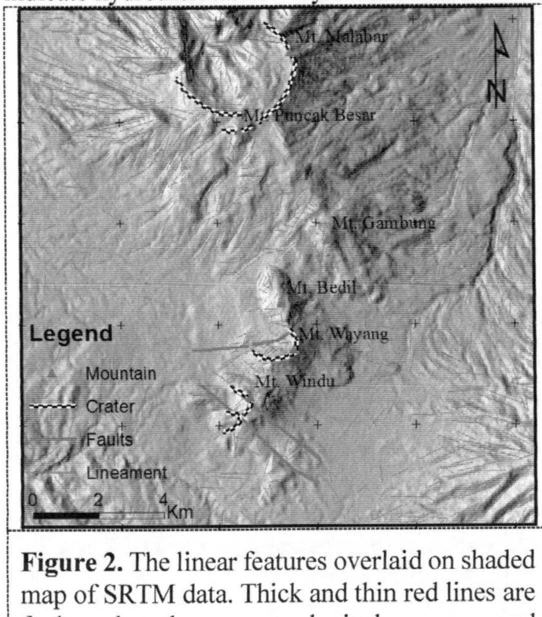

Figure 2. The linear features overlaid on shaded map of SRTM data. Thick and thin red lines are faults based on geological map and interpretation from SRTM and ALOS PALSAR data, respectively. The dotted-lines are crater features based on SRTM data.

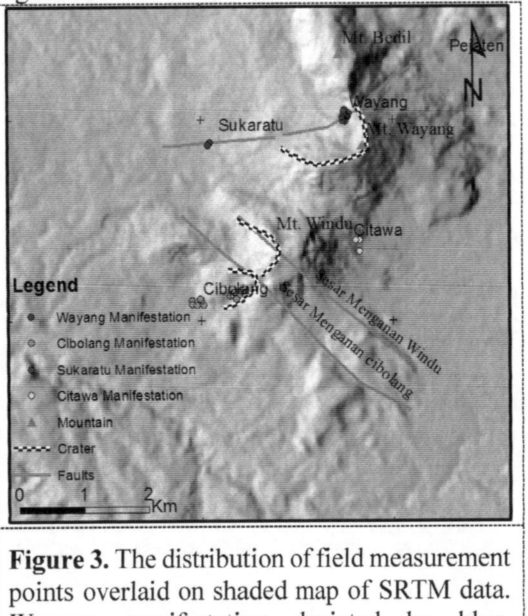

Figure 3. The distribution of field measurement points overlaid on shaded map of SRTM data. Wayang manifestation depicted by blue, Cibolang by green, Sukaratu by red and Citawa by yellow points.

The hot springs temperatures and acidity at Citawa and Sukaratu were measured about 40 – 50 °C and pH 6, respectively. The medium pH with warm temperature indicated a mixing with surface water at the hot springs. This mixing process decreases the temperature, but increase the pH. These conditions lead to interpretation that Citawa and Sukaratu are outflow zones of Wayang-Windu geothermal systems.

According to magnetic susceptibility measurements, the Wayang and Cibolang craters were less than 1.5×10^{-3} SI with average about 0.5×10^{-3} SI. Meanwhile, the areas outside from the crater showed magnetic susceptibility larger than 2 with average about 3.5×10^{-3} SI. The low magnetic susceptibility might be associated with altered rock/soil. The hydrothermal activity caused alteration that the magnetic minerals decomposed into non-magnetic minerals such clays [1].

3.3. Correlation of Surface Manifestation with SAR Dual Orbit Data

Figure 4 shows the locations of surface manifestation in ascending and descending orbits. The surface manifestations are consistent for both images that bright and dark tones are identical with rough and smooth surfaces, respectively. Interestingly, different orbits or Line Of Sight (LOS) provide different brightness level for the same surface manifestation. For descending image, the surface manifestation of altered zones at Wayang crater with ground pH about 2 is detected by dark tone with medium to low (-20 – -10 dB) backscattering coefficient. Then, the altered zones were surrounded by medium (0 – 10 dB) backscatter coefficient. For ascending image, the altered zones are identified by medium to high (10 - 20 dB) backscatter coefficient. Therefore, the altered zones at Wayang crater are presented by medium to high (ascending image) and medium to low (descending image) backscattering coefficient. The slope toward the sensor and incidence angle are supposed to be responsible to the different backscattering coefficient values. Contrary, the hot springs are difficult to be identified because of no signature anomaly in ascending nor descending orbits. The limitation spatial resolution of the images and the spotted location of hot springs were predicted responsible to the problem.

Figure 5 shows the LFD based on lifedSAR method from ALOS PALSAR data in ascending and descending orbits. The grey color bar indicates the area containing minimum (black) and maximum (white) linear features about 60% for grid cell 30×30 pixels. For ascending image, the Wayang and Cibolang craters with steaming and altered grounds were identified by medium to high density about 35 – 45 % grid. Contrary, the craters were identified by a low density less than 20 % grid in descending image. Therefore, combining both LFD provided a complete LOS and surface structures such as roughness effects from surface manifestation. The surface roughness is one parameter to identify surface characteristics related to geological features [6].

3.4. Surface Manifestation on LFD

The total LFD in ascending and descending is depicted in figure 6. The minimum and maximum density values 0 to 80 % were marked by dark blue to red, respectively. The LFD map provided information about the spatial distribution of the density of linear features [4].

Interestingly, the steaming and altered grounds are located at the green portion or medium LFD (see figure 6). Therefore, we interpreted that the surface manifestation zones were located at the medium density about 40 - 45 % grid. These phenomenon agreed with bright tone in ascending image (see Figure 5A). The Western LOS in ascending image can provide better view of Wayang and Cibolang crater because the high topography was located at the Eastern side.

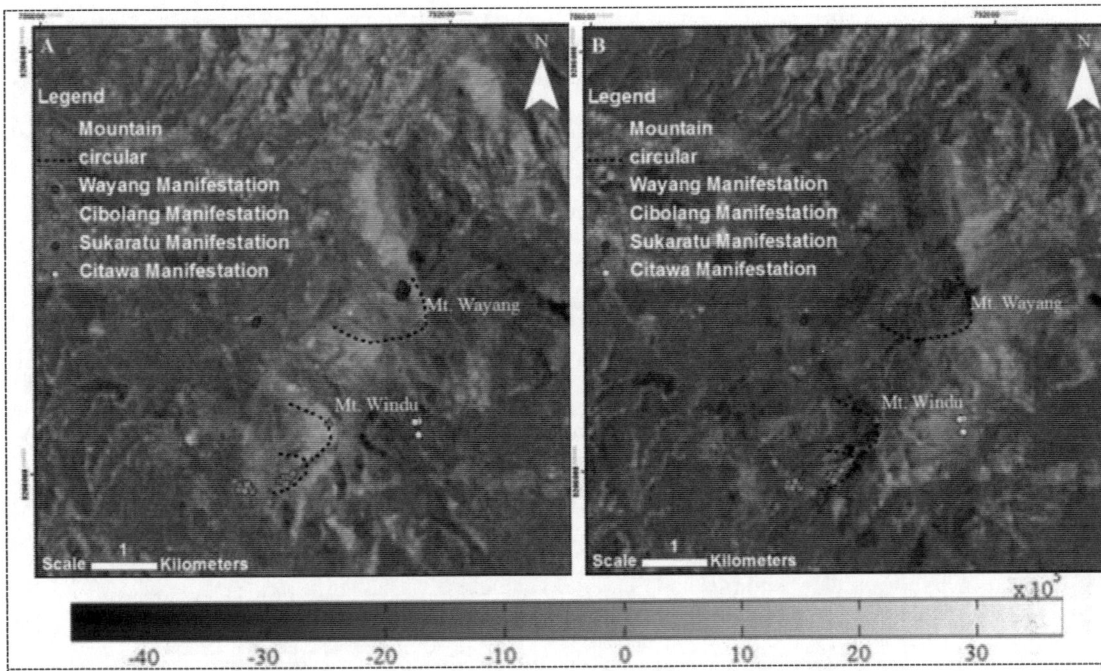

Figure 4. The backscattering intensity images of ALOS PALSAR data in Ascending (A) and Descending (B) at surface manifestation of Wayang – Windu Geothermal Field (WWGF).

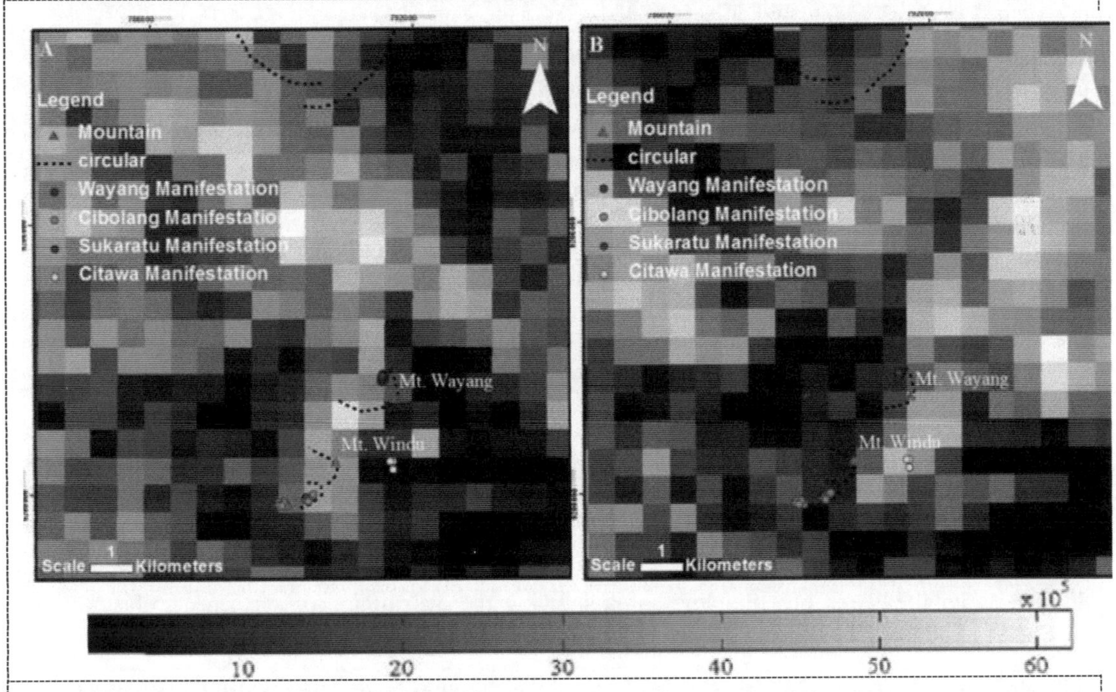

Figure 5. The Linear Features Density (LFD) map based on lifedSAR method from ALOS PALSAR data in ascending (A) and descending (B) orbits.

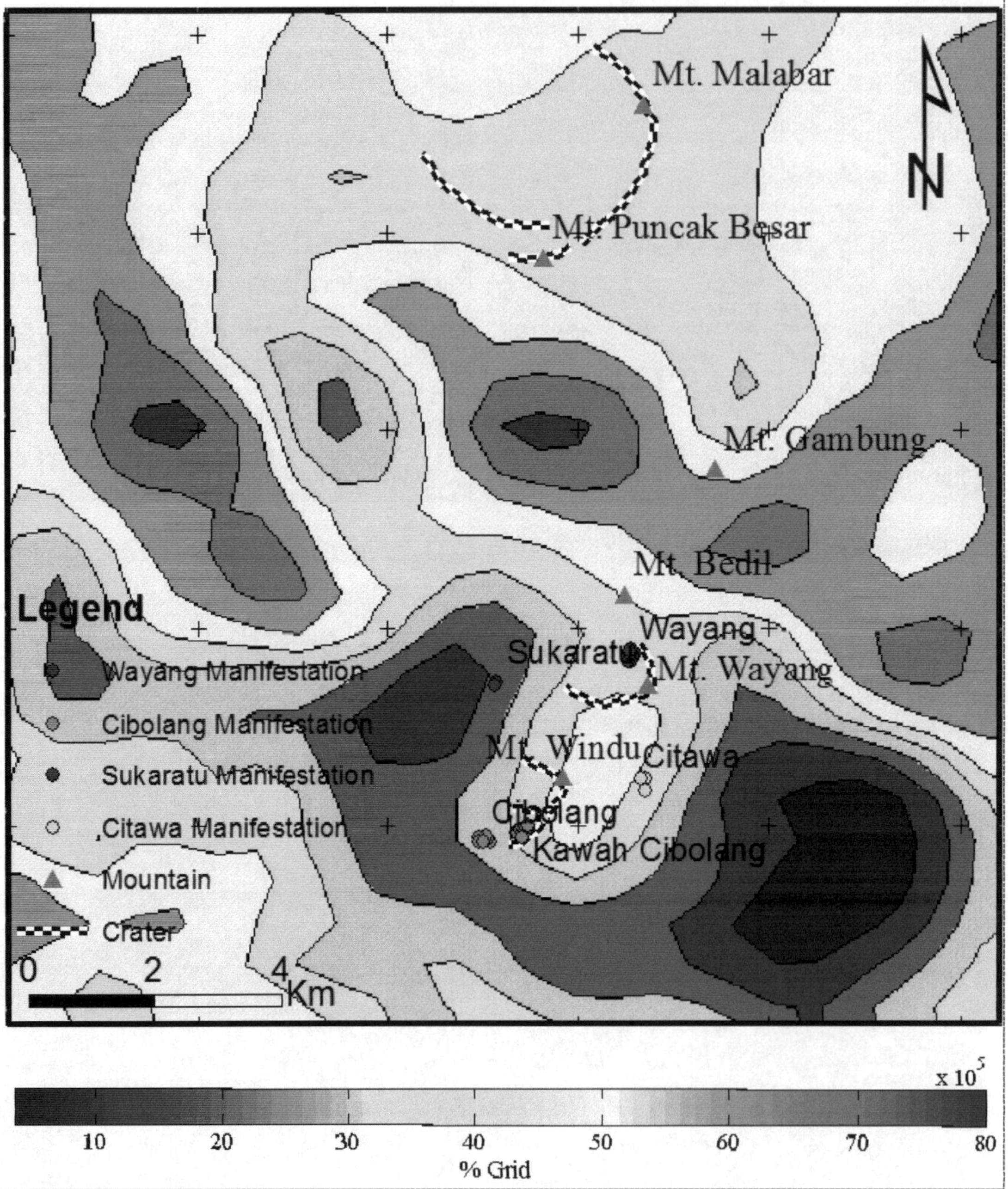

Figure 6. Total LFD map of the study area showing the distribution of surface manifestation at medium to high density (green - yellow).

3.5. Discussions

The LFD map and field observation data provided a good correlation between medium densities with the presence of surface manifestations of geothermal system at Mt. Wayang - Windu. The altered surfaces at the northern flank of Wayang crater are depicted by blue dots and the hot springs at the southern flank of Cibolang crater are depicted by dark green dots (see figure 6). We defined the low and high densities correspond to LFD value less than 30 and more than 50 % grid, respectively. The altered surfaces at the Wayang and Cibolang craters are located at medium density. The Malabar, Wayang, and Windu craters in the North and South part were located at medium – high density. The altered surfaces at Wayang crater depicted by blue dots and at Cibolang crater depicted by dark green dots north part and hot springs were also located at medium density as depicted by yellow and dark green dots south part. But, hot springs also found with low density at Sukaratu depicted by red dot at western area. The surface manifestations could be detected at Wayang and Cibolang craters succesfully at the green portion with density about 45 %. The surface manifestation such as altered and steaming grounds located at medium density, this surface manifestation probably serves as upflow zone from geothermal system. The upflow zones at the crater were composed by medium linear features with smoother surface than ridges or valleys. The medium density at the surface manifestation corresponded with linear features obtained from geological map, SRTM, and ALOS PALSAR dual orbit (see Figure 2). The manual extraction of the linear features confirmed that the densities of linear features at ridges or valleys and foothills of Mt. Malabar are corresponded to the geological structures. The activity of hydrothermal at crater causes an alteration of the minerals smoother surfaces than ridge or valley. It may indicate that the medium density of LFD affected by complex factor, but the activity of hydrothermal is the most influence to the density.

4. Conclusions

The density of linear features at Mt. Wayang Windu are less than Mt. Malabar. The Wayang crater was composed by steaming and altered surface manifestations with ground temperature and acidity about 60 °C and pH 3, the steaming ground at Cibolang crater was 90 °C and pH 3 with hot springs temperature about 40 °C at the foothill. The Sukaratu and Citawa are composed by hot springs with temperature about 40 – 50 °C and pH 6. The magnetic susceptibilty measurement at the altered surfaces of the Wayang and Cibolang crater about 0.5×10^{-3} SI, but weathered surfaces outside the crater about 3.5×10^{-3} SI. The low magnetic susceptibilty originated from decomposition of magnetic minerals due to hydrotermal activities. The ALOS PALSAR dual orbits provided variation of brightness levels for the same surfaces. The rough surface topographic was presented relatively by brighter tone than smooth surface in the backscattering images. The altered zone at the craters were presented by medium - high backscattering coefficient for ascending orbit image, but medium - low for descending orbit image. The difference originated not only from surface roughness, but also slope and incidence angle. According to the LFD for ascending image, the crater with steaming and altered grounds were identified by medium - high density about 35 – 45 % grid and descending image less than 20 % grid. The ascending and descending image were combined to reduce the difference of surface roughness, slope and incidence angle and get the better result of the LFD. Total LFD shows surface manifestation at Wayang Windu Geothermal Field (WWGF) were located at medium density in general. Therefore, the surface manifestations could be detected at Wayang and Cibolang craters succesfully at the green portion with density about 45 %.The field observations and soil measurements were used to interpret the correlation between LFD and surface manifestations. The value of magnetic susceptibility and pH are low, and temperature manifestation is high, this value have good correlation with medium to high density on LFD map. Hydrothermal activity at the crater (upflow zone) affected strongly to the transform value of magnetic susceptibility, pH and temperature and also appearence of surface manifestation.

Acknowledgments
The authors sincerely thank to the "BAGUS" SATREPS project for providing field equipments, LPPM-ITB for survey accommodation supports, and Star Energy for collaborating to this research.

References

[1] Beaton, J. D.; Burns, G. R.; Platou, J. 1968. Determination of Sulphur in Soils and Plant Materials. *Sulphur Institute Technical Bulletin 14: 56 pp.*
[2] Bogie, I., Kusumah, Y. I., & Wisnandary, M. C. (2008). Overview of the Wayang Windu geothermal field, West Java, Indonesia. Geothermics.

[3] Draniswari, W. A. (2014). Geologi dan Volkanostratigrafi Gunung Wayang – Windu dan Sekitarnya, Kabupaten Bandung, Jawa Barat. Bandung: ITB.

[4] Saepuloh, A., Susanto, A., Sumintadiredja, P., & Suparka, E. (2015). Characterizing Surface Manifestation of Geothermal System under Torrid Zone using Synthetic Aperture Radar (SAR) data. Proceeding World Geothermal Congress. Melbourne.

[5] Saepuloh, A., Urai, M., Meilano, I., & Sumintadiredja, P. (2013). Automatic Extraction and Validation of Linear Feature Density from ALOS PALSAR Data for Active Faults and Volcanoes. Proceedings of the International Symposium on Remote Sensing. Chiba, Japan.

[6] Saepuloh A., Wikantika K., & Urai M. (2015). Observing Lava Dome Roughness on Synthetic Aperture Radar (SAR) Data Case Study at Mt. Sinabung and Merapi Indonesia. Proceedings of the IEEE 5th Asia-Pacific Conference on Synthetic Aperture Radar (APSAR), Singapore.

[7] Sudarman, S., Pujianto , R., & Budiarjo, B. (1986). The Gunung Wayang Windu Geothermal Area in West Java. (pp. 141-153). in :Proceedings of the Fiftheenth Annual Convention of the Indonesian Petroleum Association

[8] Twiss, R., & Moores, E. (1995). Tectonics. New York: W.H. Freeman and Company.

Geochemical Study of Ampallas Geothermal Area, Mamuju District, West Sulawesi Province

F FAUZIYYAH[1], T R PRABOWO[2], M G J SHALIHIN[2], D I SETIAWAN[3] and A YUSHANTARTI[3]

[1] Engineering Hydrogeology, Faculty of Earth Science and Technology, ITB, Indonesia
[2] Engineering Geology, Faculty of Geology, Universitas Padjadjaran, Indonesia
[3] Pusat Sumber Daya Geologi, Ministry of Energy and Mineral Resource, Indonesia

Abstract. Ampallas is one of the areas with geothermal potential which located in Mamuju district, near from the capital city of West Sulawesi. This research was carried out to understand the characteristic of this geothermal field based on chemistry of the surface manifestation, including fluid characteristic and soil anomaly. Geothermal research in Ampallas area focused on 4 hot springs; Ampallas, Batupane, Karema, and Gantungan. With average temperature around 34 – 67^{0}C. Ampallas 1,2,3,4,7,8 hot springs water type is chloride – bicarbonate, which means it came from the reservoir while Batupane, Gantungan, Karema and Ampallas 5 are all bicarbonate type. Ampallas 1,2,3,4,7,8, Karema and Gantungan hot springs fluid plotted in partial equilibrium zone while Batupane and Ampallas 5 plotted in immature water zone. It means the Ampallas hot springs (except Ampallas-5) mixed with meteoric water right after reached the equilibrium state. It is also concluded that Ampallas 5 hot springs came from the same reservoir with Batupane, but not Gantungan and Karema hot springs. The speculative resource potential of Ampallas geothermal system is estimated around 30 MWe. But if detailed geophysical method was applied the result could be more accurate.

1. Introduction

Located inside the "ring of fire", made Indonesia rich with tectonic activities. This created many geothermal energy potentials which suitable to be developed. One of them is Ampallas in West Sulawesi. The tectonic activities and the geology settings made Ampallas as a potential area for geothermal energy development.

Ampallas composed mainly from volcanic rocks. And the tectonic activites has made a lot of faults and fractures on the rocks, hence making it a good condition for a reservoir rocks with high permeability.

The faults also made a way for the fluid onto the surface and creating manifestations such as hot springs. This geological setting has made Ampallas potential to be researched.

Ampallas located in Kaluku and Mamuju subdistrict, Mamuju district, West Sulawesi.

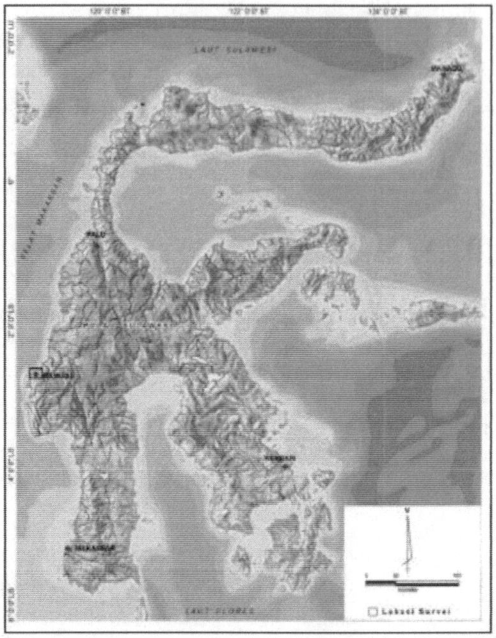

Figure 1:Ampallas research area location, Mamuju district, West Sulawesi

2. Method

This research is using both geology and geochemistry method as its main method. In addition, the geophysics method is only for preliminary survey so it acted as supporting data. In the future it needs a survey with a more detailed geophysics method.

The geology method is used for collecting data regarding the geology condition, such as lithology, morphology, tectonic activity, and surface manifestation. While the geochemistry is used to obtain fluid and soil chemistry related data, such as fluid characteristic, reservoir condition, and soil anomaly distribution.

3. Geology

3.1 Lithology

There are 14 lithology units in Ampallas research area (from the youngest); Karampuang Limestone (Qg), Manututu Andesite Dome (Qkm), Manututu Andesite (Qlm), Mamuju Limestone (Tmgm), Mamuju Conglomerate (Tmkm), Mamuju Tuff (Tmtm), Ampallas Conglomerate (Tmka), Mamuju Andesite (Tmlm), Ampallas Tuff (Tmta), Ampallas Breccia (Tmba), Membungkeng Andesite (Tmlb), Ahuni Basal (Tmla), Paweang Tuff (Tmtp), and Tandung Breccia (Tmbt).

Tandung Breccia, Paweang Tuff, Ahuni Basal, and Membungkeng Andesite were the product of tertiary volcanic activity which took place around middle Miocene – late Miocene. And in almost the same time, Ampallas Breccia together with Ampallas Tuff and Mamuju Andesite were precipitated with estimated time middle Miocene – late Miocene. These volcanic products located in the south west area. Mamuju Tuff becomes more abundant to the south west. On the other side of the research area, Ampallas Conglomerate can be found in estuary consisting of andesite, basal, and tuff fragments from the previous rocks. And more to the south west the sedimentation process took place and precipitated Mamuju Conglomerate and Mamuju Limestone.

The tectonic activity became active again around Pliocene – Pleistocene. This tectonic activity formed Manututu andesite followed by Manututu Andesite Dome. The last tectonic activity was the formation of Manututu Andesite Dome in Pleistocene, or around 0,3 – 1 million years old. The dating was estimated by using the *fission track* method. This latest volcanic product was estimated as the heat source for Ampallas geothermal system.

3.2 Geological Structures
Geological structures in Ampallas area is dominated by normal faults with northwest – southeast orientation and southwest – northeast orientation. These faults then made a depression zone in the middle of Ampallas area and believed as the one that control the presence of these hot springs.

4. Manifestation
There are 4 hot springs that became the main objects in this research; Ampallas, Batupane, Gantungan, Karema. Ampallas hot springs located in Ampallas village. It was found on the river bed and has temperature around 35 – 66^{0}C with atmosphere temperature 29^{0}C. The pH was around 6,8 – 7,2 with the electrical conductivity 551 – 1920 µs/cm. The flow rate was around 0,5 – 1 L/s.

Batupane hot spring located in Batupane village. It was found in the crack between breccia rocks and has temperature of 40,3^{0}C with atmosphere temperature of 28,9^{0}C. The pH was 8,6, with the electrical conductivity 360 µs/cm. The flow rate was 1L/s. Gantungan hot spring located in Gantungan village. It was found on the riverbed of Gantungan river and has temperature of 34,6^{0}C with atmosphere temperature of 25,9^{0}C. The pH was 9,8 with the electrical conductivity 527 µs/cm. The flow rate was 1 L/s. Karema hot spring located in Karema village. It was found as a man-made hot spring and has become a tourism object. The temperature was 48^{0}C with atmosphere temperature of 29^{0}C. The pH was 8,7 with the electrical conductivity 1074 µs/cm. The flow rate was around 0,5 L/s.

Table 1: Hot springs characteristics

	Hot Springs			
	Ampallas	Batupane	Gantungan	Karema
pH	6,8 - 7,2	8,6	9,8	8,7
Temperature (°C)	35 - 66	40,3	34,6	48
Electrical Conductivity (µs/cm)	551 - 1920	360	527	1074
Water Flow (L/s)	0,5 - 1	1	1	0,5

Beside the hot springs, there was other manifestation like altered rocks. In one of Manututu Andesite area, located in north side of Mt. Manututu foothills, the rock has been altered. This area still has a hint of sulphuric scent and on some of the joints can be seen thin layers of sulphur mineral. The host rock still can be seen clearly but the andesite has been altered in a weak intensity. Using the *portable infra-red mineral analysis* (PIMA), the clay mineral was analyzed and shown that the clay minerals area kaolinite and montmorillonite with argilic types alteration. Geographically it is located at 719441 mT and 9707717 mS on elevation 39 m asl.

5. Geochemistry

5.1 Water types

The result obtained from the lab shows that Ampallas 1,2,3,4,6,7,8 hot spring is chloride – bicarbonate type hot spring. And with near neutral pH it could be assume that Ampallas hot spring came directly from the reservoir. Unlike Ampallas hot spring, Karema, Gantungan, Ampallas 5 and Batupane hot springs are bicarbonate type hot springs. This means that the hot springs are either mixed with meteoric water or heated surface water but not from the reservoir. In other words Gantungan, Karema, and Batupane hot springs indicated outflow setting.

Table 2: Cl-SO_4-HCO_3 contents of Ampallas, Batupane, Gantungan, and Karema hot springs

Hot Springs	Cl	SO4	HCO3
Ampallas -1	513,26	1,53	596,37
Ampallas -2	492,07	0,85	606,64
Ampallas -3	532,69	1,99	609,05
Ampallas -4	174,7	12,24	360,63
Ampallas -5	1,37	78,92	283,04
Ampallas - 7	355,37	1,5	435,31
Ampallas - 8	161,09	1,88	329,89
Batupane	3,01	6,6	238,26
Gantungan	4,2	20,65	316,71
Karema	31,68	17,9	481,69

According to Na-K-Mg plotting, Ampallas (except Ampallas 5), Karema, and Gantungan hot springs located in *partial equilibrium* area. Which means it indicates that shortly after the hot springs reach the equilibrium state, they mixed with meteoric water on the surface. Batupane and Ampallas 5 hot spring located in *immature water* area, which means that the hot springs are dominated with meteoric water.

Table 3: Na-K-Mg contents of Ampallas, Batupane, Gantungan, and Karema hot springs

Hot Springs	Na	K	Mg
Ampallas -1	535,6	19,74	3,85
Ampallas -2	521,43	20,97	4,02
Ampallas -3	522,53	20,97	4,59
Ampallas -4	229,79	29,05	3,51
Ampallas -5	90,86	24,12	13,04
Ampallas - 7	370,09	16,07	5,05
Ampallas - 8	181,27	19,62	8,72
Batupane	88,47	16,27	0,19
Gantungan	141,72	1,96	0,19
Karema	316,99	16,6	0,08

And looking from the Na-K relationship in the Na-K-Mg chart, Ampallas hot springs located in 170^{0}C straight line which indicates the estimated reservoir temperature. This is similar with the estimation calculation of the reservoir, which is 169 – 173^{0}C.

5.2 Cl-Li-B Diagram

From chloride, lithium, and boron contents in the hot springs, it can be seen that almost all Ampallas hot springs have high chloride concentration. This can be assumed that Ampallas hot springs is associated with magmatic rock environment. Unlike Karema, Gantungan, and Batupane hot springs. Karema hot springs plotted between chloride and boron, while Gantungan and Batupane have higher boron contents. This indicated that Gantungan and Batupane hot springs are associated with sedimentary rock environment.

Table 4: Cl-Li-B contents of Ampallas, Batupane, Gantungan, and Karema hot springs

Hot Springs	Cl	Li	B
Ampallas -1	513,26	0,39	2,64
Ampallas -2	492,07	0,39	2,9
Ampallas -3	532,69	0,37	2,76
Ampallas -4	174,7	0,04	1,41
Ampallas -5	1,37	0,01	0,17
Ampallas - 7	355,37	0,35	2,51
Ampallas - 8	161,09	0,09	1,11
Batupane	3,01	0,02	0,6
Gantungan	4,2	0,02	0,6
Karema	31,68	0,01	1,92

And if using Cl/B ratio, which is commonly used to measure *common reservoir source* (Nicholson, 1993), it can be seen that Ampallas hot springs have relatively same Cl/B ratio. Which means that those hot springs may came from the same reservoir. Karema and Gantungan have different Cl/B ratio value from Ampallas so it can be assumed that those hot springs didn't come from the same reservoir as Ampallas hot springs, while Batupane and Ampallas-5 hot spring may have come from the same reservoir.

5.3 Isotope

In general, geothermal fluids will have a process of ^{18}O isotope ($\delta^{18}O$ *shifting)* increase, in this case is meteoric water (Craig, 1963 in Nicholson, 1993). The enrichment of ^{18}O in geothermal fluids relative to local meteoric water is caused by rock-water reactions at depth. Deuterium isotope increase can't happen in rocks because in general the deuterium concentration is too low. From the isotope analysis data using meteoric water line equation $\delta D = 8\ \delta^{18}O + 14$, it shows that fluid from Ampallas-1 hot springs is drifting away from the local meteoric water equation line. This indicates that the fluid from Ampallas-1 hot spring came straight from the reservoir.

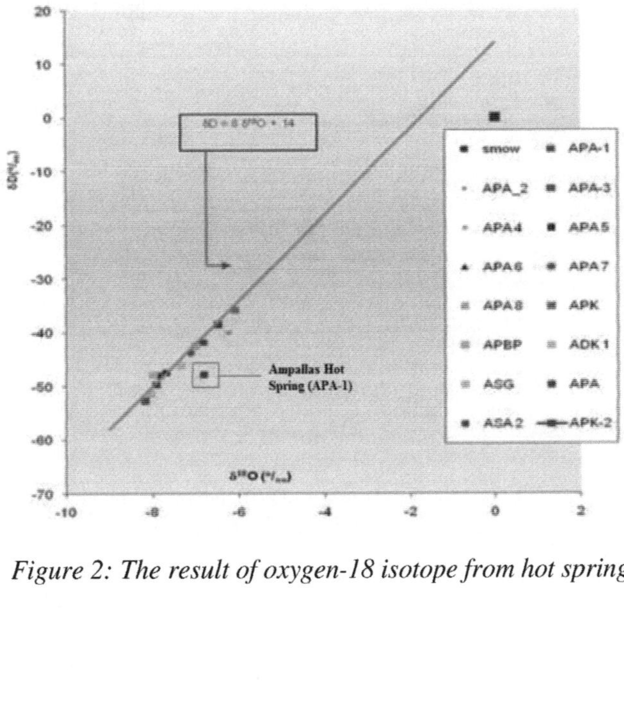

Figure 2: The result of oxygen-18 isotope from hot springs

6. Geothermal system

Ampallas geothermal system was found in tertiary volcanic and quarter volcanic rocks. This can be assumed that it associated with volcanic environment. The heat source was estimated came from Manututu Andesite Dome (0,3 – 1 million years old) as the youngest volcanic product.

The manifestation found in the research area was hot springs and altered rocks. The altered rocks show that the cap rock was composed by clay mineral as the result from the rock alteration due to contact with the hot fluids. The alteration type is argilic type. There are three hot spring locations near Mt. Manututu in Ampallas area; Ampallas hot springs (1,2,3,4,6,7,8) located in the northeast area, Ampallas-5 hot spring in the east area, and Batupane in the west area.

Fluids from Ampallas hot spring has chloride – bicarbonate type and located in partial equilibrium. This means that the fluid came from the deep and a moment after it just reach equilibrium state, it mixed with meteoric water. While Karema, Gantungan, and Batupane located in Immature water area and have bicarbonate type. This shows that they are mixed with meteoric water a lot or basically just heated up meteoric water.

The Ampallas geothermal system was estimated came from different system with Karema and Gantungan geothermal system, and Ampallas-5 and Batupane came from the same geothermal system. The altered rock found in Ampallas was the resulted from the interaction between the rocks and the hot fluid. Both altered rocks and the sulphur deposit indicate that Ampallas hot springs are located in upflow area. But from geology analysis, it is concluded that Ampallas, Ampallas-5, and Batupane hot springs came from the same reservoir. The reservoir located in a small Ampallas depression zone under Mt. Manututu. Ampallas hot springs interacted more with the youngest volcanic rocks while the other hot springs interacted more with the older volcanic rocks. And as the altered rocks and sulphur deposit indicates, Ampallas hot springs located in upflow area while Ampallas-5 and Batupane probably located in the outflow area in Ampallas geothermal system.

7. Estimated Geothermal Resource potential

Using geology and geochemistry method, this research can conclude the geothermal speculative resource potential. It was estimated that the reservoir temperature is around 169 – 173^{0}C. So Ampallas geothermal system is included in medium enthalpy geothermal system (Hochestein, 1990 in Dickson and Fanelli, 1990).

Table 5: Classification of geothermal resources by temperature. (Dickson and Fanelli, 1990)

	(a)	(b)	(c)	(d)
Low enthalpy	<90 °C	<125 °C	<100 °C	≤150 °C
Intermediate	90-150	125-225	100-200	---
High enthalpy	>150	>225	>200	>150

Source: (a) Muffler and Cataldi, 1978.
(b) Hochstein, 1990.
(c) Benderitter and Cormy, 1990.
(d) Haenel, Rybach and Stegena, 1988.

Table 6: Power per unit area

Reservoar	Temperature Limit (°C)	Power per Unti Area (MWe/km²)
Low Temperature	**< 125**	**5**
Medium Temperature	**125-225**	**10**
High Temperature	**>225**	**15**

From this research it is known that the prospect area is approximately 3 km^2. Using the equation:

$$Hel = A \times Qel \qquad (1)$$

Hel = Speculative Resource Potential (MWe)
A = Prospect Area (km^2)
Qel = Power per unit area (MWe/km^2)

it can be obtained the speculative resource potential for Ampallas geothermal system, which is about 30 MWe, although this is just the speculative potential. Which means this potential is not very accurate. By using detailed geophysics method the result can become more accurate.

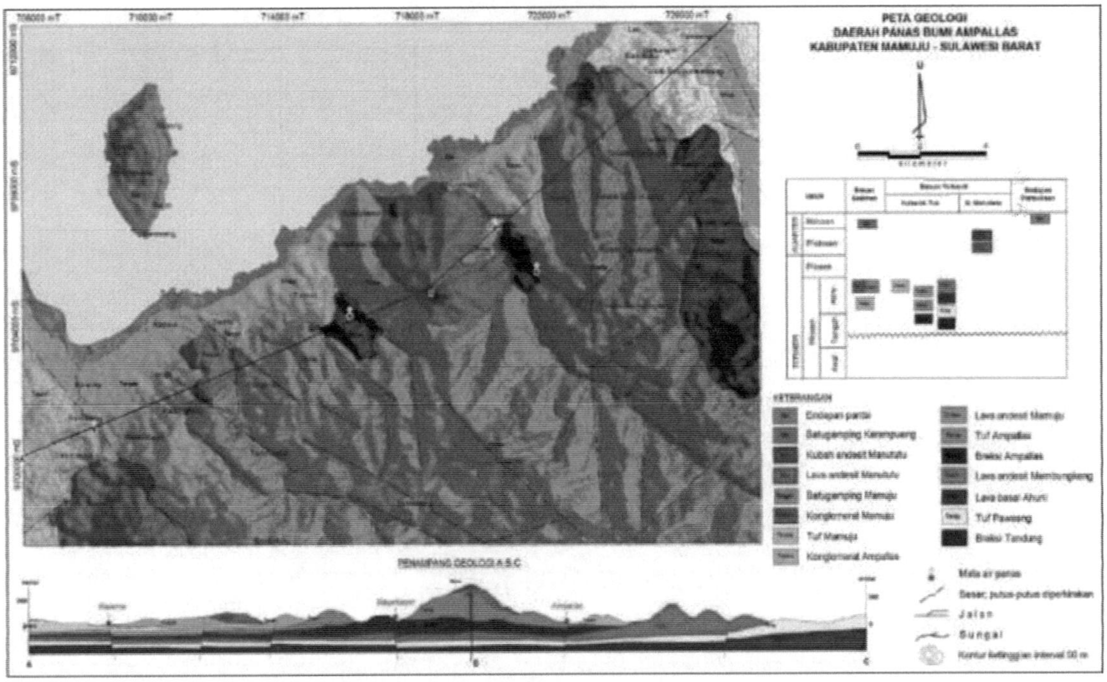

Figure 3: Geological map of Ampallas geothermal area

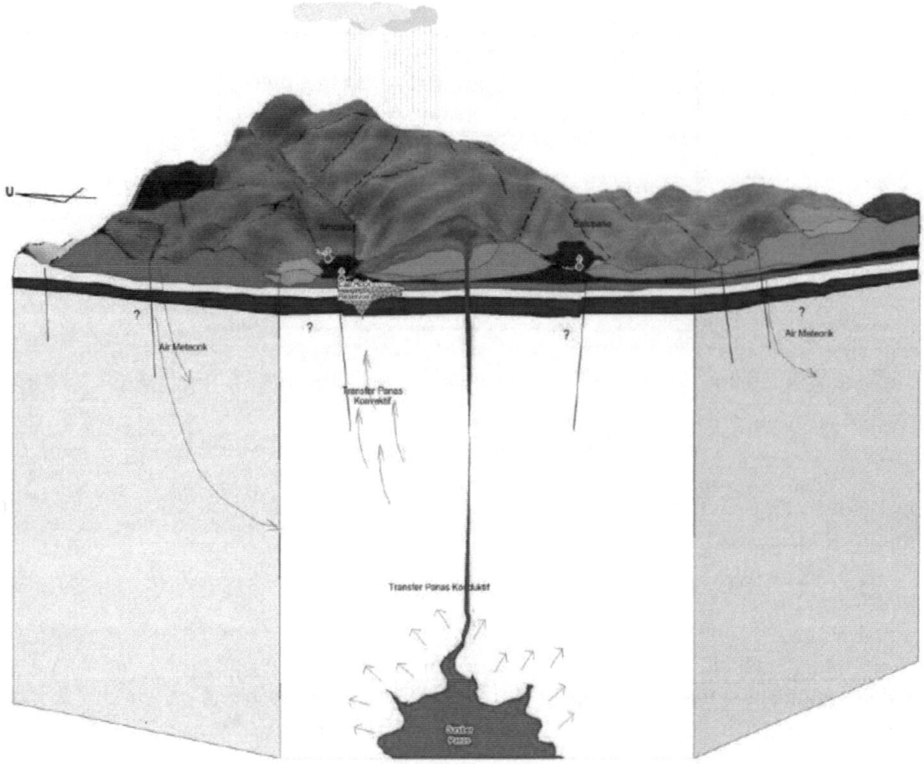

Figure 4: Tentative model of Ampallas geothermal system

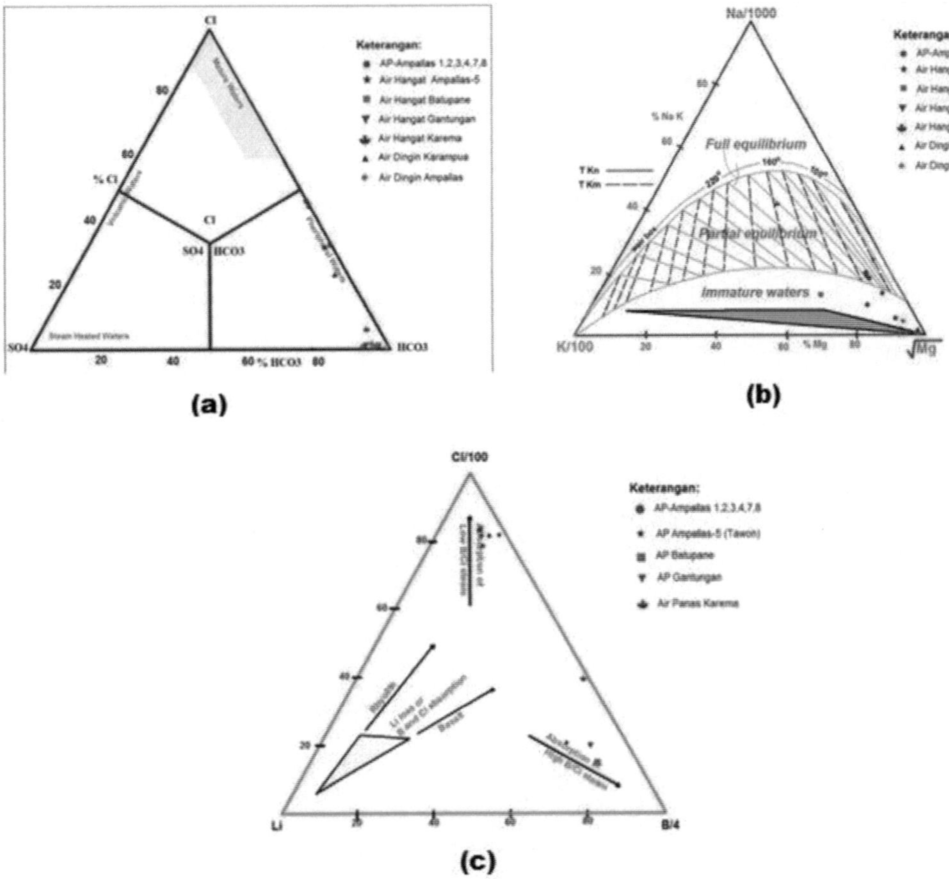

Figure 5: (a) Cl-SO$_4$-HCO$_3$ diagram, (b) Figure 6: Na-K-Mg diagram, (c) Cl-Li-B diagram

8. Conclusion

Ampallas area is consisted from 14 rock units. The Ampallas geothermal system itself is associated with young Mt. Manututu volcanic activity (0,3 – 1 million years old). There are many tectonic activities (tertiary) that made this area a potential for geothermal system by creating many faults and joints. Hot springs and altered rocks are manifestations that created in Ampallas area. There are 4 hot springs; Ampallas, Batupane, Gantungan, and Karema. Ampallas 1,2,3,4,7 and 8 hot springs water type is chloride – bicarbonate, which means it came from the reservoir while Batupane, Gantungan, Karema and Ampallas 5 are all bicarbonate type. Ampallas 1,2,3,4,7,8, Karema and Gantungan hot springs fluid plotted in partial equilibrium zone while Batupane and Ampallas 5 plotted in immature water zone. It means the Ampallas hot springs (except Ampallas-5) mixed with meteoric water right after reached the equilibrium state. It is also concluded that Ampallas 5 hot springs came from the same reservoir with Batupane, but not Gantungan and Karema hot springs. The altered rocks and sulphur deposit were found near the Ampallas hot springs. The altered mineral is clay mineral included in argilic type. Both the altered rocks and sulphur deposit indicated upflow area, so it can be concluded that Ampallas hot springs located in the upflow area, while Ampallas-5 and Batupane hot springs located in the outflow area of Ampallas geothermal system. The speculative resource potential of Ampallas geothermal system is estimated around 30 MWe. But if detailed geophysical method was applied the result could be more accurate. Ampallas geothermal area is an interesting area to be researched and to be developed. With more detailed method, like magnetotelluric, the result could be much more accurate.

Aknowledgement

We want to thank you to all who has been helping for the completion of this paper, especially Ampallas geothermal research team members whose names cannot be mention one by one. And we want to thank you for Pusat Sumber Daya Geologi for providing the data needed for this paper. Last but not least for all the people who have been always give us their supports.

References

Dickson, M. H. and Fanelli, M. 2004. Geothermal Energy. Istituto di Geoscienze e Georisorse, CNR. Pisa. Italy

Nicholson, Keith (1993), "Geothermal Fluids, Chemistry and Exploration Techniques," *Springer Verlag Inc*

Fournier, R.O. (1981), "Application of Water Geochemistry Geothermal Exploration and Reservoir Engineering, Geothermal System: Principles and case Histories," *John Willey & Songs. New York*

Ratman, N. and S. Atmawinata (1993), "Peta Geologi Regional Lembar Mamuju," *Pusat Penelitian dan Pengembangan Geologi. Bandung*

Pusat Sumber Daya Geologi (2010), "Survei Panas Bumi Pendahuluan/Regional," *Badan Geologi, Kementrian Energi dan Sumber Daya Mineral. Bandung*

Identification of linear features at geothermal field based on Segment Tracing Algorithm (STA) of the ALOS PALSAR data

Haeruddin[1], A Saepuloh[2], M N Heriawan[1] and T Kubo[3]

[1] Faculty of Mining and Petroleum Engineering, ITB, Bandung, Indonesia
[2] Faculty of Earth Sciences and Technology, ITB, Bandung, Indonesia
[3] Graduate School of Engineering, Kyoto University, Kyoto, Japan

E-mail: haedin90@gmail.com

Abstract. Indonesia has about 40% of geothermal energy resources in the world. An area with the potential geothermal energy in Indonesia is Wayang Windu located at West Java Province. The comprehensive understanding about the geothermal system in this area is indispensable for continuing the development. A geothermal system generally associated with joints or fractures and served as the paths for the geothermal fluid migrating to the surface. The fluid paths are identified by the existence of surface manifestations such as fumaroles, solfatara and the presence of alteration minerals. Therefore the analyses of the liner features to geological structures are crucial for identifying geothermal potential. Fractures or joints in the form of geological structures are associated with the linear features in the satellite images. The Segment Tracing Algorithm (STA) was used for the basis to determine the linear features. In this study, we used satellite images of ALOS PALSAR in Ascending and Descending orbit modes. The linear features obtained by satellite images could be validated by field observations. Based on the application of STA to the ALOS PALSAR data, the general direction of extracted linear features were detected in WNW–ESE, NNE–SSW and NNW–SSE. The directions are consistent with the general direction of faults system in the field. The linear features extracted from ALOS PALSAR data based on STA were very useful to identify the fractured zones at geothermal field.

Keywords: Segment Tracing Algorithm (STA), ALOS PALSAR, geothermal, extracted linear features, Wayang Windu.

1. Introduction

Indonesia is located on the world ring of fire along with many volcanoes which potentially providing the geothermal energy resources. The potency of geothermal resources owned by Indonesia reached about 13,440 MW to 28,100 MW of potential resources and 14,473 MW of reserves spread across 265 locations throughout Indonesia archipelago. However, the current production is about 1189 MW (4%) which used to generate the electricity [1]. Thus, the comprehensive understanding about the geothermal system is indispensable for continuing the development. One of area that shows the potency of geothermal energy resources is Wayang Windu field located at West Java Province.

Geological structures such as faults and their associated fracture zones are an important aspect in the exploration of geothermal systems. Faults can provide permeable paths for geothermal fluids (through zones of fractured rocks) where the study of their pattern can help often to delineate the productive parts of geothermal reservoir [2]. Geological structures could be associated with the linear features in satellite imagery.

Detection of the surface geological structure presented in the form of linear features in the image using optical sensor data hampered some problems such as the canopy of vegetation, clouds, and weather conditions. Meanwhile the direction and quantity of surface geological structures are important for geothermal exploration because it could accommodate the fluid flows and surface manifestations. Although the visual interpretation is effective to track the linear features in images, this technique is not efficient, especially for large study area or multiple datasets [3]. To overcome this problem, the satellite imagery with the microwave sensor could be an option in the detection of geological structures within a certain region. The advantage of satellite imagery is able to operate in any weather conditions and without acquisition time providing accurate surface observation.

A method for determining the linear features is Segment Tracing Algorithm (STA). The principle of the STA is to detect a line of pixels as a vector element by examining local variance of the gray levels in the digital image, and to connect retained line elements along their expected directions. The advantage of this method is able to observe the linear features of the low contrast and parallel to the sun's azimuth [4].

2. Study area

Wayang Windu geothermal field took the name after Mt. Wayang and Mt. Windu, the two small lava domes without historic eruption history [5]. The field is located at Pangalengan town, approximately 40 km to the south of Bandung city, the capital of West Java Province (Figure 1). The elevation of study area is about 1500 to 2100 m above sea level [6].

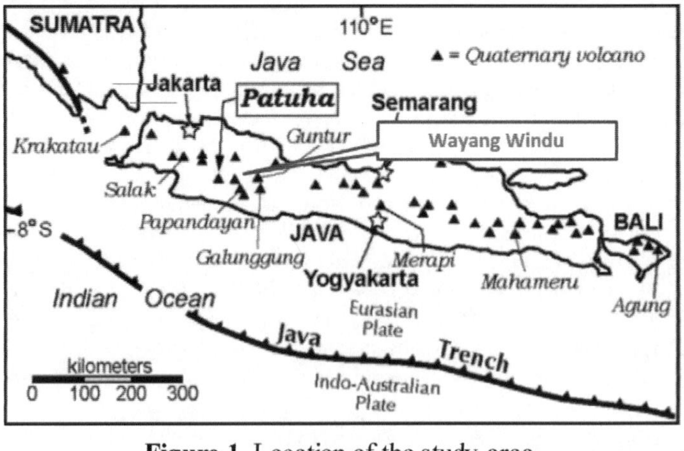

Figure 1. Location of the study area.

Several thermal manifestations including fumaroles, hot springs, mud pools and altered soil are situated along the slopes of the mountain range. Fumaroles which are located between Mt. Wayang and Mt. Windu show temperatures about 93°C to 96°C, and therefore slightly superheated if we compared to the normal water boiling point at altitudes of 2000 m. Hot springs which are located in altitudes about 1495 to 1985 m reach temperatures 41°C to 88°C. Water discharge hot springs are bicarbonate type, except for the hot springs which is located near the fumaroles has discharge acid sulfate water [7]. The geothermal system at Wayang Windu field was classified as transitional between vapor and liquid dominated systems [8].

3. Data

The Phased Array type L-band Synthetic Aperture Radar (PALSAR) onboard the Advanced Land Observing Satellite (ALOS) with two orbits mode, i.e. toward north (Ascending) and toward south

(Descending) flight directions was used in this study (see Figures 2A and B). Ascending and Descending images have a size of 800 × 800 pixels with a resolution of 30.72 m × 30.72 m.

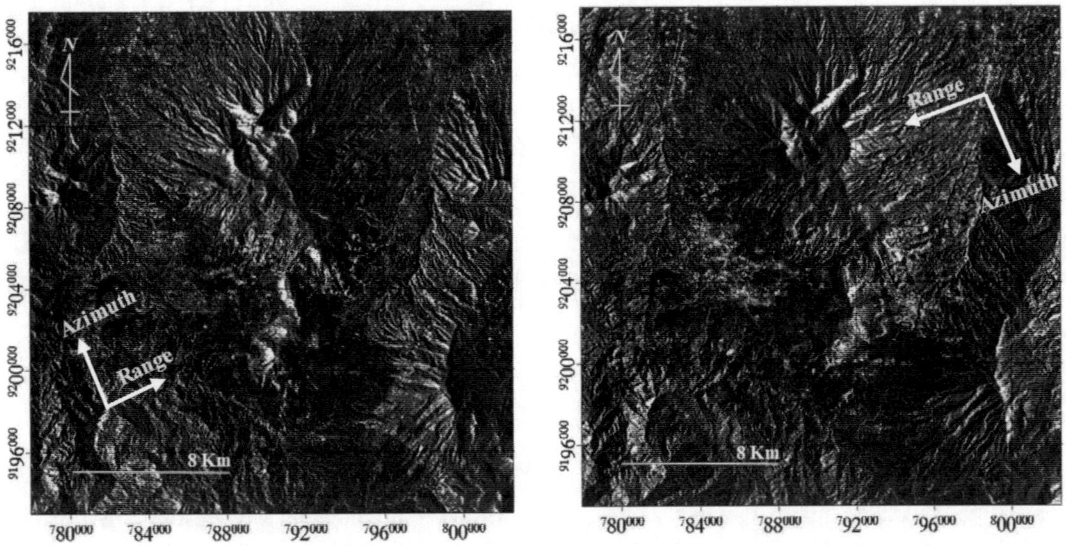

(a) (b)

Figure 2. ALOS PALSAR image with: (a) Ascending mode (hereinafter referred as Ascending image), (b) Descending mode (hereinafter referred as Descending image)

4. Methodology

Linear features represent the geological structure with a different hue contrast and relief in the images. Segment Tracing Algorithm (STA) is linear features extraction method consist of five main steps [4, 9]. The program used for the linear features extraction is developed using MATLAB syntax.

4.1. Segment Tracing Algorithm (STA)

Step 1. Local window with array size of 11 × 11 pixels is set around a centered pixel which will be judged whether it is a line element or not (Figure 3). The sixteen directions at 11.25° intervals passing through the center of the window defined to examine a local variation of gray levels along a line. The numbers shown in Figure 3 denotes the examples of line directions. The direction that minimizes the variation is assumed the valley direction and expressed by a symbol k_{min}.

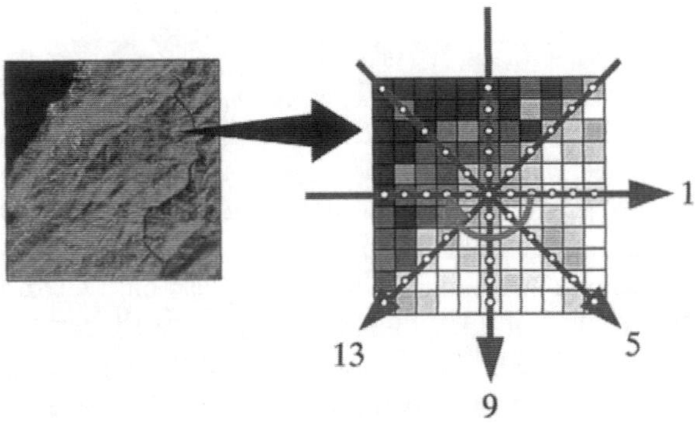

Figure 3. Local window with array size of 11 × 11 pixels and scan lines at pi/16 radian intervals used in the Segment Tracing Algorithm (STA).

Step 2. Let a gray level at location x by $z(x)$. Along the direction perpendicular to k_{min}, that is k_{max}, squared secondary differentiation for gray levels, λ, is calculated by:

$$\lambda = \left(z(x)\right)^{-1}\left(\frac{d^2 z(x)}{dx^2}\right)^2 \tag{1}$$

Expressing the mean and standard deviation of λ by m and σ, a dynamic threshold T defined as:

$$T = m + \mu\sigma, \mu = \phi_1\left(\left(1 + \sin\theta\right)^{-1} - 0.5\right) \tag{2}$$

where θ is the included angle between the sun's azimuth (s_p) and k_{max}, and ϕ_1 is a constant. The line elements that lie in the closer direction of s_p have a lower threshold level. If the value calculated at the centered pixel of the window is above the threshold, the pixel is retained as a line element (p).

Step 3. Judging whether p represents a ridge or valley feature was performed and p was eliminated when it is assumed to be distributed at a ridge.

Step 4. This routine linked to the derived line elements. The distance between p and connectable pixels was examined within the distance D from p by the following formula:

$$D = \phi_2\left(1 + \sin\theta *\right)^{-1} + \theta_3 \tag{3}$$

where $\theta *$ is included angle between s_p and k_{min}, then ϕ_2 and ϕ_3 are constants. The value D is a dynamic threshold where two pixels lying parallel to s_p and having a larger distance which could be linked.

Step 5. A centerline for the line elements, which have similar directions and intersect each other is obtained from the (x, y) coordinates of the line elements using Principal Component Analysis.

4.2. Interpretation

After linear features were obtained, the Rosette diagram was used to determine the major direction of linear features in the study area, either for Ascending or Descending images. Interpretation of linear features was then validated by using the field data (fault system at study area) in order to know the relationship between the linear features extraction and the actual field conditions.

5. Result and discussion

Segment Tracing Algorithm (STA) was used to extract the linear feature structures which interpreted as fault or fracture based on ALOS PALSAR data with Ascending and Descending modes (see Figures 4 and 5). In the Rosette diagram, ratios of the counts of individual linear feature to the total amount of all linear features for Ascending and Descending image along each direction was calculated. In the Ascending image, WNW–ESE direction is the most dominant direction than the others. In the Descending image, linear feature direction more varied in all directions compared with linear features of Ascending image.

The number of extracted linear features in Ascending image was larger than in Descending image. The number of linear features in the Ascending and Descending images is 10,167 and 9842 respectively (see Table 1). In this case, the range direction affects the appearance of linear features in the image, i.e. toward East (Ascending) and West (Descending) directions (see Figure 2).

Composite Rosette diagram was obtained by a combination of linear features in Ascending and Descending image, where counts of individual linear features to total amount of all linear features for Ascending and Descending image along each direction were also calculated. Extracted linear features from ALOS PALSAR (Ascending and Descending) image showed the three major directions, i.e. N305°E, N335°E and N15°E. WNW–ESE is the most dominant direction than the others, where it was counter-clockwise according to the increasing order of fractures abundance (see Figure 7a).

In this study, we use the field data which formed permeable fault system in Wayang Windu showed 27 faults. These structures influenced the geothermal fluid path to the surface [10]. The overlay of linear features and fault system was conducted to know the position of faults and linear features in the image.

Subset area was arranged to determine the relationship between extracted linear features of imagery with the field data (fault system). The boundary of subset area was arranged based on the maximum and minimum coordinates of the faults (see Figure 6).

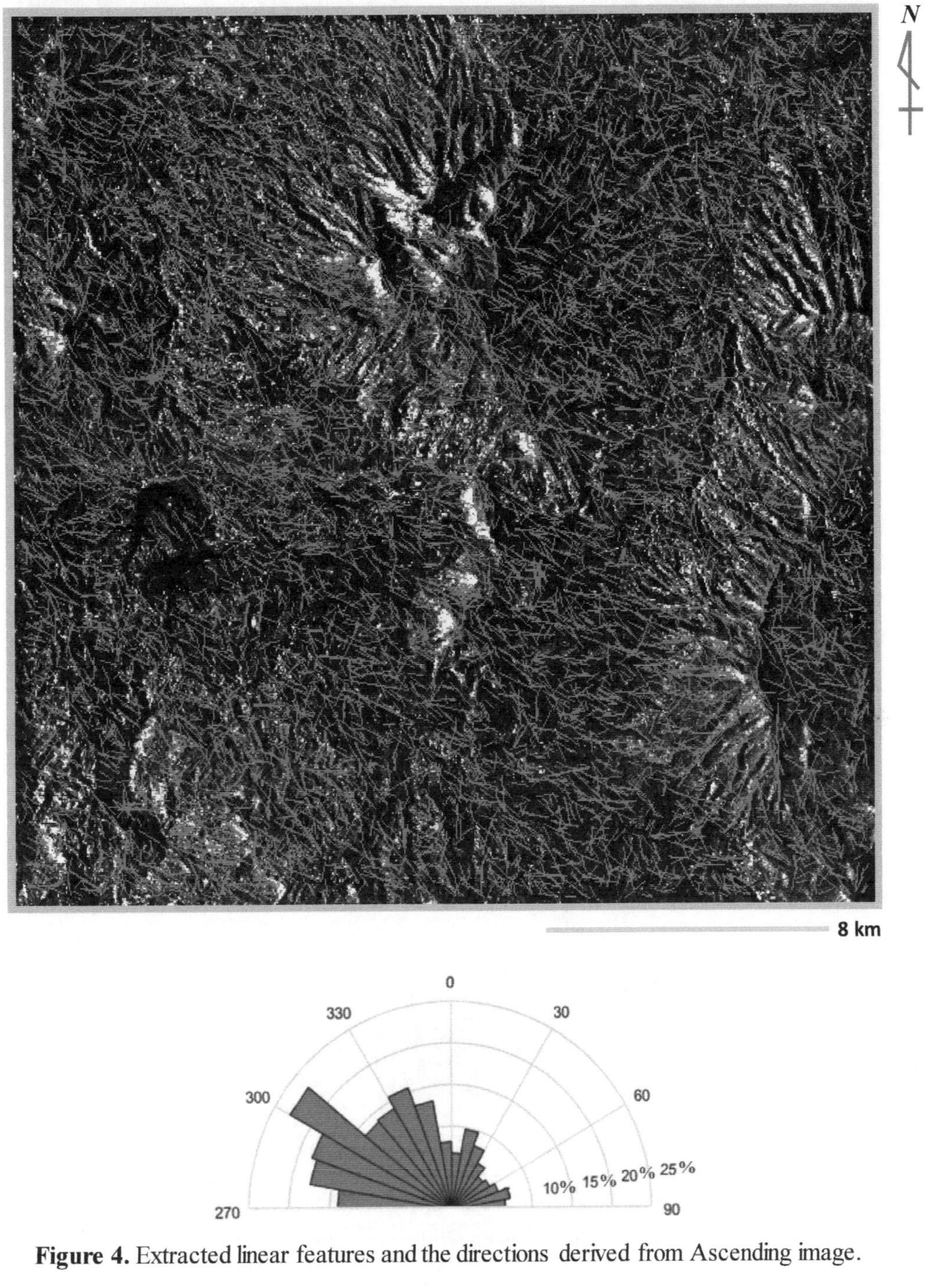

Figure 4. Extracted linear features and the directions derived from Ascending image.

Figure 5. Extracted linear features and the directions derived from Descending image.

At subset area, the number of linear features in the Ascending and Descending images is 1706 and 1602, respectively. The major direction of linear features is N285°–305°E, N335°E and N25°E. In this case, WNW–ESE direction is dominant, while NNW–SSE and NNE–SSW directions are subordinate (see Figure 7b). The major direction of fault system at Wayang Windu field is N295°E, N330°E, and N35°E (see Figure 7c).

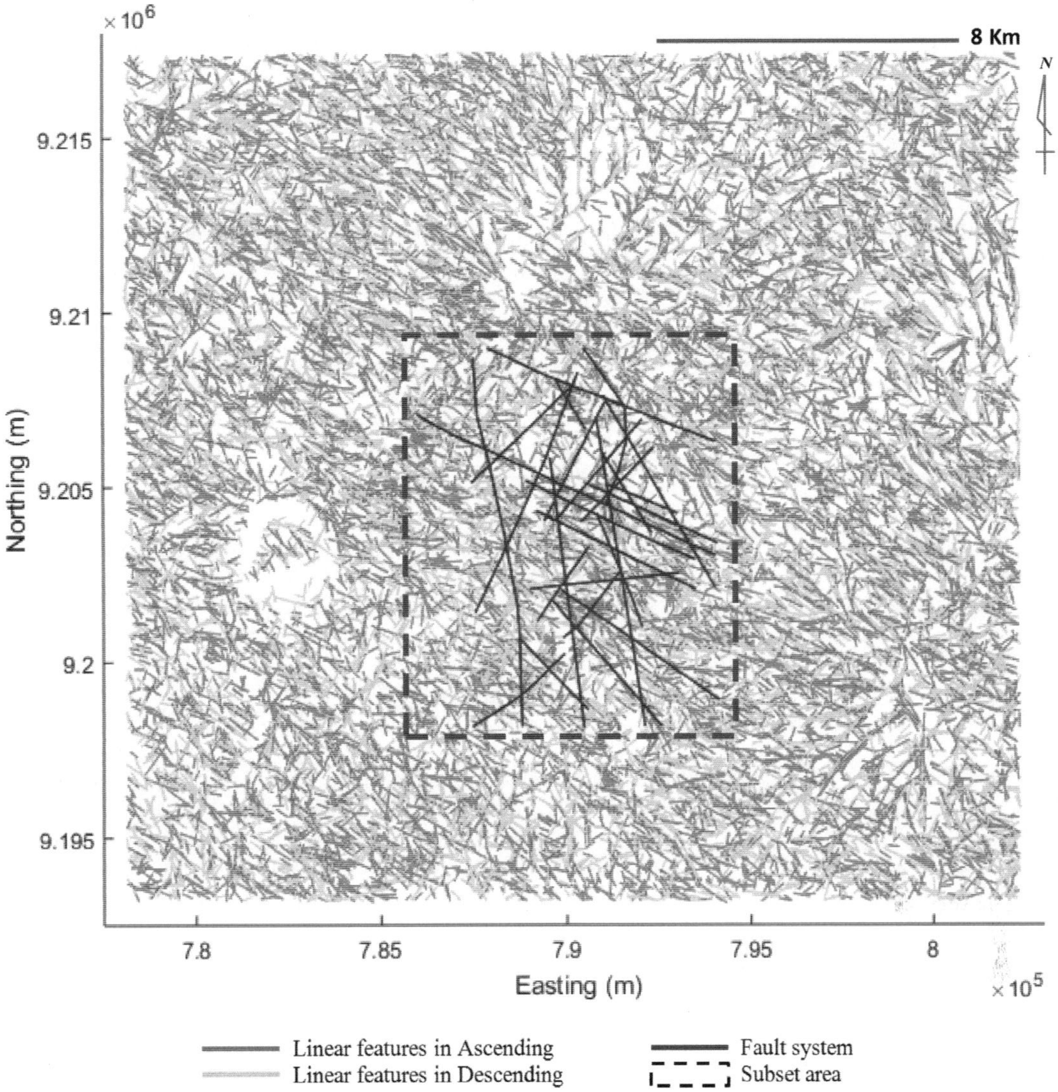

Figure 6. Overlaying between extracted linear features and fault system at Wayang Windu field [10].

Table 1. Summary of extracted linear features at Wayang Windu field

Image	Orbit mode	Count
Full image	Ascending	10,167
	Descending	9842
Subset area	Ascending	1706
	Descending	1602

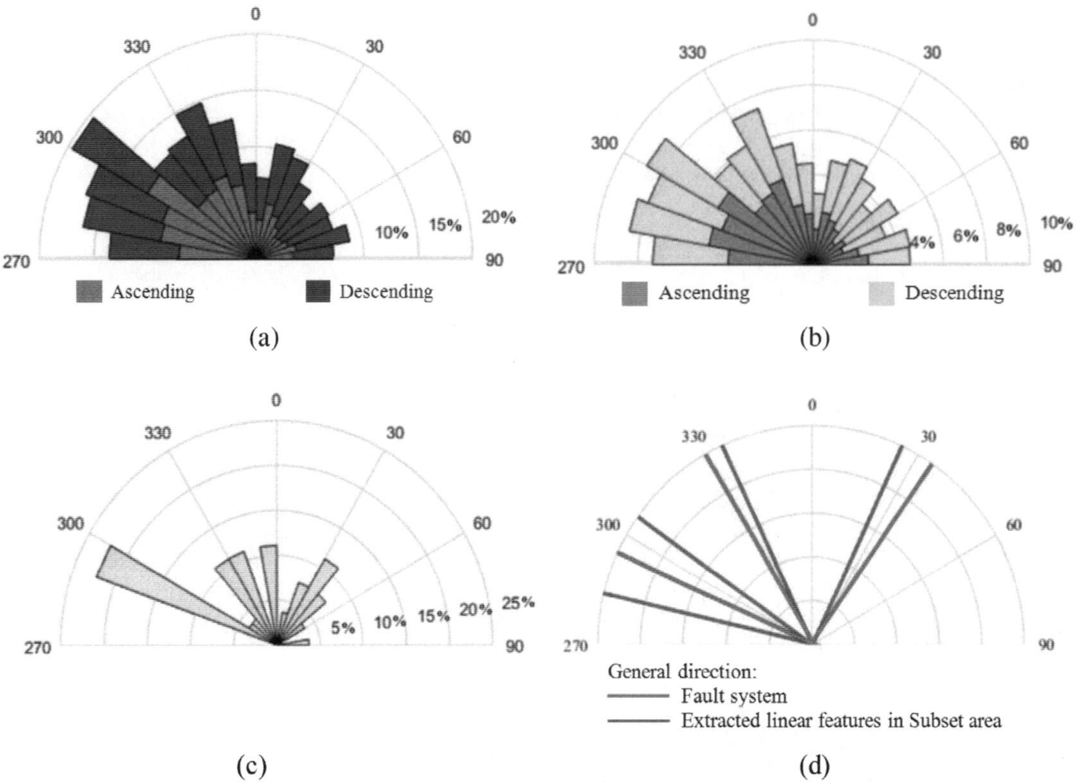

Figure 7. (a) The direction of linear features in Ascending and Descending images (full image), (b) The direction of linear features in Ascending and Descending images (subset area), (c) The direction of faults in Wayang Windu, (d) Comparing the major direction between fault system and extracted linear features in subset area.

Based on the general direction of fault system and extracted linear features in subset area, the different direction between WNW–ESE and NNE–SSW is about 10°, and the NNW–SSE is about 5° (see Figure 7d). The WNW–ESE direction is dominant, while NNW–SSE and NNE–SSW directions are subordinate. Those directions showed appropriateness between extracted linear features and the field data. Thus, extracted linear features in these directions are more representative for the fault system in Wayang Windu. In general, it can be said that the major direction of extracted linear features and fault system were similar.

Extracted linear features using Segment Tracing Algorithm (STA) for ALOS PALSAR (Ascending and Descending) data can help to explain phenomena in the field. In other words, these extracted linear features could support to identify the fractured zones in the geothermal field.

In order to determine the areas that have the geothermal potency, we need to know the geological structures more intensively. In the geothermal system, the fluid will flow upward through permeable zones which are generally derived from the geological structure, so that the higher density of structure, the greater level of permeability. For this purpose, Linear Feature Density (LFD) could be used as the further work. In order to obtain better results, LFD could be modeled spatially using geostatistical approach by taking into account the spatial aspects of extracted linear features.

6. Conclusion

Extracted linear features derived from Ascending and Descending images produced 10,167 and 9842 data, respectively. The major direction of linear features from the imagery was WNW–ESE. Extracted linear features were validated by the fault system in the field. At subset area, the number of linear features in Ascending and Descending were 1706 and 1602, respectively. In addition, the general direction of extracted linear features and fault system were similar as: WNW–ESE, NNW–SSE, and NNE–SSW. The directions showed appropriateness between extracted linear features and the field data. Thus, extracted linear features from ALOS PALSAR dual orbit (Ascending and Descending) using Segment Tracing Algorithm (STA) could support to identify fractured zones in Wayang Windu geothermal field.

Acknowledgement
The authors sincerely thank to the "BAGUS" SATREPS project for providing field equipment, LPPM-ITB for survey accommodation supports, and Star Energy for collaborating to this research.

References
[1] Surana T, Prastawa A, Yasser R, Priandana E R, Zaky T, Mayasari A P and Endang 2010 *Rancang Bangun Sistem PLTP Tipe Turbin Kondensasi 5 MW dalam Rangka Menunjang Kemandirian Industri Nasional di Sektor Kelistrikan* (Jakarta: BPPT)
[2] Soengkono S 1999 Te Kopia Geothermal System (New Zealand) - The Relationship Between Its Structure and Extent *Geothermics* **28** 767
[3] Saepuloh A 2013 Potential Use of Synthetic Aperture Radar (SAR) Data for Geothermal Exploration *Proc. Int. Conf. on Geol. Eng.* (Yogyakarta-Indonesia)
[4] Koike K, Nagano S and Ohmi M 1995 Lineament Analysis of Satellite Images Using a Segment Tracing Algorithm (STA) *Comp. & Geosc.* **21** 1091
[5] Hochstein M P and Sudarman S 2008 History of Geothermal Exploration in Indonesia from 1970 to 2000 *Geothermics* **37** 220
[6] Purnanto M H and Purwakusumah A 2015 Fifteen Years (Mid-Life) of Wayang Windu *Proc. World Geoth. Cong. 2015* (Melbourne – Australia)
[7] Prasetio R, Wiegand B A, Malik D and Sauter M 2015 Stable Isotopes Study of Wayang Windu Geothermal Field, Indonesia *Proc. World Geoth. Cong. 2015* (Melbourne – Australia)
[8] Bogie I, Kusumah Y I and Wisnandary M C 2008 Overview of the Wayang Windu Geothermal Field, West Java, Indonesia *Geothermics* **37** 347
[9] Koike K, Kouda R and Ueki T 2001 Characterizing Fracture Systems of Kyushu, Southwest Japan Through Satellite-Image Derived Linear Features Superimposed on Topographic and Lithologic Data *Bul. of the Geol. Surv. of Japan* **52** 405
[10] Star Energy Geothermal Ltd 2015 unpublished report

Repeat Absolute and Relative Gravity Measurements for Geothermal Reservoir Monitoring in the Ogiri Geothermal Field, Southern Kyushu, Japan

J Nishijima[1], C Umeda[2], Y Fujimitsu[1], J Takayama[3], N Hiraga[3], and S Higuchi[3]
[1] Department of Earth Resources Engineering, Faculty of Engineering, Kyushu University, Fukuoka, Japan
[2] Department of Earth Resources Engineering, Graduate School of Engineering, Kyushu University, Fukuoka, Japan
[3] Nittetsu Mining Co., Ltd., Tokyo, Japan

E-mail: nishijima@mine.kyushu-u.ac.jp

Abstract. Repeat hybrid microgravity measurements were conducted around the Ogiri Geothermal Field on the western slope of Kirishima volcano, southern Kyushu, Japan. This study was undertaken to detect the short-term gravity change caused by the temporary shutdown of production and reinjection wells for regular maintenance in 2011 and 2013. Repeat microgravity measurements were taken using an A-10 absolute gravimeter (Micro-g LaCoste) and CG-5 gravimeter (Scintrex) before and after regular maintenance. Both instruments had an accuracy of 10 μgal. The gravity stations were established at 27 stations (two stations for absolute measurements and 25 stations for relative measurements). After removal of noise effects (e.g., tidal movement, precipitation, shallow groundwater level changes), the residual gravity changes were subdivided into five types of response. We detected a gravity decrease (up to 20 μgal) in the reinjection area and a gravity increase (up to 30 μgal) in the production area 1 month after the temporary shutdown. Most of the gravity stations recovered after the maintenance. The temporal density changes in the geothermal reservoir were estimated based on these gravity changes.

1. Introduction

Repeat microgravity measurements are used for the assessment of mass balance of ground water and geothermal fluid storage changes. The repeat microgravity measurement has been shown to be an effective method to monitor the balance of production and reinjection in a geothermal reservoir [1, 2, 3]. The production and reinjection of geothermal fluid cause mass movement and redistribution, and these mass changes cause a measurable gravity change on the ground surface. It is possible that the mass balance of the geothermal reservoir could be an index of sustainability of a geothermal power plant.

The results in the previous studies included the gravity change at the reference gravity station, and so the absolute gravimeter was introduced to evaluate this change. The combined absolute and relative gravity measurement is called the hybrid gravity measurement [4], and this type of measurement is applied to volcanological and geothermal monitoring [5, 6].

The Ogiri geothermal power plant (30 MW) is located in the southern part of Kyushu, Japan (Figure 1). Geothermal exploration was started in 1973 with various geophysical surveys and drilling, and operations commenced in 1996. Nittetsu Mining Co., Ltd. operates geothermal steam production, and

Kyushu Electric Power Co., Inc. operates the electric power generation. The production area is located in the eastern part of the geothermal field where the depths of the production wells are about 1000–1500 m. The amount of production is about 10.2 Mt/year, and about 79.5% of production (hot water) is reinjected into the western part of the field [7].

Little research has focused on the short-term gravity changes associated with the shutdown of production and reinjection wells [8]. The present study was undertaken to detect the influence of the shutdown on the production and reinjection of geothermal fluid. Hybrid gravity measurements were collected during the regular maintenance activities in 2011 and 2013. The effect of shallow groundwater level changes is estimated to extract the geothermal reservoir changes. This paper reports the results of the 2013 measurements.

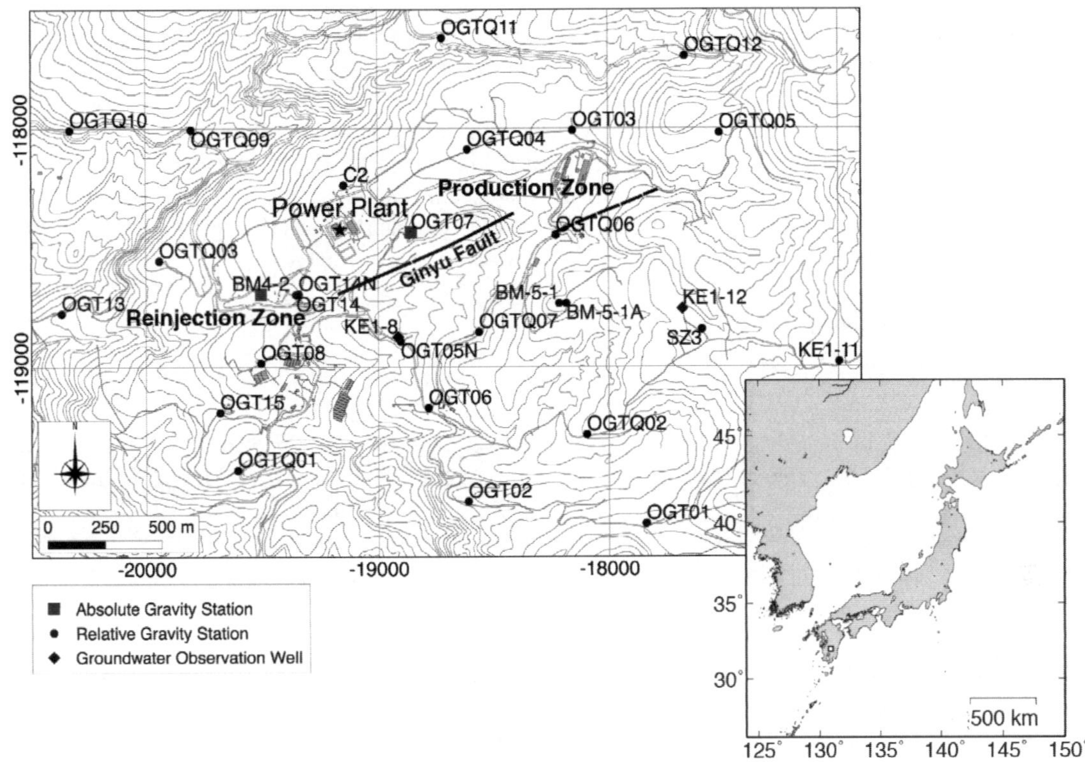

Figure 1. Location of the gravity stations at the Ogiri Geothermal Field, Kagoshima prefecture, Japan. Red circles and green squares show the relative and absolute gravity sampling stations, respectively.

2. Gravity measurement

2.1. Absolute gravity measurement

The A10 is a portable absolute gravimeter manufactured by Micro-g LaCoste Inc. It can operate on a 12-V DC power supply, much like a vehicle battery. The A10 consists of a laser, interferometer, long period inertial isolation device, and an atomic clock to accurately measure the position of the test mass. A test mass is dropped vertically in a vacuum chamber and then allowed to fall an average distance of 7 cm. The raw gravity data are processed with "g" version 9 software, which is designed to work with the Micro-g LaCoste absolute gravimeter to acquire and process gravity data. The software needs the input of some parameters, including the location of the site (latitude, longitude, altitude) and geophysical corrections. We can correct the effect of the tidal movement, ocean load, barometric pressure, and polar motion in acquiring the gravity data [9].

There were two gravity stations for the absolute gravity measurement. The setting of the measurement is shown in Table 1. It took about 30 min for the measurement, and the error of measurements is about 10 μgal.

Table 1. A10 absolute gravimeter settings.

Drop interval	1 sec
Number of drop/set	100 drops
Set interval	6 min
Number of sets	10

2.2. Relative gravity measurement

The relative gravity measurements were conducted using a Scintrex CG-5 gravimeter. Measurements were taken March, April, and May in 2013, at an interval of 1 month. There are 25 gravity stations around the Ogiri geothermal power plant. The two-way measurement method was used to evaluate instrumental drift and precision. The reference station was set at BM 4-2, which monitored the gravity change using an A10 absolute gravimeter. The measured gravity data were reduced to the necessary corrections, such as the earth tide, the height of the instrument, and drift corrections. The earth tidal corrections, including the ocean tidal effect, were calculated using GOTIC2 [10] instead of the instrument's internal program. The errors of each observation were estimated within 10 μgal.

3. Results and discussion

3.1. Absolute gravity changes

Figure 2 shows the results of the absolute gravity measurements from 2011 to 2013. A gravity decrease (up to 19 μgal) was observed just after the regular power plant maintenance in April 2011, in the reinjection area (BM4-2). The gravity recovered by May 2011, and again by May 2013 after the maintenance. These changes seem to reflect the storage change in the reinjection area.

In the production area (OGT-7), small gravity changes were observed just after regular maintenance, but gravity was basically stable from 2011 to 2013.

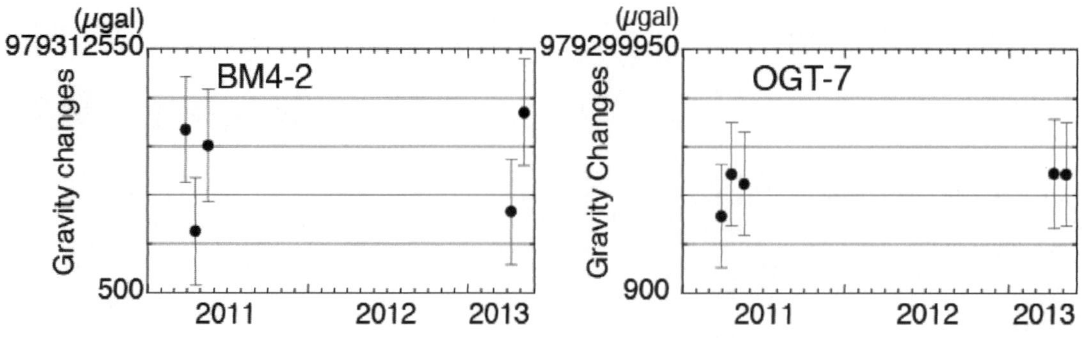

Figure 2. Absolute gravity data at the Ogiri Geothermal Field. BM4-2 is located in the reinjection area, and OGT-7 is located in the production area.

3.2. Effect of shallow ground water level changes

The effect of shallow ground water level changes can be estimated using the equation in Figure 3, assuming an infinite slab aquifer with a porosity of 10%.

The gravity effect that is caused by the ground water level change is estimated using the data from observation wells KE1-8 and KE1-12, shown in Figure 1. Figure 4 shows the ground water level change and converted gravity changes in KE1-8 and KE1-12. The ground water level changes were less than 1 m during these periods. These water level changes cause about 4 μgal, and the observed gravity changes cancelled out this effect.

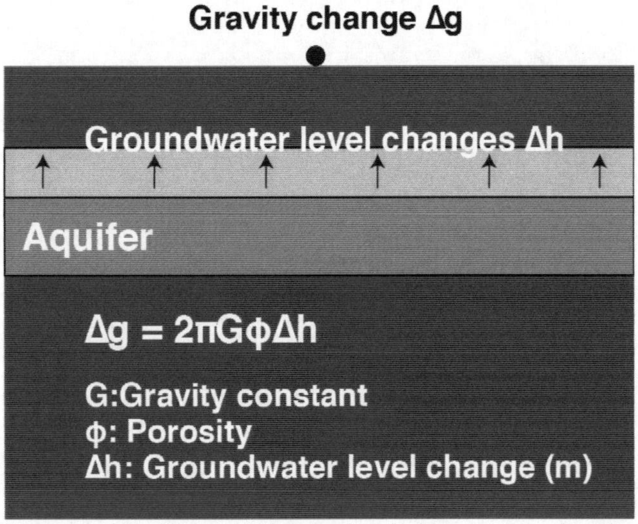

Figure 3. Gravity effect caused by the ground water level change, assuming the infinite slab aquifer.

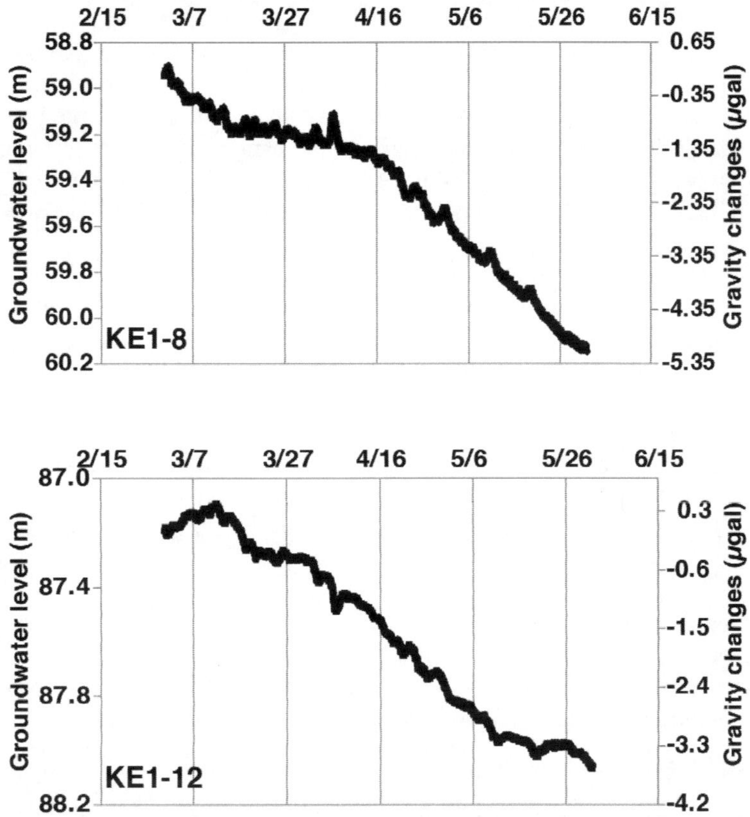

Figure 4. Ground water level changes from the beginning of March to the end of May 2013. The right axis shows the gravity changes converted from ground water level changes.

The effect of soil water content changes was estimated by GWATER[E] [11], using 10 m-mesh digital elevation maps (DEMs) and the precipitation at the power plant. The estimated result is depicted in Figure 5. Although the effect of this method was less than several μgal, it was calculated at each gravity station and thus removed this effect from the observed gravity changes.

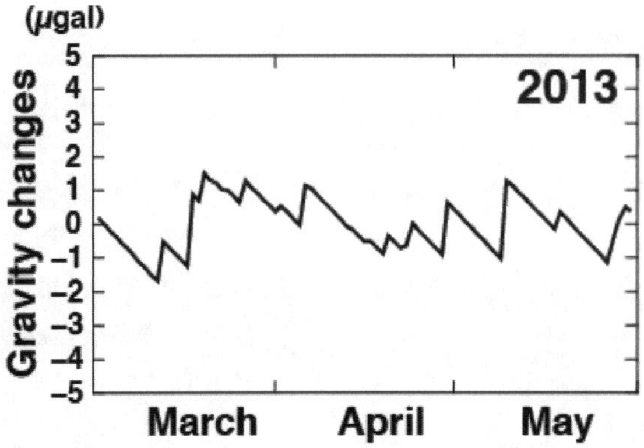

Figure 5. The effect of soil moisture changes from March to May 2013.

3.3. Gravity changes before and after the shutdown of production and reinjection

According to the results of the leveling surveys, the vertical ground movements are less than a few mm/year. Assuming a normal free-air gradient of –308.6 μgal/m, ground movement causes less than 3 μgal. This effect is very small in comparison with the observed gravity change. Consequently, the effect of vertical ground movement is negligible on the observed gravity in the short term.

The residual gravity changes (due to reservoir effects), taken as the difference between the observed and the calculated gravity effect of ground water level changes at each observation station, can be classified into four categories according to the characteristics of the trend (Figure 6). These categories are distributed in three areas as described below.

- Northern and Southern Area (Type A)
  This category is situated in the outside of the production and reinjection zone. Although the gravity changes are within 10 μgal, which is the nominal accuracy announced by the manufacturer, slight gravity changes can be seen before and after maintenance.

- Western Area (Type B and C)
  This category is located in the reinjection zone, in the western part of the observation area. Gravity decreases from the onset of stopping the reinjection, and then recovers after restarting the reinjection (Type B). However, the southern area of the reinjection zone kept the low gravity value (Type C).

- Eastern Area (Type D and E)
  This category is typical of stations located in the eastern production zone along the Ginyu fault, in the eastern part of the observation area. An increase of residual gravity was seen after the shutdown of the geothermal fluid production, and gravity decrease was observed after production was restarted (Type D). But in the southeastern part of the observation area the high gravity value persisted (Type E).

The data suggest there were increases of residual gravity (up to 30 μgal) in the production zone and decreases of residual gravity (up to 20 μgal) in the reinjection zone just after the production and reinjection stopped. Figure 7 shows the contour map of gravity changes from March to April 2013, before and after stopping the production and reinjection in the Ogiri geothermal power plant. The center of residual gravity decrease is located just to the southwestern part of the geothermal field. The center of this change is located in the western part of Ginyu fault. This gravity decrease indicated the reinjected hot water flowed outside the observation area. In the production area, a gravity increase (up to 30 μgal) was observed in the eastern part of the observation area. These data corresponded to the borehole pressure changes in the reservoir, so this gravity response indicated the recharge came from outside of the observation area.

Figure 8 shows the contour map of gravity changes from April to May 2013. The gravity changes recovered in this period. Some stations still remained at their former level in the southern area of the Ginyu fault. The center of the residual gravity increase is located in the reinjection area of the geothermal field. The gravity decrease is detected in the production area on the Ginyu fault.

These residual gravity changes are consistent with the pressure changes in the geothermal reservoir. Especially for the gravity stations along Ginyu fault, the response of the effect of stopping and restarting of the production and reinjection was quick. Thus, the effects of field operations can be isolated, even for fields with relatively low production rates, like the Ogiri Geothermal Field.

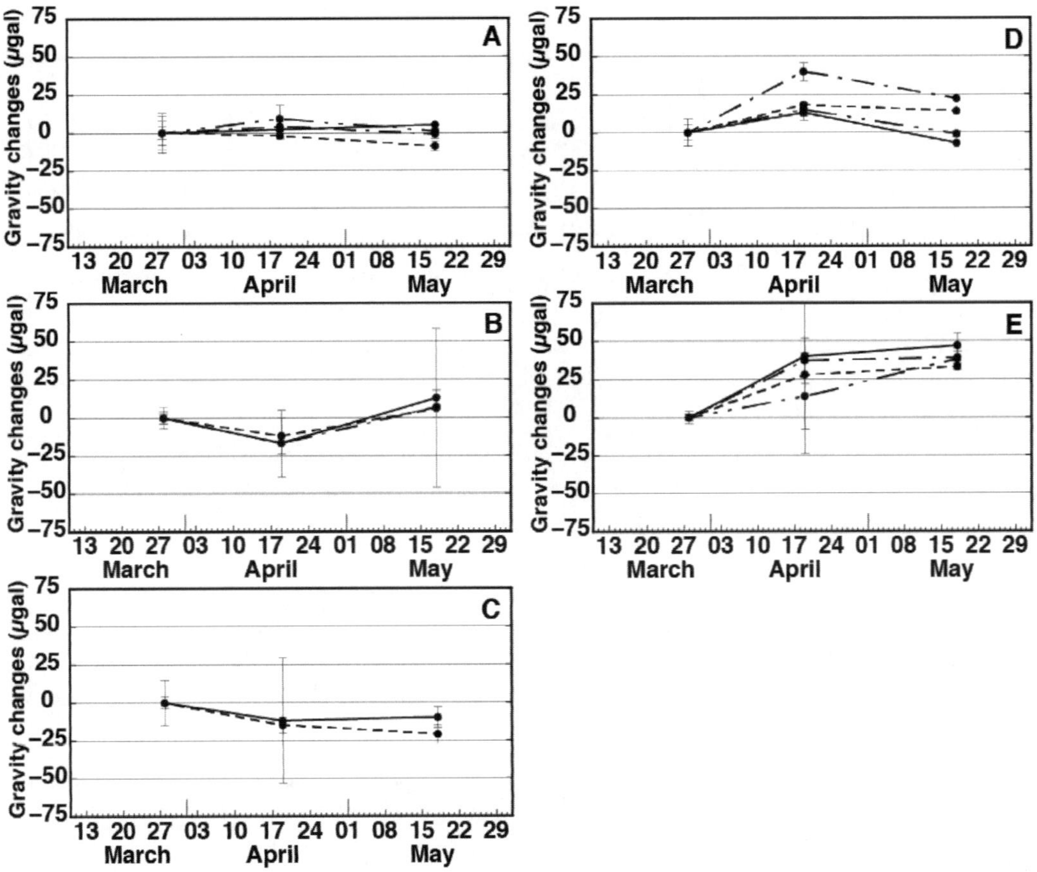

Figure 6. The effect of soil moisture changes from March to May 2013.

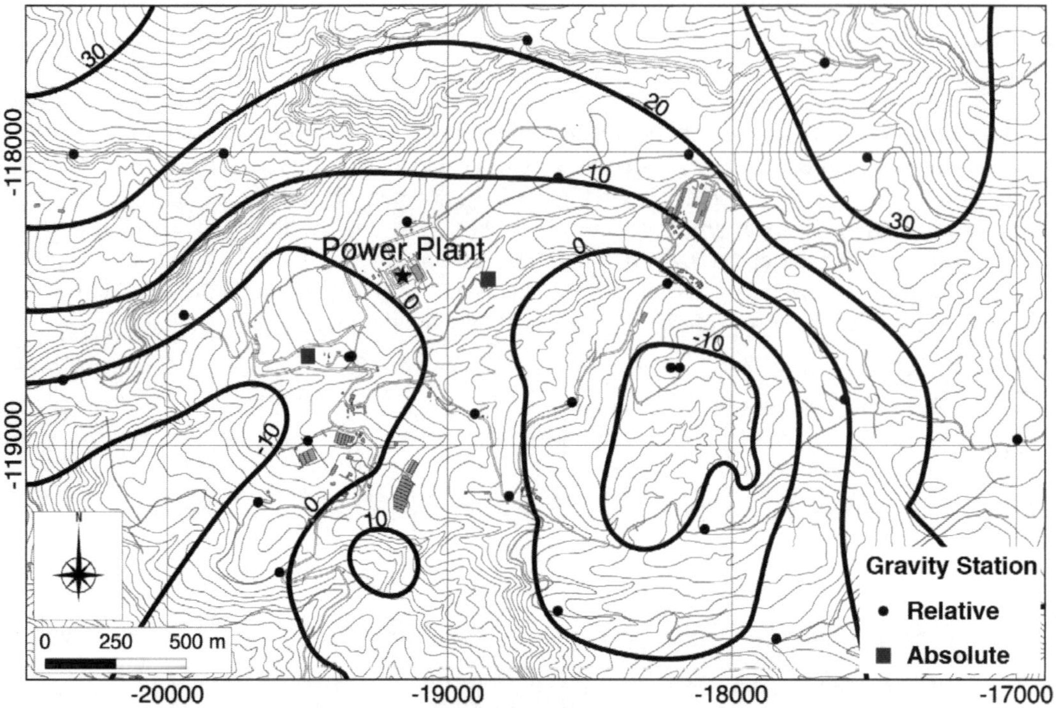

Figure 7. Distribution of gravity changes (μgal) from March to April 2013.

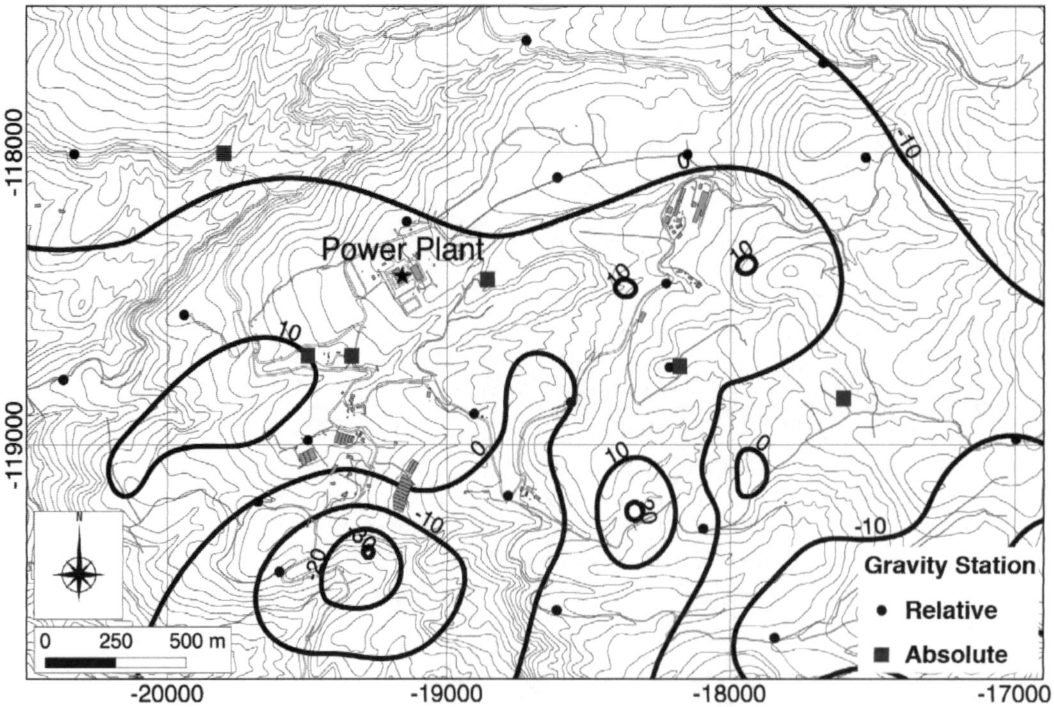

Figure 8. Distribution of gravity changes (μgal) from April to May 2013.

4. Conclusion

Hybrid microgravity monitoring was undertaken to detect the influence of a shutdown on the regular maintenance period around the Ogiri geothermal power plant. Noises that were caused by soil moisture changes and shallow groundwater level changes were removed. Based on these results, residual gravity increases of up to 30 μgal were detected in the production zone, and residual gravity decreases of up to 20 μgal were detected in the reinjection zone. These residual gravity changes are consistent with the changes in the geothermal reservoir. Thus, the effects of field operations can be isolated, even for fields with relatively low production rates, like Ogiri. This study indicates that repeated hybrid gravity measurement is an effective method not only to monitor the long-term mass balance but also to detect short-term reservoir behaviors.

References

[1] Allis R G and Hunt T M 1986 Analysis of exploitation induced gravity changes at Wairakei geothermal field *Geophysics* **51** pp 1647-1660.

[2] Nordquist G, Protacio J A P and Acuna A 2004 Precision gravity monitoring of the Bulalo geothermal field, Philippines: Independent checks and constraints on numerical simulation *Geothermics* **33** pp 37–56.

[3] Nishijima J, Saibi H, Sofyan Y, Shimose S, Fujimitsu Y, Ehara S, Fukuda Y, Hasegawa T and Taniguchi M 2010 Reservoir monitoring using hybrid micro-gravity measurements in the Takigami Geothermal Field, Central Kyushu, Japan *Proceedings World Geothermal Congress 2010*.

[4] Okubo S, Satomura M, Furuya M, Sun W, Matsumoto S, Ueki S and Watanabe H 2002 Grand design for the hybrid gravity network around the Mt. Fuji volcano *International Symposium on Geodesy, Kanazawa, Abstracts* pp 39–40.

[5] Furuya M, Okubo S, Sun W, Tanaka Y and Oikawa J 2003 Spatiotemporal gravity changes at Miyakejima volcano, Japan: Caldera collapse, explosive eruptions and magma movement: *Journal of Geophysical Research* **108** pp 1–17.

[6] Sugihara M and Ishido T 2008 Geothermal reservoir monitoring with a combination of absolute and relative gravimetry *Geophysics* **73** WA37-WA47.

[7] Thermal and Nuclear Power Engineering Society 2015 The situation and a trend of geothermal power generation 2014 pp 22–23 (in Japanese).

[8] Nishijima J, Oka D, Higuchi S, Fujimitsu Y, Takayama J and Hiraga N 2015 Repeat microgravity measurements using absolute and relative gravimeters for geothermal reservoir monitoring of the Ogiri Geothermal Power Plant, South Kyushu, Japan *Proceedings World Geothermal Congress 2015*.

[9] Micro-g LaCoste Inc. 2008 *A10 Portable Gravimeter User's Manual* pp 7-8.

[10] Matsumoto K, Sato T, Takanezawa T and Ooe M 2001 GOTIC2: A program for computation of oceanic tidal loading effect: *Journal of the Geodetic Society of Japan* **47** pp 243–248.

[11] Kazama T, Yamamoto K, Fukuda Y and Iguchi M 2014 The importance of hydrological disturbance corrections for relative gravity data: A case study at Sakurajima volcano, southern Kyushu, Japan *Journal of the Geodetic Society of Japan* **60** pp 73-89 (in Japanese with English abstract).

The Effectiveness of Hydrothermal Alteration Mapping based on Hyperspectral Data in Tropical Region

R R D Muhammad[1] and A Saepuloh[1,2]

[1]Geothermal Technology Study Program, Faculty of Mining and Petroleum Engineering, Institut Teknologi Bandung, Indonesia
[2]Geology Engineering Study Program, Faculty of Earth Sciences and Technology, Institut Teknologi Bandung, Indonesia

E-mail: reynorivelino@gmail.com

Abstract. Hyperspectral remote sensing could be used to characterize targets at earth's surface based on their spectra. This capability is useful for mapping and characterizing the distribution of host rocks, alteration assemblages, and minerals. Contrary to the multispectral sensors, the hyperspectral identifies targets with high spectral resolution. The Wayang Windu Geothermal field in West Java, Indonesia was selected as the study area due to the existence of surface manifestation and dense vegetation environment. Therefore, the effectiveness of hyperspectral remote sensing in tropical region was targeted as the study objective. The Spectral Angle Mapper (SAM) method was used to detect the occurrence of clay minerals spatially from Hyperion data. The SAM references of reflectance spectra were obtained from field observation at altered materials. To calculate the effectiveness of hyperspectral data, we used multispectral data from Landsat-8. The comparison method was conducted by comparing the SAM's rule images from Hyperion and Landsat-8, resulting that hyperspectral was more accurate than multispectral data. Hyperion SAM's rule images showed lower value compared to Landsat-8, the significant number derived from using Hyperion was about 24% better. This inferred that the hyperspectral remote sensing is preferable for mineral mapping even though vegetation covered study area.

1. Introduction
Hyperspectral remote sensing has the potential to provide the detailed physico-chemistry (mineralogy, chemistry, morphology) of the earth's surface. This capability is useful for mapping and characterizing the distribution of host rocks, alteration assemblages, and minerals. Furthermore, minerals can be uniquely identified based on their spectra [1], [2], [3].

Contrary to the multispectral sensors, the hyperspectral measures targets with high spectral resolution. In particular, the 2.0 to 2.5 μm (SWIR) spectral range covers spectral features of hydroxyl-bearing minerals, sulfates, and carbonates common to many geologic units and hydrothermal alteration [3].

2. Study Area
As an objective of this study, the effectiveness of hyperspectral remote sensing in tropical region was targeted. The Wayang Windu Geothermal Field (WWGF) in West Java, Indonesia was selected as the study area due to the existence of surface manifestation and dense vegetation environment. WWGF is

located about 40 km southwest from Bandung, surrounded by tea plantation with elevation about 1.700 to 2.100 meter above sea level. The land cover of study area is depicted by **Figure 1**. There are four portions detected by observing its color which are dark green, green, white-grey, and bright-white. Based on field observation, the dark green zone correlated with dense vegetation, green is medium vegetation, white-grey is rare vegetation such as plantation/farm-field, and bright-white is cloud covers. Mt. Malabar, Mt. Bedil, Mt. Wayang, and Mt. Windu were clearly observed at dense vegetation area as presented by 1 – 4, respectively.

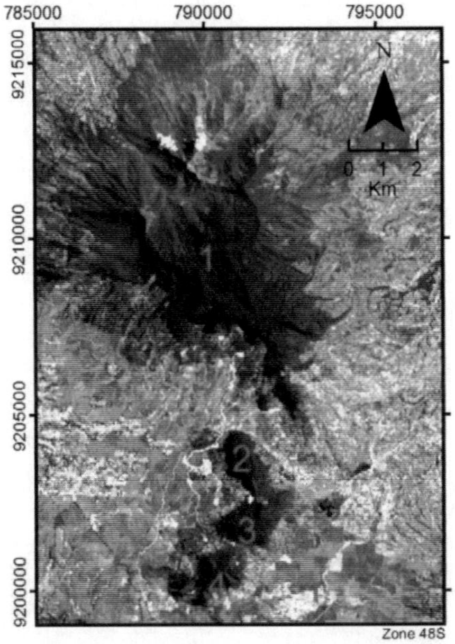

Figure 1. True Color Composite (TCC) of Landsat 8 reflectance data for R,G,B= Band 4, 3, 2. Mt. Malabar, Mt. Bedil, Mt. Wayang, and Mt. Windu presented by no. 1 to 4 were clearly observed at dense vegetation area (dark green portion).

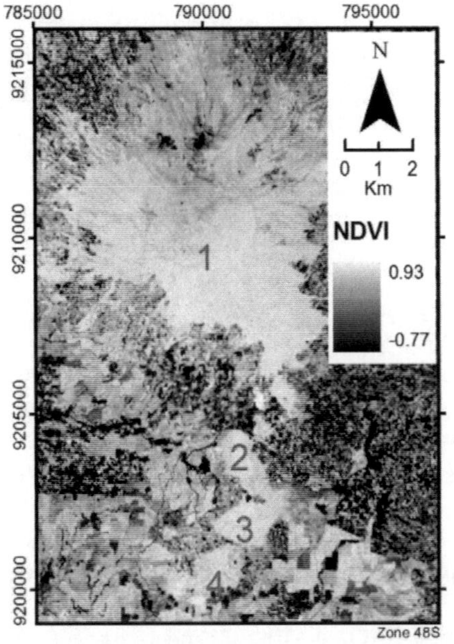

***Figure 2.** NDVI Map of study area. Confirming the vegetation covers at study area especially at Mt. Malabar, Mt. Bedil, Mt. Wayang, and Mt. Windu presented by no. 1 to 4 with dense vegetation covers as presented by NDVI value about 0.8.*

The NDVI (Normalized Difference Vegetation Index) was applied to the Landsat 8 reflectance data to calculate the vegetation cover quantitatively. This calculation was applied after atmospheric correction process as explained later. The NDVI is an index of plant "greenness" or photosynthetic activity. The green-less and unhealthy vegetation reflects more visible red light and less near infrared (NIR). Contrary, the green and healthy vegetation absorb most visible red and reflect high portion of near-infrared [5]. The NDVI is calculated by the following equation:

$$NDVI = \frac{(NIR-Red)}{(NIR+Red)} = \frac{(Band\ 5-Band\ 4)}{(Band\ 5+Band\ 4)} \qquad (1)$$

Applied on reflectance values of each band mentioned

Brighter pixel in NDVI means high value related to healthy vegetation and darker pixel means lower value related to unhealthy vegetation (Figure 2). The value range of NDVI is -1 to 1 where healthy vegetation generally falls between values of 0.20 to 0.80 [6]. The range NDVI value in this study area is -0.77 to 0.93 with average 0.72 (**Figure 3**). The average value shows that the study area has dense vegetation environment. In particular, confirming the visual analyses, Mt. Malabar, Mt. Bedil, Mt. Wayang, and Mt. Windu were observed at dense vegetation area with NDVI value about 0.80.

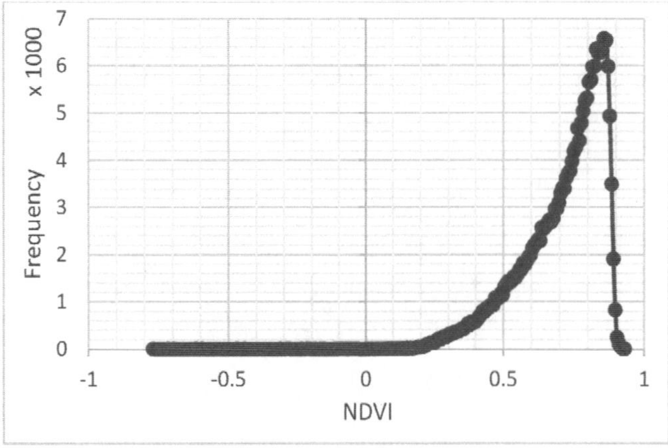

Figure 3. *Histogram of NDVI, the minimum, maximum, and mean values are -0.77, 0.93, and 0.72. The Skewness is located on right side of the data distribution, which is around the 0.20 – 0.80 threshold for healthy vegetation [6].*

3. Data

The satellite data used in this study are Landsat 8 data with acquisition time on September 10th 2013 (**Figure 1**) and Hyperion data with acquisition time on June 25th 2012 (**Figure 4**). There are 242 bands of Hyperion data in total, but 129 bands are not available. For producing reliable reference data, we excluded the uncalibrated bands and selected only 113 bands with high signal to noise ratio. The remote sensing data used in this study were provided by the http://earthexplorer.usgs.gov/.

Ten rock samples were collected at and around Mt, Malabar and identified as altered rocks by analyzing its spectra. Altered rocks are commonly identified with spectra absorption at 2.100 and 1.400 nm. The reflectance spectra of those samples were measured from visible to short-wave infrared regions (0.2 to 2.6 μm) using a SolidSpec-3700 spectrometer (Shimadzu) [1].

4. Methods

4.1. Atmospheric Correction

Remote sensing data on optical region carry the influence of some external factors, which may degrade the spectral features quality of ground objects. These factors are: (1) effects of the solar irradiance curve, and (2) atmosphere and topography. The effect of the solar irradiance curve arises from the fact that solar radiation intensity peaks at 0.48 μm and the radiation intensity drops off towards longer wavelengths, therefore the effects of solar irradiance is not uniform throughout the image. Atmospheric effects arise due to the fact that the image data are collected over a wide wavelength range, which includes atmospheric windows as well as atmospheric absorption and scattering. Topographic effects arise due to local landscape orientation [7].

In this study, The Fast Line-of-sight Atmospheric Analysis of Spectral Hypercubes (FLAASH) module in ENVI 5.2 software was applied to Hyperion and Landsat 8 data. The FLAASH is a first-principles atmospheric correction modeling tool for retrieving spectral reflectance from hyperspectral and multispectral radiance images. FLAASH can accurately compensate for atmospheric effects, it corrects wavelengths in the visible through near-infrared and shortwave infrared regions, up to 2.5 μm [8]. The parameters used for the calculation were Tropical Atmospheric Model, 40 km for Initial Visibility, and Rural Aerosol Model.

4.2. Spectral Angle Mapper

The Spectral Angle Mapper (SAM) method was used to detect the occurrence of clay minerals spatially. The SAM determines the similarity between target spectra (t) and each of the reference spectra ($\boldsymbol{r}$). This method treats both the unknown and reference spectra as vectors and calculates the spectral angle between them (**Figure 5**), a smaller angle means a closer match between those two spectra. The SAM is applied by the following equation [10]:

$$\alpha = cos^{-1}\left[\frac{\vec{t}*\vec{r}}{\|\vec{t}\|*\|\vec{t}\|}\right] \quad (2)$$

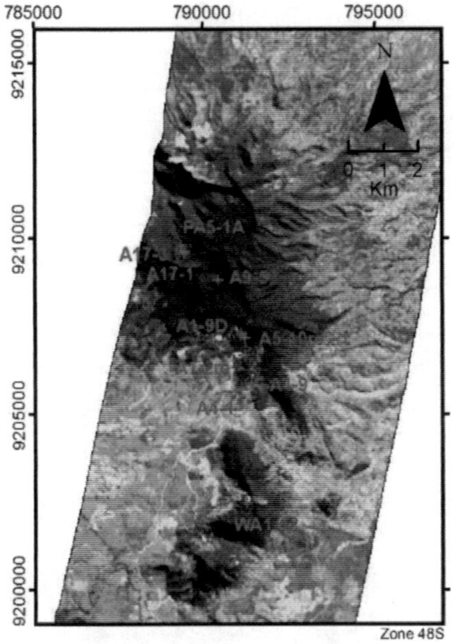

Figure 4. True Color Composite (TCC) of Hyperion reflectance data for RGB= Band 29, 20, 12. Locations of the measured field spectra were plotted in red plus symbol.

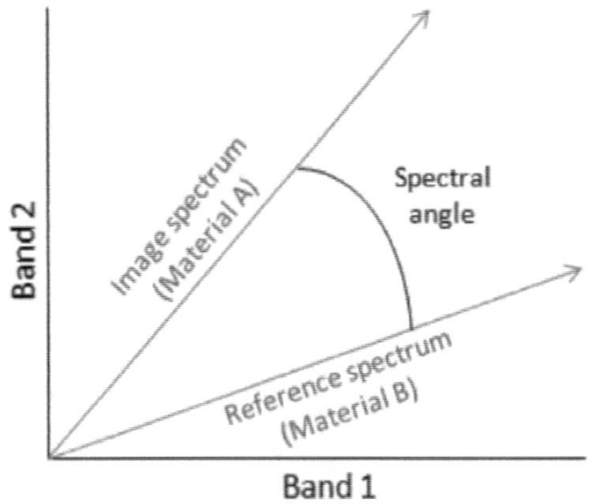

Figure 5. Illustration of Spectral Angle Mapper (SAM) method. A smaller spectral angle means a similarity between target (image) and reference spectra [9].

The SAM method was applied to both Hyperion and Landsat 8 data with ten references spectra obtained from field observation at ten altered material locations (

Figure 6). Following ten input spectra from spectrometer measurements, the SAM method produced ten rule images from Hyperion and Landsat 8 data.

5. Result

Pixel spectra from both Hyperion and Landsat 8 reflectance data were compared visually. Generally, there was a reflectance increase from 400 to 800 nm, then decrease in wavelength longer than 800 to 2.400 nm. The reference spectra showed the same reflectance increase, in particular from 200 to 1.000 nm. Then decrease in wavelength longer than 1.000 to 2.600 nm. Some local maxima were found at 1.300, 1.500, 1.650, and 2.100 nm for Hyperion spectra.

Compared to reference spectra, reflected pattern at 1.300, 1.650, and 2.100 nm were found to be similar in Hyperion spectra. The absorption pattern at 1.200, 1.350, and 1.750 nm were found to be similar, either. Relatively, compared to reference spectra, Landsat 8 data only showed similarity between 400 to 800 nm. Thus, visual analysis showed that Hyperion spectra showed higher similarity to reference spectra than Landsat 8 (**Figure 7**).

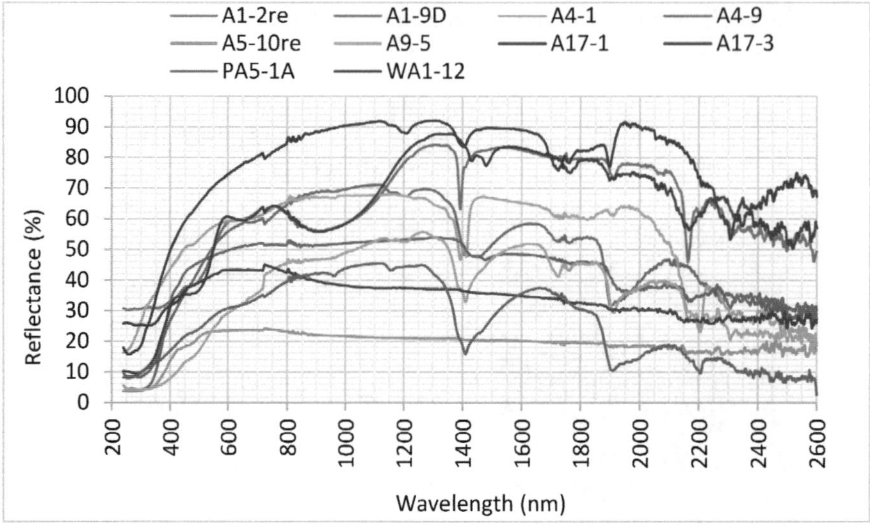

Figure 6. *Reference Spectra of SAM at ten observed locations, identified as altered materials as there was absorption at 1.400, 1.900, and 2.100 nm. The x-axis is the spectra wavelength ranging from 240-2.600 nm and y-axis is reflectance (%).*

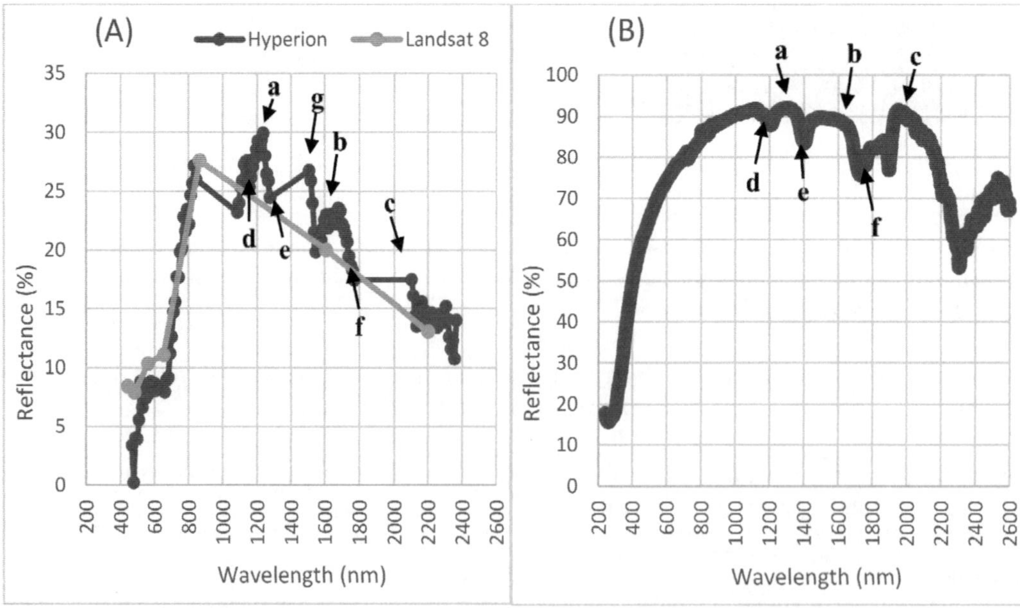

Figure 7. *Spectra of Hyperion and Landsat 8 reflectance data (A) and measured spectra of alteration rock (B) at A1-2re in Figure 4. Generally, there was a reflectance increase from 400 to 800 nm, then decrease in wavelength longer than 800 nm. Some local maxima were found at 1.300, 1.500, 1.650, and 2.100 nm for Hyperion spectra (a, g, b, and c). Hyperion spectra showed higher similarity to reference spectra than Landsat 8 spectra, as shown by reflection (a, b, and c) and absorption (d, e, and f) pattern.*

Quantitative analyses were conducted by pixel to pixel comparison between Hyperion and Landsat 8 SAM rule image. To compensate the spatial uncertainty, 3x3 pixels around the observed location were used for analyses. Therefore, each observed location was composed by nine pixels of rule image value to be compared. Pixel-1 is located at the top left, then pixel-9 at the bottom-right position (**Error! Reference source not found.**).

Lower rule image value means a high similarity of the reference to the target spectra, or high accuracy to detect target. Individually, most Hyperion SAM's rule images at observed locations showed lower value compared to Landsat 8 SAM's rule images, except for three pixels which were pixel-5 at location **A17-1**, and pixel-3 and pixel-6 at location **WA1-12 (Table 2).**

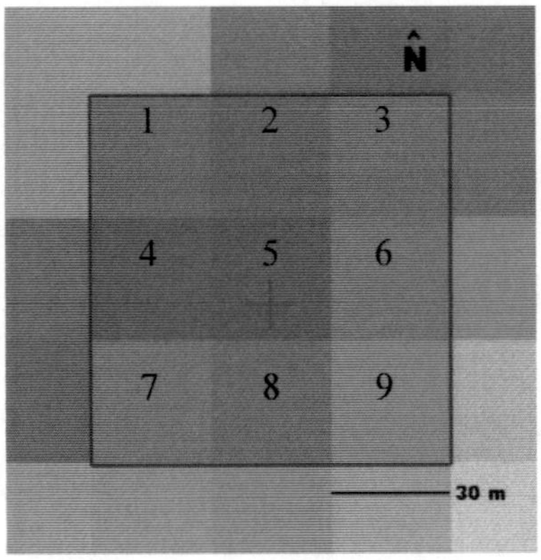

*Figure 8. Pixels value counting at SAM's rule image for Hyperion data at location **A1-2 re** (red plus symbol). To compensate the spatial uncertainty, 3×3 pixels window (blue rectangle) were used. Each observed location had nine pixels for comparison purpose.*

Table 1. *Pixels value of Hyperion and Landsat 8 SAM's rule images at ten observed locations. Lower rule image value means a high similarity to spectra target. Most Hyperion SAM's rule values showed lower than Landsat 8 SAM's, except three pixels location in highlighted yellow.*

Location		A1-2 re	A1-9D	A17-1	A17-3	A4-1	A4-9	A5-10	A9-5	PA5-1	WA1-12
Hyperion	**1**	0.47	0.45	0.67	0.59	0.51	0.60	0.70	0.49	0.46	0.38
Hyperion	**2**	0.45	0.52	0.62	0.55	0.53	0.60	0.71	0.49	0.46	0.22
Hyperion	**3**	0.47	0.56	0.55	0.48	0.54	0.60	0.72	0.48	0.45	0.24
Hyperion	**4**	0.45	0.46	0.66	0.60	0.49	0.57	0.68	0.50	0.46	0.43
Hyperion	**5**	0.44	0.45	0.62	0.53	0.49	0.57	0.70	0.49	0.46	0.30
Hyperion	**6**	0.48	0.51	0.57	0.43	0.50	0.57	0.70	0.48	0.46	0.36
Hyperion	**7**	0.48	0.45	0.65	0.58	0.44	0.56	0.69	0.50	0.46	0.53
Hyperion	**8**	0.47	0.43	0.65	0.53	0.43	0.56	0.71	0.49	0.47	0.47
Hyperion	**9**	0.50	0.50	0.65	0.46	0.44	0.57	0.71	0.48	0.46	0.51
Landsat 8	**1**	0.68	0.64	0.74	0.67	0.70	0.80	0.81	0.71	0.63	0.69
Landsat 8	**2**	0.54	0.73	0.62	0.55	0.67	0.77	0.78	0.71	0.63	0.29
Landsat 8	**3**	0.52	0.74	0.87	0.73	0.71	0.77	0.86	0.70	0.63	0.09
Landsat 8	**4**	0.72	0.56	0.80	0.64	0.61	0.76	0.80	0.71	0.64	0.73
Landsat 8	**5**	0.59	0.65	0.70	0.40	0.62	0.78	0.78	0.70	0.64	0.33
Landsat 8	**6**	0.54	0.71	0.86	0.81	0.69	0.70	0.86	0.70	0.64	0.29
Landsat 8	**7**	0.77	0.55	0.84	0.70	0.58	0.74	0.84	0.71	0.66	0.76
Landsat 8	**8**	0.70	0.62	0.84	0.59	0.55	0.74	0.85	0.72	0.65	0.71

	9	0.67	0.71	0.86	0.79	0.65	0.74	0.87	0.70	0.67	0.70

Averaging was applied within pixels window for each location. This comparison showed that Hyperion SAM's rule values at ten observed location were lower than Landsat 8. Obtaining the effectiveness of the two data, significant number (SN) was calculated and defined by the differential pixel means value ($\Delta\overline{P}$) and divided by Landsat 8 pixel means value. The calculation could be presented as follows:

$$SN = \frac{\overline{P}_{Landsat\ 8} - \overline{P}_{Hyperion}}{\overline{P}_{Landsat\ 8}} \qquad (3)$$

The Hyperion SAM's rule images showed better results for targeting altered rocks with its significant number ranging 15-31% and 24% average (**Table 2**). There are two craters among the sample locations, which are Burung Crater (**A1-9D**, **A5-10 re**) and Wayang Crater (**WA1-12**). The Lowest Hyperion SAM's Rule images is located at **WA1-12**, agree with the Lowest on Landsat 8's. This is relatively acceptable because of its lack vegetation cover as shown with NDVI value of 0.27. In general, higher NDVI value will result in higher rule images value. As a result, Hyperion data is 25% better than Landsat 8 at **WA1-12.**

The correlation between SN and NDVI value has not been clearly enough since SN at **WA1-12** is both lower and higher compared to Burung Crater (**A1-9D**, **A5-10 re**) with high NDVI value ~1. Another result that certify the justification is that the lowest SN is located at **A17-3** even its NDVI value showed relatively low (0.37).

Table 2. *Pixel means value of Hyperion and Landsat 8 SAM's rule images. Hyperion SAM's rule images showed better results for targeting clay mineral with its average significant number 24%.*

Location	Hyperion	Landsat 8	Significant Number	Description	NDVI
A1-2 re	0.47	0.64	27%	Altered Rock	0.67
A1-9D	0.48	0.66	27%	Burung-1 Crater	0.90
A17-1	0.63	0.79	21%	Altered Rock	0.69
A17-3	0.53	0.65	19%	Altered Rock	0.37
A4-1	0.49	0.64	24%	Altered Rock	0.68
A4-9	0.58	0.76	24%	Altered Andesite	1.01
A5-10 re	0.70	0.83	15%	Burung-2 Crater	1.02
A9-5	0.49	0.71	31%	Altered Andesite	1.02
PA5-1	0.46	0.64	28%	Altered Rock	0.92
WA1-12	0.38	0.51	25%	Wayang Crater	0.27
AVERAGE	0.52	0.68	24%		

6. Conclusions

Hyperspectral data showed better absorption similarity to the reference spectra than multispectral data. Hyperion with narrow spectral bands delivered the relative similar reflection and absorption pattern with the reference spectra. Based on pixels window comparison, most Hyperion SAM's rule images showed lower values. The significant number using Hyperion was about 24% better than Landsat 8. This inferred that the hyperspectral remote sensing is preferable for mineral mapping even though vegetation covered study area.

The Lowest Hyperion SAM's Rule images is located at **WA1-12**, agree with the Lowest on Landsat 8's. In general, higher NDVI value will result in higher rule images value. The correlation between SN and NDVI value has not been clearly enough since SN at **WA1-12** is both lower and higher compared to Burung Crater (**A1-9D**, **A5-10 re**). Another result that certify the justification is that the lowest SN is located at **A17-3** even its NDVI value showed relatively low.

Acknowledgement
The authors sincerely thank to "Beneficial and Advanced Geothermal Use System (BAGUS)" project of the "Science and Technology Research Partnership for Sustainable Development (SATREPS)" for providing license of ENVI 5.2 software, Geothermal Capacity Building Program (GEOCAP) and LPPM-ITB for providing field observation support, and Arif Susanto, M.T. for collecting and sharing the rock samples at and around Mt, Malabar. The remote sensing data used in this study were provided by the http://earthexplorer.usgs.gov/.

References

[1] Saepuloh, A., Koike, K., Omura, M., Iguchi, M., & Setiawan, A. (2010). SAR- and gravity change-based characterization of the distribution pattern of pyroclastic flow deposits at Mt. Merapi during the past ten years. Bulletin of Volcanology, Vol. 72, No.2, 221-232.

[2] Satpathy, R., Singh, V. K., Parveen, R., & Jeyaseelan, A. T. (2010). Spectral Analysis of Hyperion Data for Mapping the Spatial Variation of AL+OH Minerals in a Part of Latehar & Gumla District, Jharkhand. Journal of Geographic Information System 2, 210-214.

[3] Kruse, F. A. (2003). Mineral Mapping with AVIRIS and EO-1 Hyperion. JPL Airborne Geoscience Workshop. Pasadena, California, USA.

[4] Ganesh, B. P., Aravindan, S., Raja, S., & Thirunavukkarasu, A. (2012). Hyperspectral Satellite Data (Hyperion) Study on Banded Magnetite Quartzite in Tamil Nadu, India. Arab J Geosci.

[5] Suryantini, Wibowo, H., Rahman, K. R., & Woldai, T. (2013). Application of Normalized Different Vegetation Index (NDVI) Method to Identify Thermal Anomaly Area from Remote Sensing. The 13th Indonesia International Geothermal Convention & Exhibition. Jakarta.

[6] Weier, J., & Herring, D. (1999). Measuring Vegetation (NDVI & EVI). Earth Observatory, NASA, USA.

[7] Gupta, R. P. (2003). Remote Sensing Geology, 2nd edition. New York: Springer Publication.

[8] Exelis Visual Information Solutions. (2014). ENVI Classic Tutorial: Using ASTER Data with FLAASH.

[9] Exelis Visual Information Solutions. (2014). Hyperspectral Analysis: SAM and SFF Tutorial.

[10] Kruse, F. A., Lefkoff, A. B., Boardman, J. W., Heidebrecht, K. B., Shapiro, A. T., Barloon, P. J., & Goetz, A. F. (1993). The Spectral Image Processing System (SIPS) Interactive Visualization and Analysis of Imaging Spectrometer Data. Remote Sensing Environment 44, 145-163.

IOP Conf. Series: Earth and Environmental Science **42** (2016) 012006 doi:10.1088/1755-1315/42/1/012006

Numerical Simulation for Natural State of Two-Phase Liquid Dominated Geothermal Reservoir with Steam Cap Underlying Brine Reservoir

Heru Berian Pratama, Nenny Miryani Saptadji

Geothermal Master Degree Program, Institut Teknologi Bandung.
Jalan Ganesha No.10, Bandung, Indonesia.

hb.pratama@gmail.com

Abstract. Hydrothermal reservoir which liquid-dominated hydrothermal reservoir is a type of geothermal reservoir that most widely used for power plant. The exploitation of mass and heat from the geothermal fluid will decrease the pressure in the reservoir over time. Therefore the pressure drop in the reservoir will have an impact on the formation of boiling zones or boiling will increase. The impacts are an increase in the fraction of steam, dryness, in the reservoir and with good vertical permeability will form a steam cap underlying the brine reservoir. The two-phase liquid dominated reservoir is sensitive to the porosity and difficult to assign average properties of the entire reservoir when there is boiling zone in some area of the reservoir. These paper showed successful development of two-phase liquid dominated geothermal reservoir and discussed the formation of steam cap above brine reservoir through numerical simulation for state natural conditions. The natural state modeling in steam cap shows a match with the conceptual model of the vapor-dominated developed. These paper also proofed the presence of transition zone, boiling zone, between steam cap and brine reservoir.

1. Introduction

Numerical model of geothermal system developed by conceptual model that develops from geosciences data and the reservoir engineering data is important to describe the condition in the reservoir. The first step is to match the natural state. The natural state of a geothermal system refers to the condition of the system before any form of exploitation. The relevant data are the natural temperatures and pressure and the amount of surface discharge as both heat and mass. The research of two-phase liquid-dominated reservoir are less than both single-phase (liquid) reservoir and vapor-dominated reservoir. These reservoirs is interesting and difficult to model because it contains a vapor-dominated zone underlain by a liquid-dominated region [1]. It has similiarity with Wayang Windu – Indonesia [2]. The main objective of this work is to develop of two-phase liquid-dominated geothermal reservoir with numerical study and achieve their natural conditions. The model should be described for all major physical process such as mass transport, heat transfer, boiling and condensation process that take place in these system.

2. Conceptual Model

The model for this study is based on liquid dominated that has a steam cap above liquid reservoir area such. Therefore the reservoir model needs a conceptual model of synthetic liquid-dominated reservoir with steam cap underlying brine reservoir. The conceptual model in these synthetic model developed based on four geothermal fields such as Wairakei, Tongonan, Awibengkok and Wayang Windu at production stage.

IOP Conf. Series: Earth and Environmental Science **42** (2016) 012006 doi:10.1088/1755-1315/42/1/012006

The synthetic conceptual model of liquid-dominated with steam cap at top reservoir shown in Figure 1 and has a reservoir characteristics shown in Table 1. These studies used six exploration wells (XXA-1, XXB-1, XXC-1, XXD-1, XXE-1 and XXF-1) and is shown in Figure 1 and Figure 2. These wells delineate the reservoir. The pressure and temperature profile of XXX-1 and XXB-1 show a high temperature well, whereas the rest of the wells have a medium temperature well.

Table 1. The Characteristics of synthetic geothermal reservoir.

Area of Reservoir	Proven Area (km^2)	Temperature (°C)	Pressure (bar)	Thickness (m)
Steam Cap	13	240	34	500-1000
Brine Reservoir	23	240-320	55	1400-1500

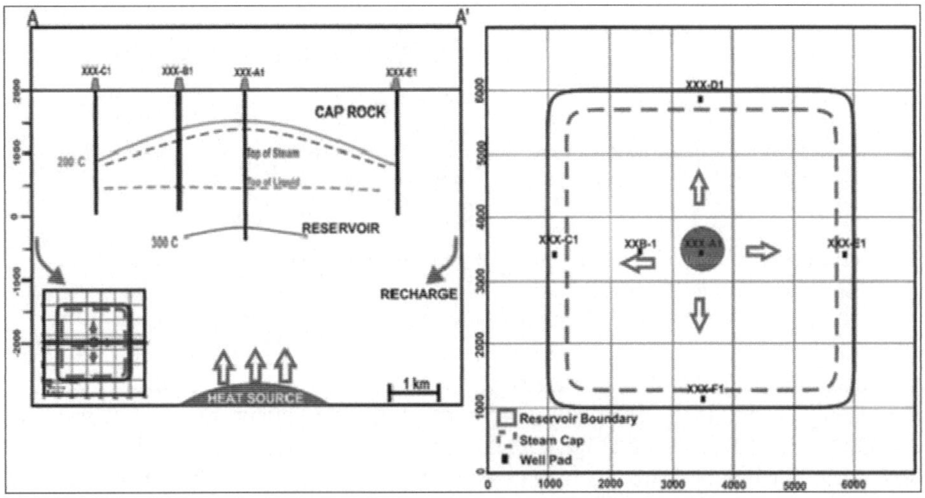

Figure 1. The conceptual model of synthetic liquid-dominated geothermal reservoir with steam cap underlying brine reservoir

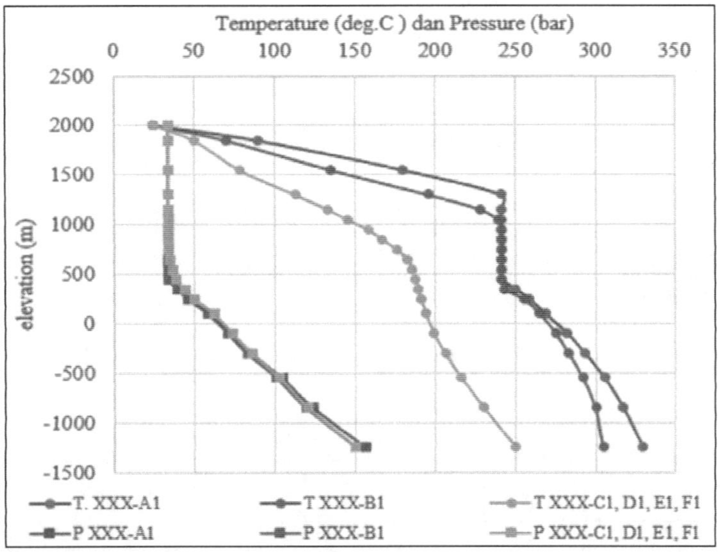

Figure 2. Pressure and temperature profile at two-phase liquid dominated geothermal wells with steam cap above brine reservoir.

3. Computer Model

The methodology of this study is shown in Figure 3. The first step is reviewed liquid-dominated geothermal reservoir then build a synthetic conceptual model. The second is developing computer model in the simulator with applying EOS, mesh type and geometry model. The third step is applying both of boundary conditions and rock properties into computer model, then running the model to achieve their natural conditions based on pressure and temperature matching. The data input is very tricky because there are a lot of unknown information in subsurface Material data should be updated to gain natural condition of model computer hence the reservoir model computer can represent their natural conditions.

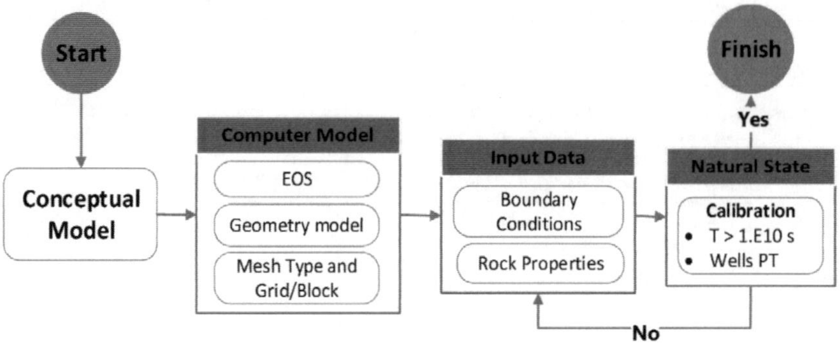

Figure 3. Methodology of natural state liquid-dominated geothermal reservoir with steam cap underlying brine reservoir.

3.1. *Grid System*

The Computer model has 7x7 km2 with reservoir area around 4.5x4.5 km2. Therefore the mesh type applied into a computer model is rectangular and a distributed parameter approach is used. The refine mesh placed in the productive area that have a high temperature and good permeability. The fine mesh has propose of the block is only trough by one well. The fine mesh is 200x200 m and the course is 1000x1000 m and placed on the edge of the computer model as boundary conditions. Those grid-blocks shown in Figure 4.

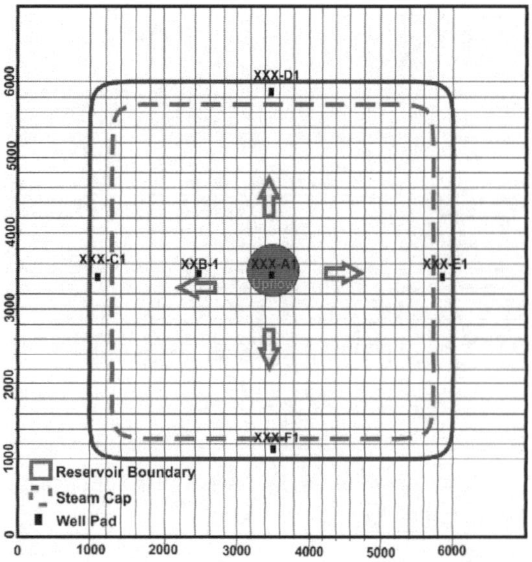

Figure 4. Cross section of the top section which is divided by a grid.

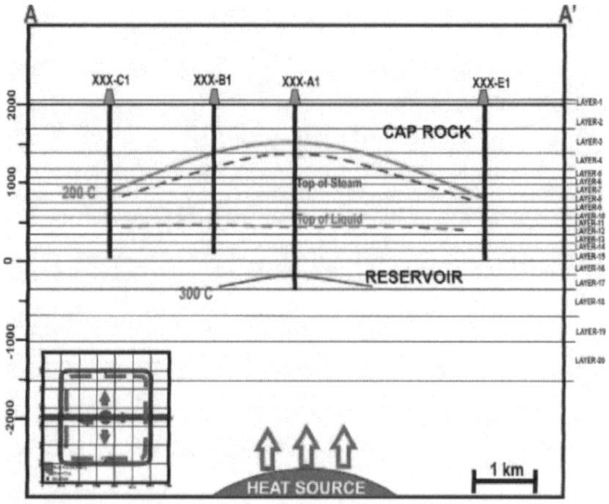

Figure 5. Distribution of layer of the conceptual model based on productive area.

The vertical extend of the model based on proposed synthetic conceptual model. It's divided into 20 layers and the thinner placed (100 m) into the area where a steam cap estimated in the top of reservoir because of lower density than brine otherwise the thickness of liquid reservoir is 200 m. The thicker layer (500 m) placed into the bottom of the model as boundary conditions for put heat source, therefore there is no change of pressure and temperature. Those shown in Figure 5.

3.2. Rock Properties

The characteristics data that represent geothermal system such as a heat source, cap rock and reservoir have been assigned in the computer model and set to reach the natural state condition. The properties of the rocks that represent geological conditions in the model is shown in Table 2. These material data assigned to each block in the computer model shown in Figure 6. Material data that has a good permeability assign into a productive area and material data that has tight permeability used as outside boundary.

Table 2. Material data for computer model.

Material Type Legend	Rock Density (kg/m^3)	Porosity	Permeability (m^2)	
			XY	Z
Atmosphere	2600	0.99	1E-10	1E-12
Ground Water	2500	0.02	2E-18	2E-18
Caprock	2600	0.05	1E-18	1E-18
Boundary1	2600	0.001	1E-19	2E-19
Boundary2	2600	0.01	1E-20	1E-20
Heat source	2650	0.07	1E-14	1E-15
Reservoir1	2500	0.25	1E-13	5E-14
Reservoir2	2550	0.2	8E-14	4E-14
Reservoir3	2600	0.15	6E-14	3E-14
Reservoir4	2600	0.15	5E-14	2E-14
Reservoir5	2600	0.1	3E-14	1E-14
Reservoir6	2600	0.1	9E-15	6E-15
Reservoir7	2600	0.1	7E-15	3E-15
Reservoir8	2600	0.09	5E-15	2E-15
Reservoir9	2500	0.05	3E-17	1E-17

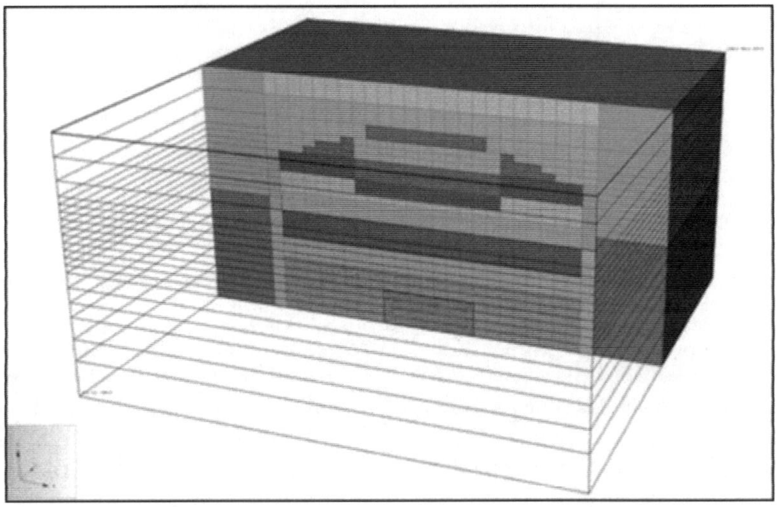

Figure 6. 3D block model at computer model.

Each block is connected is connected with each other block therefore the diversity of permeability, porosity, water saturation in the reservoir and fluid, both laterally and vertically, were taken into account. The magnitude of the pressure, temperature, water and steam saturation in each block and the mass flow rate from one block to another block for a range of time is calculated by using the simulator. Thus the results of the calculation can illustrate the changes in pressure and temperature versus depth, both in well and in other places as well as other changes in reservoir characteristics versus time.

3.3. Initial and Boundary Conditions

The boundary conditions such as outside, heat source and the atmosphere was assigned in this model. The objective is to give initial conditions from a model therefore the simulator can calculate every thermodynamic properties in each block. Each block a boundary condition used 1E+30 for volume factor, hence it give stationary of pressure and temperature versus time. The assumptions of boundary condition are:

1) The top of layer is atmosphere layer. In this layer, each block is considered to have similar properties with typical atmosphere condition (1 atm and 25°C).
2) The Outside layer is the boundaries representing a surrounding environment.
3) The heat source layer is located in the bottom of model. The thickness is 500 m.

4. Natural State

Pressure and temperature profile at natural state is shown at Figure 7. It shows a perfectly match between a model output and actual well data. The pressure and temperature in steam cap close to steam-static pressure and slightly above temperature saturation. The pressure and temperature in brine reservoir have a hydrostatic pressure and temperature below temperature saturation. It is obvious there are transition zones between steam cap and brine reservoir.

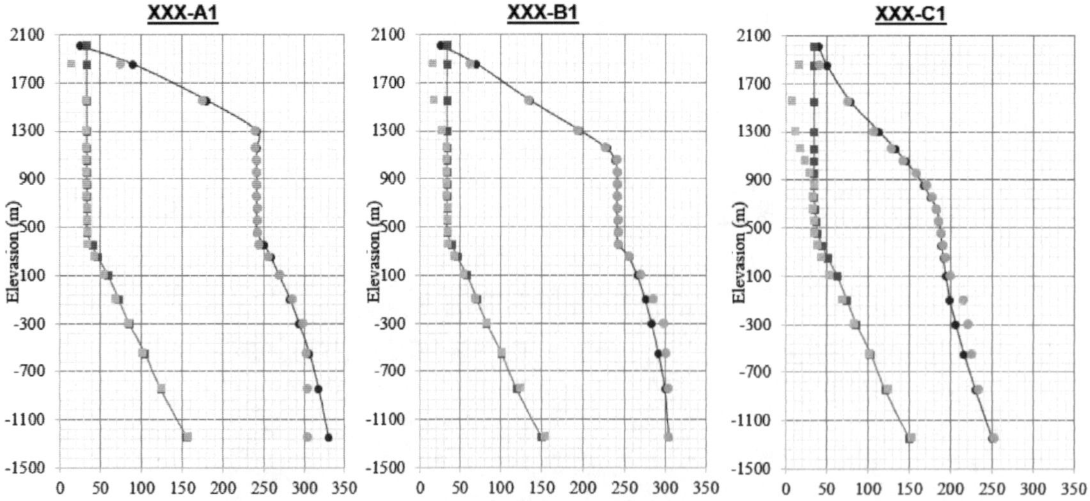

Figure 7. Matching pressure and temperature data between model computer and actual well data.

The cross-section of the natural state model of liquid dominated with a steam cap at the top reservoir is shown at Figure 8. This cross section of natural state has similarity with a conceptual model of a vapor dominated system of Kamojang Geothermal Field proposed by Hochstein [3] but have a thinner steam cap. This natural state of the steam cap has a similarity with a conceptual model of vapour dominated proposed by White et al [4] and enhanced by D'amore and Trusdell [5].

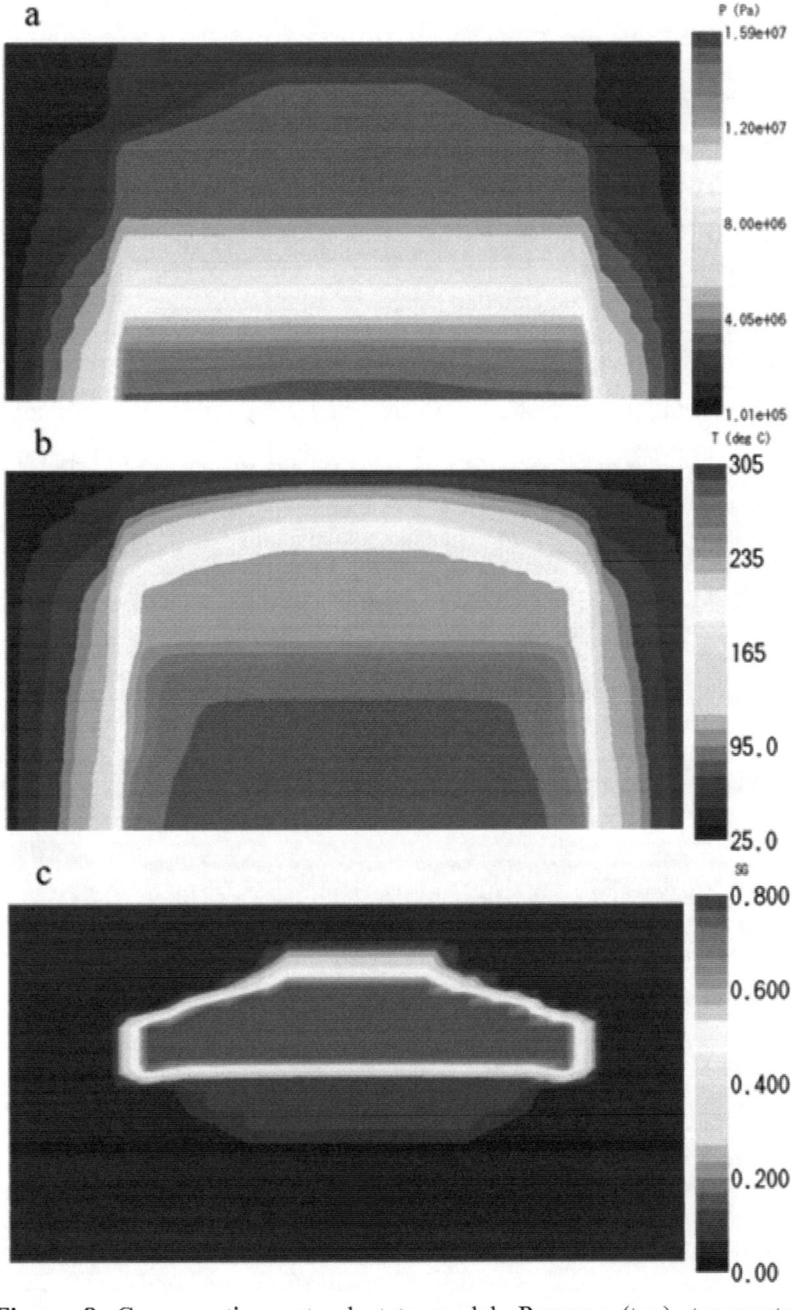

Figure 8. Cross section natural state model. Pressure (top), temperature (middle) and gas saturation (bottom) simulation results (Y-axis = 3500).

The conductive heat transfer occurs from heat source into the reservoir and convective entire steam cap reservoir (Figure 9a). Hot water is heated by hot reservoir rock and will go up because high temperature fluid has a lower density. Throughout the heat transfer, the boiling zone will segregate vapour and liquid phase. The greater boiling zone, the greater of a steam cap will be formed. The steam has formed will be moved laterally through the reservoir and spread, a steam cap formed (Figure 9b and Figure 9c). The steam saturation formed at steam cap is 80% and close to value of 85% of vapour dominated geothermal field.

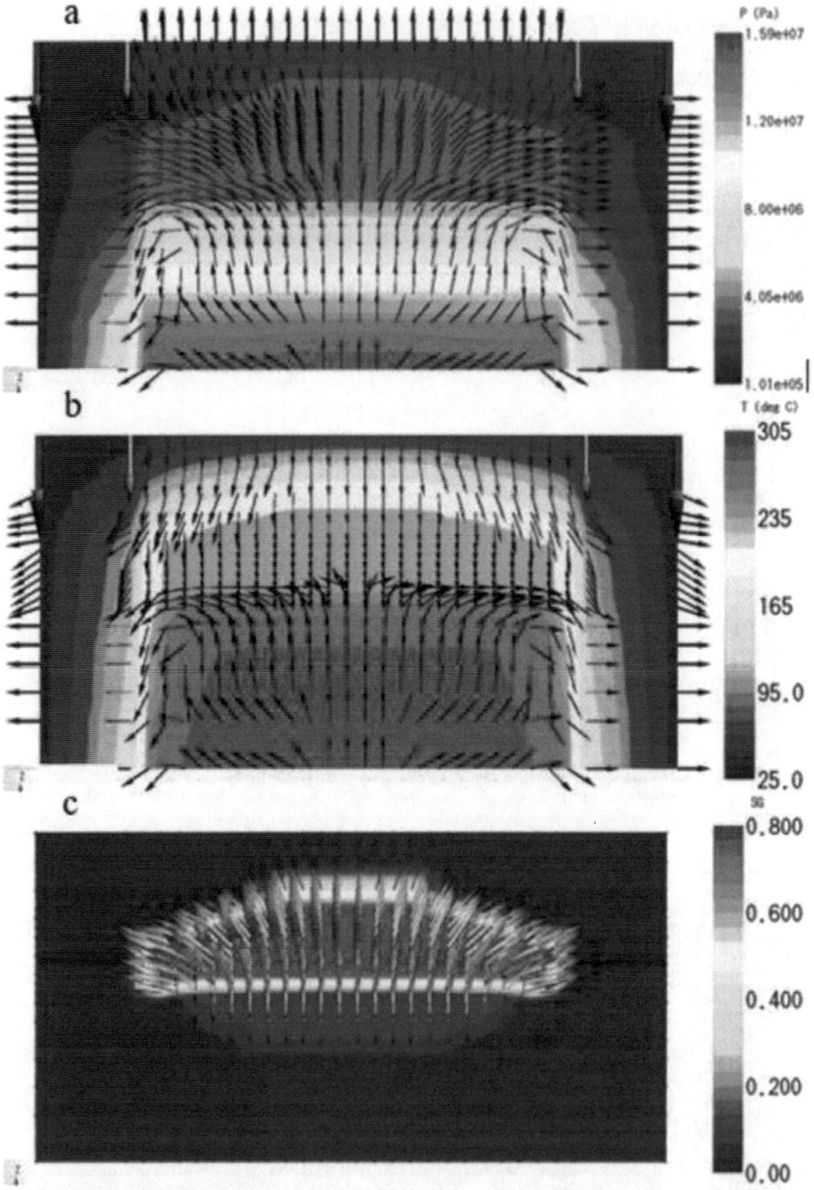

Figure 9. Heat transfer pattern (top), the mass transfer of water (middle) and the mass transfer of steam (bottom) on cross section natural state model (Y-axis = 3500).

The steam mass transfer is hindered by impermeable layers above the top reservoir and the heat loss from steam take place. Therefore steam will be condensed and move downward by the influence of gravity. The concept of steam and condensate movement is counterflow heat transfer. The steam move upward while the condensate move downward. This counterflow heat transfer occurs continuously hence the pressure gradient will show the steam gradient pressure (34 bar) and uniform temperature 245°C throughout the whole steam cap.

At the natural state mode, the highest permeability found on top of the reservoir and surrounding by poor permeability at cooler rock outside the steam reservoir. Therefore the natural state model output has similarities with the conceptual model from D'Amore and Truesdell [5].

Studies of the reservoir mineralogy indicate that all of the four vapor-dominated fields ware at one time liquid dominated [6] and have somehow "boiled dry" to produce current state. The previous study [7] shown the expansion steam cap with vertical extends occur in the liquid dominated with moderate production. If these production applied in the long time, hence it will create a thicker steam cap. Therefore the liquid dominated reservoir with the steam cap underlying brine reservoir will evolve to vapor-dominated geothermal reservoir.

5. Conclusion

The model of liquid dominated reservoir with steam cap underlying brine reservoir successfully developed and it showed a good match with actual data after tested and validated. There is an evidenced of transition zone between steam cap and brine reservoir. The steam cap has a similarity with the model of vapor-dominated reservoir model.

References

[1] O'Sullivan, M.J., Karsten, P., Lippmann, M.J., 2000, Geothermal Reservoir Simulation: The State-of-Practice and Emerging Trends, World Geothermal Congress 2000, Kyusu-Tohoku, Japan.

[2] Mulyadi dan Ashat, A. 2011, Reservoir Modelling of the Northern Vapor-Dominated Two-Phase Zone of the Wayang Windu Geothermal Field, West Java, Indonesia, Workshop on Geothermal Reservoir Engineering 36th, Stanford University, Stanford, California SGP-TR-191.

[3] Hochstein, M. P., 1990, Classification and assessment of geothermal resources. In ''Small Geothermal Resources: A Guide to Development and Utilization'' (M. H. Dickson and M. Fanelli, eds.). UNITAR/UNDP Centre for Small Energy Resources, Rome.

[4] White, D.E., Muffler, L.J.P., Truesdell, A.H., 1971. Vapor-dominated hydrothermal systems compared with hot-water systems. Econ. Geol.

[5] D'Amore, F., Truesdell, A.H., 1979, Models for steam chemistry at Larderello and The Gesyers. Proceedings, 4th Workshop on Geothermal Reservoir Engineering, Stanford University. Grant, M.A., Donaldson I.G., Bixley P.F (1st edition 1982, 2nd edition 2011): *Geothermal Reservoir Engineering*, Academic Press, Oxford.

[6] Allis, R, 2000, Insights on the Formation of Vapor-Dominated Geothermal Systems, World Geothermal Congress 2000, Kyusu-Tohoku, Japan.

[7] Pratama, H.B. and Saptadji, (2015), Study of Production-Injection Strategies of Synthetic Geothermal Reservoir Liquid-Dominated Model with Numerical Simulation, 37th NZGW, Rotorua, New Zeland.

Performance analysis of mineral mapping method to delineate mineralization zones under tropical region

M H Wakila[1], A Saepuloh[2], M N Heriawan[1], A Susanto[2]

[1]Faculty of Mining and Petroleum Engineering, ITB, Bandung, Indonesia
[2]Faculty of Earth Sciences and Technology, ITB, Bandung, Indonesia

E-mail: wakilamuh.hardin@yahoo.com

Abstract. Geothermal explorations and productions are currently being intensively conducted at certain areas in Indonesia such as Wayang Windu Geothermal Field (WWGF) in West Java, Indonesia. The WWGF is located at wide area covering about 40 km^2. An accurate method to map the distribution of heterogeneity minerals is necessary for wide areas such as WWGF. Mineral mapping is an important method in geothermal explorations to determine the distribution of minerals which indicate the surface manifestations of geothermal system. This study is aimed to determine the most precise and accurate methods for minerals mapping at geothermal field. Field measurements were performed to assess the accuracy of three proposed methods: 1) Minimum Noise Fraction (MNF), utilizing the linear transformation method to eliminate the correlation among the spectra bands and to reduce the noise in the data, 2) Pixel Purity Index (PPI), a designed method to find the most extreme spectrum pixels and their characteristics due to end-members mixing, 3) Spectral Angle Mapper (SAM), an image classification technique by measuring the spectral similarity between an unknown object with spectral reference in n-dimension. The output of those methods were mineral distribution occurrence. The performance of each mapping method was analyzed based on the ground truth data. Among the three proposed method, the SAM classification method is the most appropriate and accurate for mineral mapping related to spatial distribution of alteration minerals.

Keywords: mineral mapping, geothermal manifestations, Minimum Noise Fraction, Pixel Purity Index, Spectral Angle Mapper.

1. Introduction

The remote sensing technology is an effective and efficient tools to identify physical properties of rocks at surface such as alteration [1, 2], moisture and texture [3], and roughness [4],[5]. The main advantage to use the remotely sensed data is to reduce the field investigation time or to define the prospect for geophysical surveys at wide coverage area [6].

Mineral mapping method is useful in geothermal potential expansion to determine the alteration mineral spatially. We selected the Wayang Windu Geothermal Field (WWGF) because the area is composed by various surface manifestations with heterogeneity of mineral compositions covering wide areas.

Therefore, precise and accurate mapping method is necessary to determine the distribution of these minerals. This study is aimed to determine the most precise and accurate methods for mineral mapping under tropical conditions for geothermal exploration purposes.

We used Hyperion data in this study. Hyperion is a hyperspectral sensor covering the 0.4 to 2.5 μm spectral range with 242 spectral bands at approximately 10 nm spectral resolution and 30 m spatial resolution from a 705 km orbit [7]. Hyperion is a pushbroom instrument, capturing 256 spectra each with 242 spectral bands over a 7.5 km-wide swath perpendicular to the satellite motion with 160 km path length [8]. The system was utilized by two spectrometers; a visible/near infrared (VNIR) spectrometer (approximately 0.4 – 1.0 μm) and a short-wave infrared (SWIR) spectrometer (approximately 0.9 – 2.5 μm).

2. Study Area

The WWGF is composed by the two small lava domes without eruption history [9]. The field is located at Pengalengan, approximately 35 km to the south of Bandung, the capital of West Java, Indonesia (see Figure 1). The elevation lies between 1500 – 2100 m above sea mean level [10]. The geothermal systems of WWGF are associated with the dominance of steam fluid caps southern part of Mt. Malabar, a large stratovolcano composed of lava, breccia, lava and dacitic tuffs, and andesite.

There are about 9 thermal manifestations including fumaroles, hot springs, mud pools and altered ground situated along the southern slopes of the Mt. Malabar. Fumaroles are located between Mt. Wayang and Mt. Windu with temperatures between 93°C to 96°C and indicated slightly superheated compared to the normal water boiling point at altitudes 2000 m. Hot springs located in altitudes between 1495 m and 1985 m with temperatures from 41°C to 88°C. Water discharge as hot springs are bicarbonate type, except for hot springs located near the fumaroles as acid sulphate water [11].

The lithology at WWGF consists of four main units, from the young to old units: Malabar-Bedil andesite, Kancana lavas, Malabar-Tilu volcanic, and the younger Wayang volcanic units (see Figure 2).The geothermal system at WWGF is classified as transition between vapor and liquid dominated systems [13].

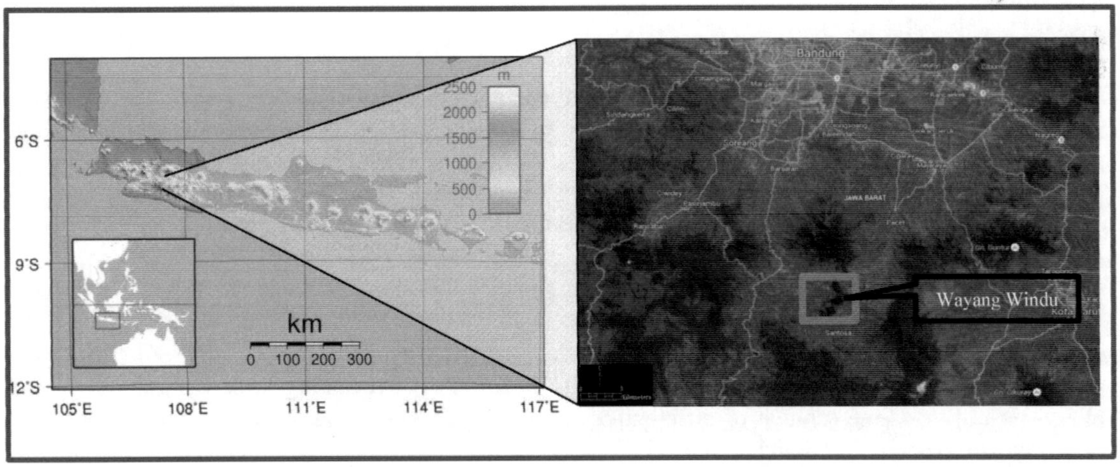

Figure 1. Study area at WWGF, West Java presented by red rectangular overlaid on Landsat -8 imagery.

IOP Conf. Series: Earth and Environmental Science **42** (2016) 012007 doi:10.1088/1755-1315/42/1/012007

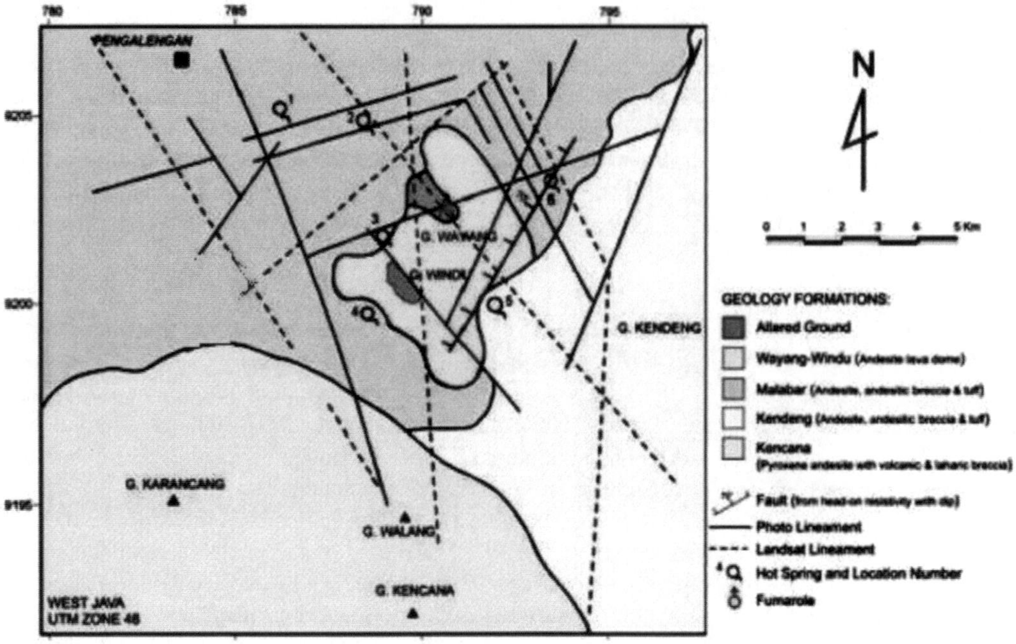

Figure 2. Geological map of Wayang Windu Geothermal Field [14].

3. Methods

There are three methods of mineral mapping that will be compared to the ground truth (field survey) data: Minimum Noise Fraction (MNF), Pixel Purity Index (PPI), and Spectral Angle Mapper (SAM).

3.1. Minimum Noise Fraction (MNF)

The MNF transformation is a linear transformation related to principal components that orders the data according to signal-to-noise-ratio [15]. The MNF transformation is used to separate the data space into two parts: one associated with large eigenvalues and coherent eigenimages, and a second with near-unity eigenvalues and noise-dominated images [16]. In general, the higher numbered MNF bands contain progressively lower signal-to-noise. MNF is also a linear transformation method for eliminating the correlation between bands and for reducing the noise in the data [17].

3.2. Pixel Purity Index (PPI)

Based on MNF results, the lower order MNF bands are usually excluded and the higher order bands are selected for further processing. For the Pixel Purity Index (PPI), processing is designed to locate the most spectrally extreme (unique or different or "pure") pixels [18]. The most spectrally pure pixels typically correspond to mixing endmembers. The PPI is computed by repeatedly projecting n-dimensional scatterplots onto a random unit vector.

The extreme pixels in each projection are recorded and the total number of times each pixel is marked. A PPI image is created in which the digital number of each pixel corresponds to the number of times that pixel was recorded as extreme. A threshold is interactively selected using the histogram and used to select only the purest pixels in order to keep the number of pixels to be analyzed to a minimum. These pixels are used as input to an interactive visualization procedure for separation of specific endmembers.

3.3. Spectral Angle Mapper (SAM)

SAM is a supervised classification algorithm, which utilizes spectral angular information for the classification of hyper-spectral data [19]. It treats each pixel in a hyper-spectral image as an n-dimensional vector, where n equals the number of spectral bands. The algorithm measures similarity of a target to a reference spectrum by calculating spectral angles between them (see Figure 3). A smaller angle represents a closer match to the reference spectrum. The angle between a target spectrum vectors *a,* reference spectrum vector *b*, can be calculated by:

$$\alpha = cos^{-1}\left(\frac{a \cdot b}{\|a\| \cdot \|b\|}\right) \qquad (1)$$

where ‖‖ is a norm function [20], Eq. (1) can also be written as:

$$\alpha = cos^{-1}\left\{\frac{\sum_{i=1}^{n} a_1 \cdot b_1}{\left[\sum_{i=1}^{n} a_1^2\right]^{1/2}\left[\sum_{i=1}^{n} b_1^2\right]^{1/2}}\right\}, \qquad (2)$$

where *n* is the number of spectral bands, a_1 denotes value of the target spectrum at i^{th} band and b_1 denotes value of the reference spectrum at i^{th} band.

This measure of similarity is insensitive to gain factors because the angle between two vectors is invariant with respect to the lengths of the vectors. As a result, laboratory spectra can be directly compared to remotely sensed apparent reflectance spectra, which inherently have an unknown gain factor related to topographic illumination effects.

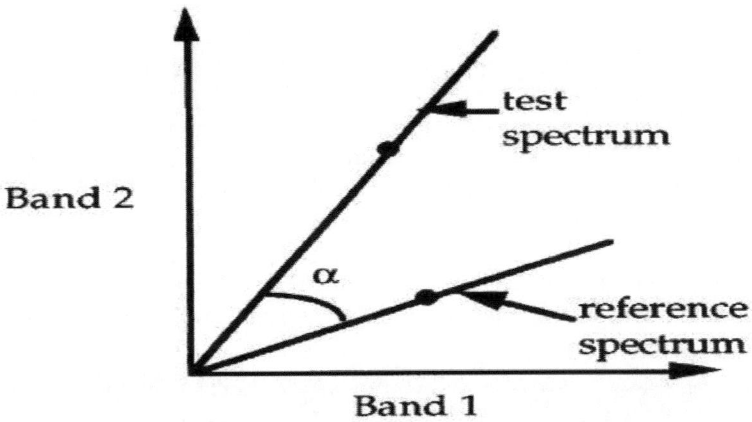

Figure 3. Plot of a reference spectrum and test spectrum for a two-band image. The same materials with varying illumination are represented by the vectors connecting the origin (no illumination) and projected through the points representing the actual spectra [19].

4. Result and Discussion

The results of each method are compared with the existing ground truth in order to determine which method is the most precise and accurate for mapping minerals (alterations). The table below is the result of ground truth:

There are 19 observation points, and 10 point that showed the presence of geothermal manifestations which characterized by lithology in the form altered rock and altered andesite, that points: (A1-2), (A1-9D), (A4-1), (A4-9), (A5-10), (A9-5), (A17-3), (A17-1), (PA5-1A), and (WA1-12).

Table 1. The result of ground truth data and the point of observation and lithology.

No.	Waypoint	Lithology
1	DR10-2	Andesite porphyri
2	A1-2	Altered rock
3	A1-9D	Altered rock
4	A4-1	Altered rock
5	A4-9	Altered andesite
6	A5-10	Altered rock
7	A7-5	Volcanic breccia
8	A9-5	Altered andesite
9	A14-2	Pyroclastic breccia
10	A17-3	Altered rock
11	PA1-4	Andesite
12	PA2-7	Brecciated andesite lava
13	PA3-7	Volcanic breccia
14	PA5-1A	Altered rock
15	WA1-12	Altered rock
16	WA3-1	Travertin
17	A17-1	Altered rock
18	A7-3	Volcanic breccia
19	WA1-10B	Andesite

4.1. MNF Result

The parameters used in analyzing the MNF Transform are input image, output noise statistics, and output MNF stats. If these parameters have been met, then eigenvalues can be obtained. Eigenvalues are values of the transformation shown the dimensional distinctions based on testing of digital values. Eigenvalue is the discriminant level of the band. The greater eigenvalue means the greater of the difference level, or the noise is getting smaller. Figure 4 shows that there are 147 bands with respective eigenvalues, where the results of MNF shows that the initial bands has large eigenvalues.

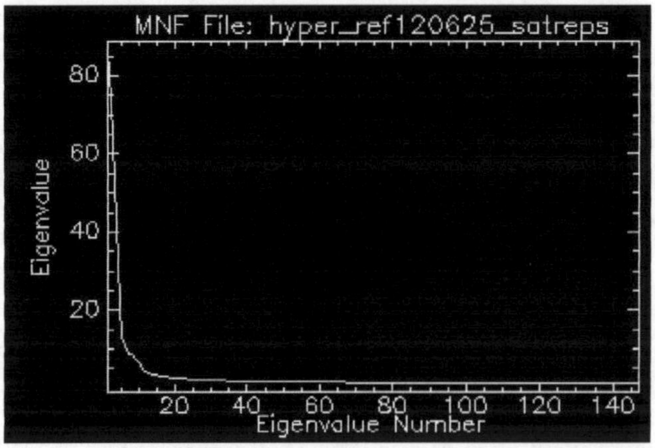

Figure 4. MNF plot with the total eigenvalue of each band.

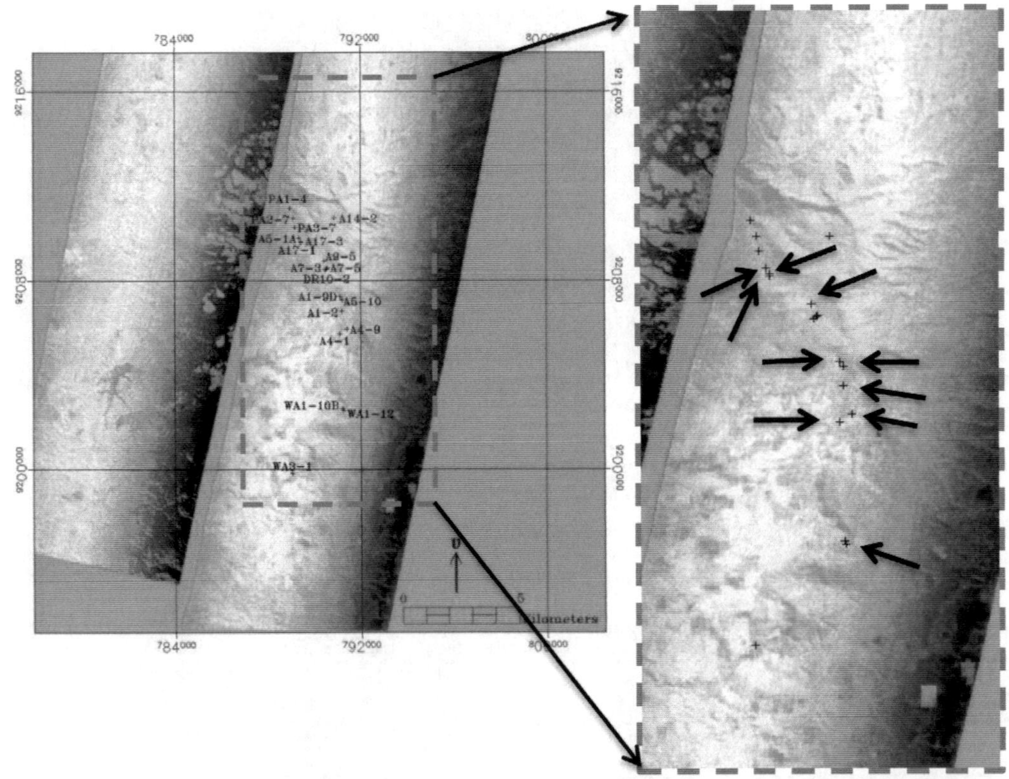

Figure 5. The MNF in color composite R,G,B for (Band 1,2,3), with sample locations in "+"symbol. The black arrows indicating a manifestation point.

MNF analysis is applied by selecting three bands with the highest eigenvalue (MNF-1, 2, 3). To analyze it, the plot surface manifestation of the composite MNF, and observe the hue and toned appearance. Based on the results of existing ground truth, the location of alteration zones was indicated by altered rocks at (A1-2), (A1-9D), (A4-1), (A4-9), (A5-10), (A9-5), (A17-3), (A17-1), (PA5-1A), and (WA1-12).

An subset image in Figure 5 shows that the altered rock samples are located at a relatively bright hues and greenish color. The appearance of the brightness and color was identified as an anomaly originated from alteration zones.

4.2. PPI Result

Pixel purity index (pixel Purity Index) is used to find the purity of a pixel on multispectral and hyperspectral imagery. Pure pixel is a pixel that is truly representative of the object to be homogeneous. Mixed pixel is a pixel representing several objects at once.

There are two parameters in pixel purity index plot, ie parameter y are number of pixels, and the parameters x are number of iterations. The number of iterations of itself means the ability an image to read a number of pixel values at once. From PPI curve can be seen that the value of iterations performed 10.000 times. It is seen that the more iterations performed, the index value is also higher purity (see Figure 6).

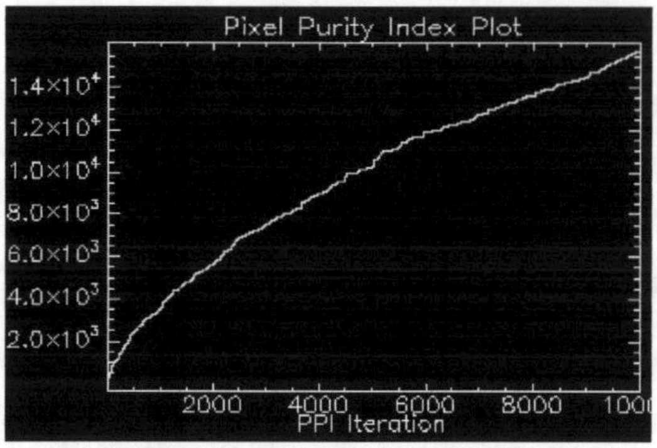

Figure 6. PPI iteration vs. total cumulative pixel

Figure 7 shows the result of the mapping based on the brightness of the pixel (PPI outout), where the bright pixels represent spectral extremes (pure) and dark pixels represent the spectral less pure. The points that indicate geothermal manifestations were located on the pixel with a relatively dark (black).

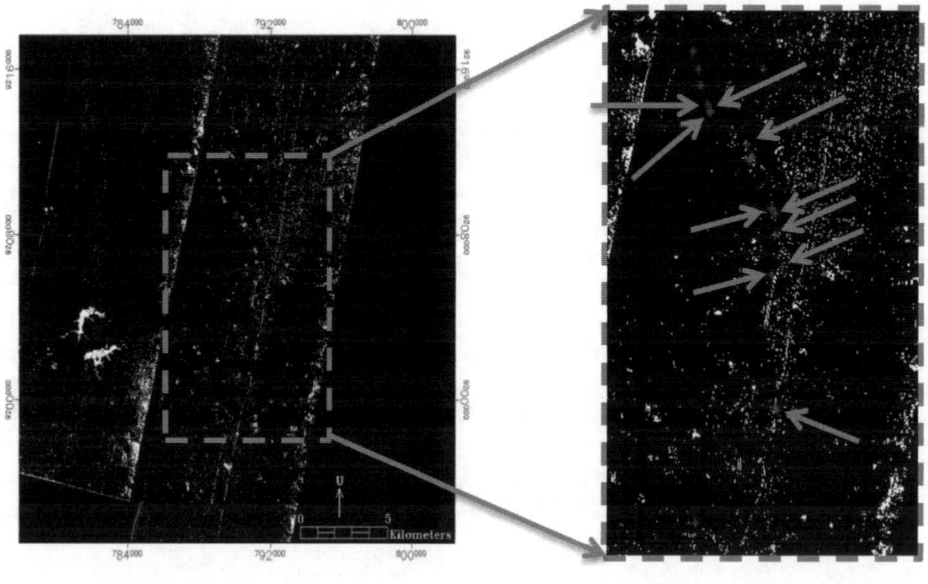

Figure 7. The PPI image indicating bright and darks pixels represent more and less spectrally extreme (pure), respectively. The red arrows are a manifestation points.

4.3. SAM Result

SAM process dividing the objects in the image into 48 classes or 48 different spectra. The points manifestations (10 points) was then plotted into the SAM map and can be seen that: three points are in class 46 that point are: (A1-9D, A1-2, A4-1), three point were in class 29: (PA5-1A, A9-5, A5-10), two points are in class 35 that are: (A17-3, A4-9), one point is in class 27, that is point (A17-1) and one point is in class 23, that is point (WA1-12).

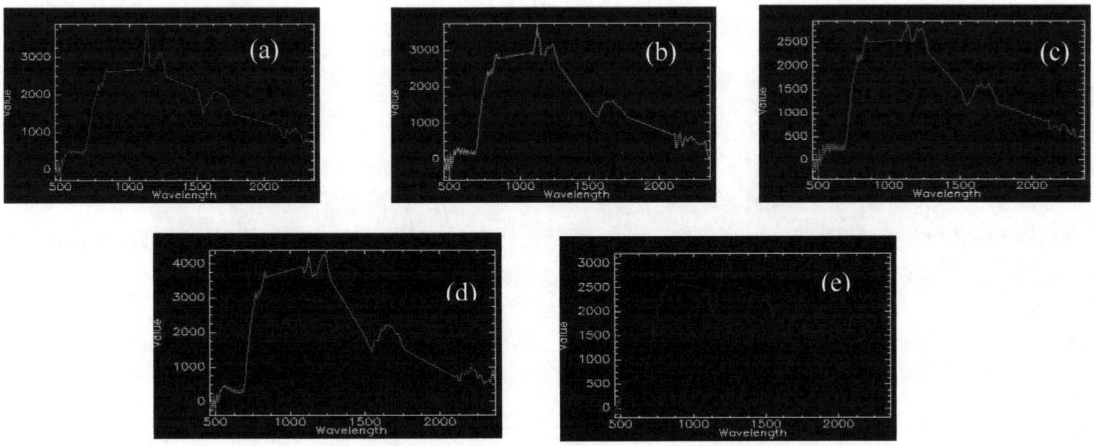

Figure 8. Plot spectrum target with (a). Spectrum class 46; (b). Spectrum class 29; (c). Spectrum class 35; (d). Spectrum class 27; and (e). Spectrum class 23.

All spectra class above (see Figure 8) will be matched with the reference spectra of clay minerals from spectra library ENVI and USGS (Illit, kaolinite and montmorillonite) to determine the minerals contained in the class.

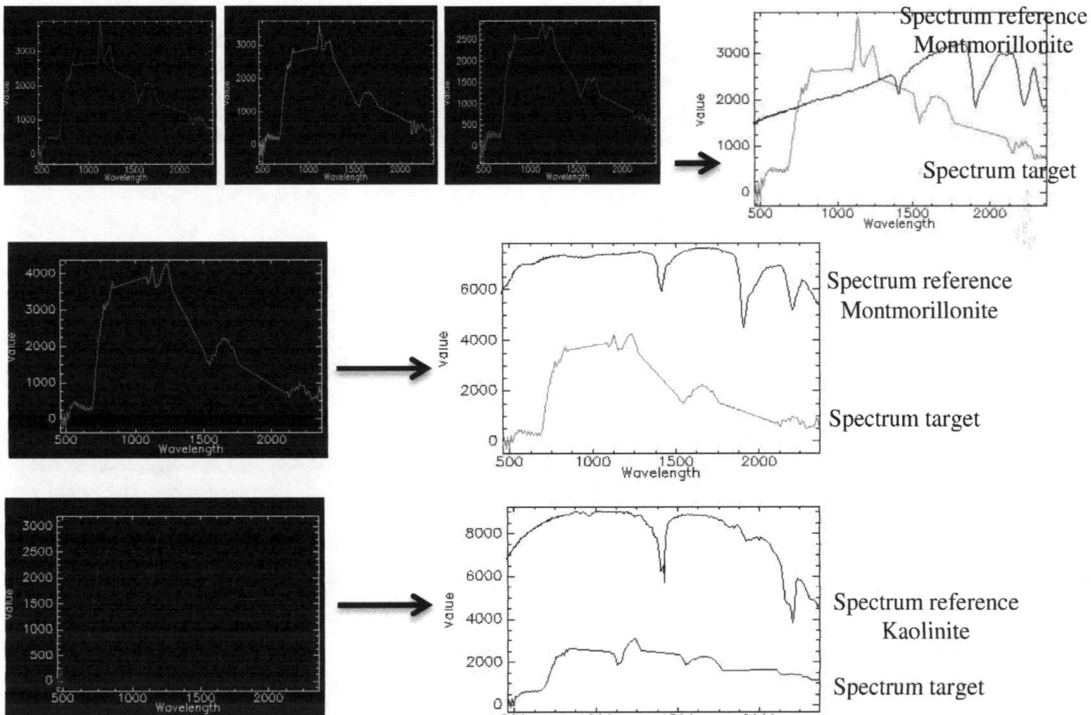

Figure 9. Matching proscess between target spectrum to reference spectra

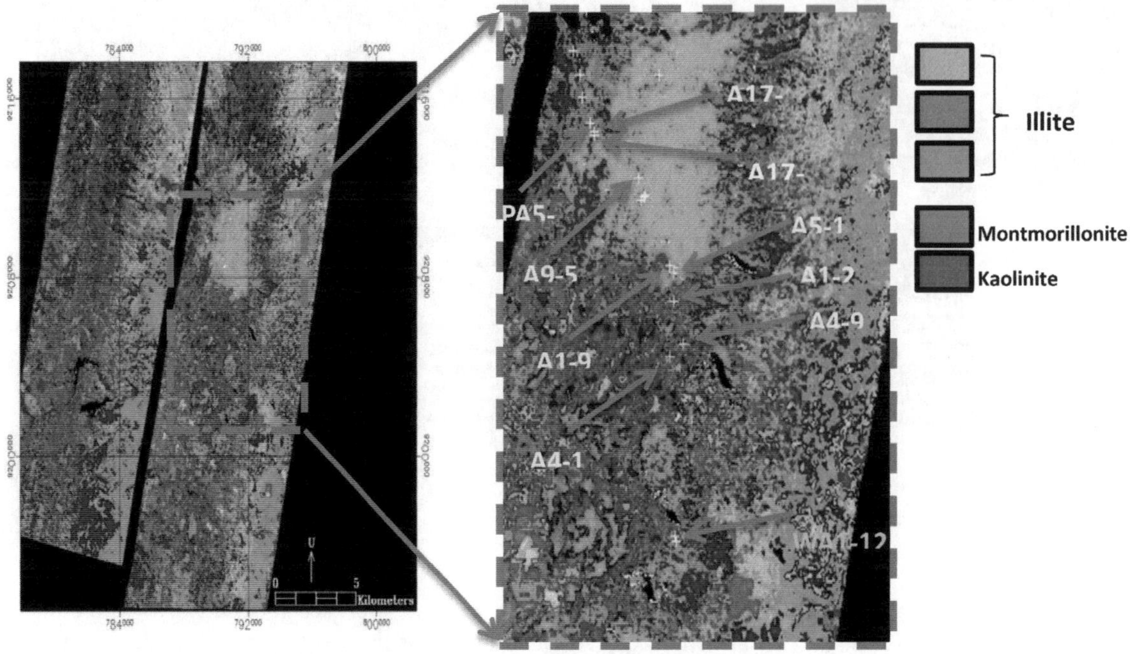

Figure 10. The SAM map with a description of each class spectrum from alteration minerals.

Figure 9 shows that the spectrum of the target (class 46, 29 and 35) has a relatively similar pattern, so we interpreted that the three spectra is same object or target. After matching proscess, we obtained that the spectra originated from illite mineral. Pattern of the spectrum in class 27 also resemble the pattern of minerals montmorillonite and class 23 is kaolinite. Figure 10 shows that the light blue, brown, and pink were identified as illite, the green is montmorillonite mineral, and dark blue is kaolinite mineral.

Conclusion

The MNF and PPI methods showed that the appearance of the target at the ground surfaces was presented by color and gray. The MNF method showed that all altered rock samples were located at bright hues and greenish color. For PPI method, the samples were located at dark pixel. According to the SAM method, there were eight out from ten rock samples were identified by illite mineral, and the other two samples were montmorillonite and kaolinite minerals. The SAM method could identify the minerals at low vegetation covers. Therefore, among the three methods, the SAM method is the most suitable for altered minerals mapping. The similarity spectra from reference and spectra was used as basis of mineral identification spatially.

Acknowledgment

The authors thank to Beneficial and Advanced Geothermal Use System (BAGUS) project of the Science and Technology Research Partnership For Sustainable Development (SATREPS) for providing equipment, LPPM-ITB for financial support.

References

[1]. Ninomiya, Y., Fu, B., and Cudahy, T.J., 2005. Detecting lithology with Advanced Spaceborne Thermal Emission and Reflection Radiometer (ASTER) multispectral thermal infared "radiance-at-sensor" data. *Remote Sens Environ, 99, pp. 127- 139. Corrigendum: Remote Sens Environ, 101, p. 567.*

[2]. Saepuloh A., Koike K., Sumintadireja, P., (2007). "Accurate detection and mapping of altered minerals by optical remote sensing for an active volcano in the Torrid Zonel", *Proceeding of 5th International Workshop on Earth Science and Technology*, Kyushu University, Japan.

[3]. Saepuloh A., Wijaya K., Sumintadireja P. (2011a), Detecting soil layer condition soon after Merapi eruption 2010 using ALOS/PALSAR datal, *Proceeding of the Annual Meeting of Science and Technology Studies 2011 (AMSTECS-2011)*, ISSN: 2088-2041, Tokyo, Japan.

[4]. Mushkin, A., Gillespie, A. R. (2005), Estimating sub-pixel surface roughness using remotely sensed stereoscopic data,‖ *Remote Sensing Environment*, 99, 75–83.

[5]. Saepuloh A., Koike K., Omura M. (2011b), Applying Bayesian decision classification to Pi-SAR polarimetric data for detailed extraction of the geomorphologic and structural features of an active volcanol, *IEEE Geoscience and Remote Sensing Letters (GRSL)*, doi. 10.1109/LGRS.2011.2174611, October 2011.

[6]. Saepuloh, A., Urai, M., Sumintadireja, P., Suryantini, (2012). "Spatial Priority Assessment of Geothermal Potentials using Multi-Sensor Remote Sensing Data and Applications". *Proceedings, 1st ITB Geothermal Workshop 2012*, Bandung

[7]. Pearlman, J, Stephen Carman, Paul Lee, Lushalan Liao and Carol Segal, 1999, "Hyperion Imaging Spectrometer on the New Millennium Program Earth Orbiter-1 System": *In Proceedings, International Symposium on Spectral Sensing Research (ISSSR)*, Systems and Sensors for the New Millennium, published on CDROM, International Society for Photogrammetry and Remote Sensing (ISPRS).

[8]. Kruse, F.A., Boardman, J.W., Huntington, J.F., Mason, P., Quigley, M.A., (2003). "Evaluation and Validation of EO-1 Hyperion for Geologic Mapping": *in Special Issue*, TGARSS, IEEE (in press).

[9]. Hochstein, M.P., Sudarman, S.: History of geothermal exploration in Indonesia from 1970 to 2000, *Geothermics, 37*, (2008), 220- 266.

[10]. Prugovecki, E.,1982. Quantum Mechanics in Hilbert Space. Academic Press, London,UK. Purnanto, M.H., Purwakusumah, A., (2015). "Fifteen Years (Mid-Life) of Wayang Windu". *Proceedings World Geothermal Congress 2015*. Australia: Melbourne.

[11]. Bogie, I., Kusumah, Y.I., Wisnandary, M.C., 2008, Overview of the Wayang Windu Geothermal Field, West Java, Indonesia. *Geothermics vol. 37, pp. 347-365.*

[12]. Alzwar, M., Akbar, N., Bachri, S., 2004, "Geological map of the Garut and Pamengpeuk quadrangle, Jawa". 2nd edition, Geological Research and Development Center, Indonesia.

[13]. Prasetio, R., Wiegand, B.A., Malik, D., Sauter, M., (2015). "Stable Isotopes Study of Wayang Windu Geothermal Field, Indonesia". *Proceedings World Geothermal Congress 2015*. Australia: Melbourne.

[14]. Sule R.M., Nugraha D.A., Widiantoro S., "The Development and Implementation of Microseismic Method for Geothermal Reservoir Monitoring and Monitoring of CO_2 Injection Into Subsurface Acquifer During CO_2-Sequestration Activity, LPPM-ITB, Bandung.

[15]. Green, A.A., Berman, M., Switzer, P., Craig, M.D., (1988). "A Transformation for Ordering Multispectral Data in Terms of Image Quality with Implications for Noise Removal", *Geoscience and Remote Sensing*, IEEE Transactions on, vol. 26, pp. 65-74.

[16]. Boardman, J.W., Kruse, F.A., (1994). "Automated Spectral Analysis: A Geological Example using AVIRIS Data, Northern Grapevine Mountains, Nevada"*: in Proceedings, Tenth Thematic Conference*, Geologic Remote Sensing, 9-12 May 1994, San Antonio, Texas, pp. I-407–I-418.

[17]. Lixin, G., Weixin, X., Jihong, P., (2015). "Segmented Minimum Noise Fraction Transformation for Efficient Feature Extraction of Hyperspectral Images". *Elsevier: Pattern Recognition 48(2015)3216–3226.* China: Shenzhen.

[18]. Boardman, J.W., Kruse, F.A., Green, R.O., (1995). "Mapping Target Signatures via Partial Unmixing of AVIRIS Data: in Summaries", *Fifth JPL Airborne Earth Science Workshop*, JPL Publication 95-1, v. 1, pp. 23–26.

[19]. Kruse, F.A., Lefkoff, A.B., Boardman, J.W., Heidebrecht, K.B., Shapiro, A.T., Barloon, P.J., Goetz, A.F.H., (1993). "The Spectral Image Processing System (SIPS) Interactive Visualization and Analysis Imaging Spectrometer Data". *Elsevier Science Publishing Co. Inc.,* Remote Sens. Environ. 44:145-163, Avenue of the Americas: New York.
[20]. Prugovecki, E.,1982. Quantum Mechanics in Hilbert Space. Academic Press, London,UK.

Geology Structure Identification based on Polarimetric SAR (PolSAR) Data and Field Based Observation at Ciwidey Geothermal Field

R A Pradipta[1], A Saepuloh[1,2], and Suryantini[1,2]

[1]Geothermal Technology Study Program, Faculty of Mining and Petroleum Engineering, Institut Teknologi Bandung, Indonesia
[2]Geology Engineering Study Program, Faculty of Earth Sciences and Technology, Institut Teknologi Bandung, Indonesia

E-mail: ratna_amalia_pradipta@yahoo.com

Abstract. Geological structure observation is difficult to be conducted at Quaternary volcanic field due to the classical problem at tropical region such as intensive erosion, dense vegetation covers, and rough terrain. The problem hampers the field observation especially for geological structures mapping. In order to overcome the problems, an active remote sensing technology based on Polarimetric Synthetic Aperture Radar (PolSAR) data was used in this study. The longer wavelength of microwave than optical region caused the SAR layer penetration higher than optics. The Ciwidey Geothermal Field, Indonesia was selected as study area because of the existence of surface manifestations with lack information about the control of geological structures to the geothermal system. Visual interpretation based on composite polarization modes was applied to identify geological structures at study area. The color composite Red-Green-Blue for HV-HH-VV polarizations provided highest texture and structural features among the other composite combination. The Linear Features Density (LFD) map was also used to interpret the fractures zones. The calculated LFD showed high anomaly about 3.6 km/km^2 with two strike directions NW-SE and NE-SW. Interestingly, the surface geothermal manifestation agreed with the low anomaly of LFD. The geological structures consisted of ten faults were successfully detected and mapped. The faults type mainly are oblique-slip with strike directions NE-SW and NW-SE.

1. Introduction

Study area is a geothermal field located around crater lake and tourism spot Kawah Putih, Ciwidey, Bandung with the average elevation about 2.000 meter above sea level (m.a.s.l). The land use in Ciwidey area consists of farm land, estate crops (i.e, tea, strawberry, coffee, and other vegetable crops), tourism area, and conservation forest. Mt. Patuha and Mt. Kendeng are two volcanoes which contribute to produce Quaternary volcanic rocks in study area (Figure 1).

There are two geothermal working areas, which are operated by PT. Geo Dipa Energi and PT. Yala Teknosa Geothermal. Patuha Geothermal Field operated by PT. Geo Dipa Energi has completed exploration activities and Unit-1 of Patuha Geothermal Power Plant with a capacity of 55 MWe has operated since July, 2014 [1].

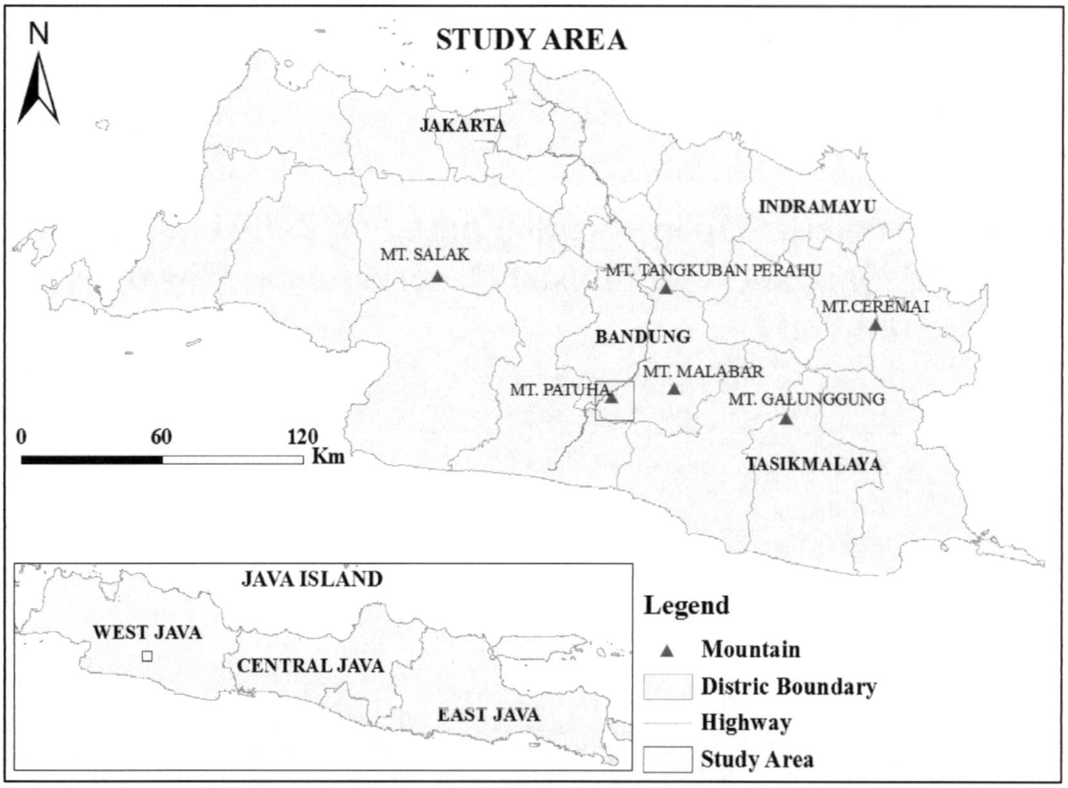

Figure 1. Study area located at Ciwidey Geothermal field about 46 km Southwest from Bandung is presented by light blue box.

Geological, Geochemical, and Geophysical surveys were able to delineate the study area widely [2]. Gravity and resistivity (MT and TDEM) survey had been conducted in the Patuha field. Based on their report, gravity lineament and electrical discontinuity were estimated for consideration of geological structure in this field [3]. In addition, geologic structures were also identified by cliff, rim features, waterfall, river stream lineament/morphology, continuity of eruption centre, and hotspring which may exist along the fault line [4]. Both methods only derived structure lineament without its kynematics and dynamics analysis, while those analyses have big role in permeability zone identification in this study area.

However, there was still minor information about surface geology structure in study area. Therefore, geology structure mapping need to be conducted in detail. Detailed distribution of geology structure could be used to predict the permeability of the surface/subsurface layers. The correct information about geological structures were required to interpret thermal fluid and conceptual model of geothermal system as well as prospecting geothermal well target.

This paper discussed about surface geologic structure identification using active remote sensing technology and geological mapping. We applied visual analyses of linear features based on Synthetic Aperture Radar (SAR) data. The Polarimetric SAR, termed as PolSAR is superior to detect Geomorphologic and Structural Features (GSF) under canopy vegetation and rough terrain in volcanic field [5]. In addition, the superiority of the microwave sensors is their capability to penetrate canopy vegetation and cloud cover [6].

2. Geology

Study area is located at Quarternary volcanic field in Java, Western part of Indonesia [7]. Historical geology of Java Island has been formed since early Tertiary through a convergent tectonic margin between the Indian-Australian Plate and the southeastern margin of Eurasian Continental Plate called the Sundaland [8]. The regional geology of study area is dominated by Quaternary volcanic products, which cover most of the Patuha field [9]. The volcanic products are determined by andesitic lavas and pyroclastics from Mt. Patuha and Mt. Kendeng. Majority of geology structures identified at Patuha is normal fault with strike direction at NW-SE and minority at E-W and N-S [4].

Rocks penetrated by geothermal wells at Patuha were dominated by lavas, tuffs, and breccias. The existence of eruption center formed volcanic cone and crater structure, apparently has been controlled by normal faults [3]. Fractures within the volcanic axis probably control intrusion at depth of the magmatic heat source for the Patuha geothermal system, as well as the eruption of magma to the surface. Structures in this zone also control fracturing system in the geothermal reservoir and serve as the path of the discharge of fumaroles to the surface [10].

Surface manifestations in study area are composed by fumaroles, thermal springs, cold gas discharges, mud pool, cold acid water, and steaming ground. Fumaroles are located at Cibuni Crater, Putih Crater, and Ciwidey Crater (Figure 2).

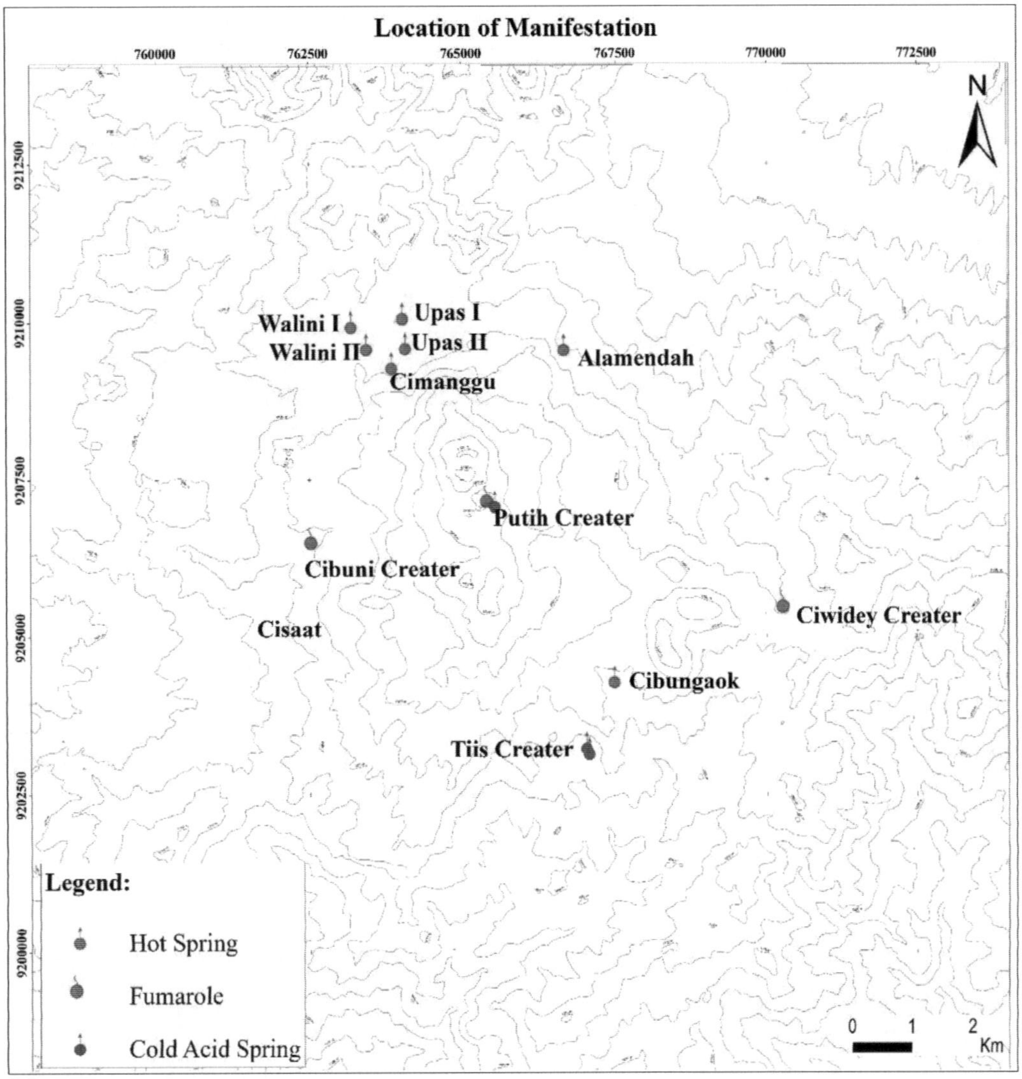

Figure 2. Location of manifestations in study area. Hot/ thermal springs and fumarol showed by red color and cold water showed by green color.

IOP Conf. Series: Earth and Environmental Science **42** (2016) 012008 doi:10.1088/1755-1315/42/1/012008

3. Remote Sensing

The Phased Array L-band type Synthetic Aperture Radar (PALSAR) full polarimetric (HH, HV, VV, and VH) with acquisition date on 19th April 2010 data onboard the Advanced Land Observing Satelite (ALOS) data was used in this study. The objective are to obtain Geomorphologic and Structural Features (GSF) and to interpret preliminary targets of permeability zone at study area. Visual analysis of GSF and Linear Features Density (LFD) based on color composite of SAR polarimetric data were applied to detect targets at ground surface. In order to predict the surface permeability zones targetted in this study, we correlated the fault at the field to the GSF and LFD.

3.1. Visual Analysis of PolSAR

Visual interpretation based on composite polarization modes was applied to identify geological structures at study area. The color composite of R-G-B for polarization HV-HH-VV provided highest texture and structural features among the others composite combinations. Comparing the histogram of each polarization data, there are two classifications: (1) high peak – high steepness and (2) low peak – low steepness. The second class, consists of HV and VH, one of them is choosen as red composite because its relatively more variety value compared to the first class. The green and blue composite are filled with the others straight polarization mode (HH and VV). The reason for this is because, based on the scatter plot, the combination between cross and straight polarization mode shows more variety value compared to the combination between cross and cross polarization mode.

This analysis was used as guidance for field observation and also used as detection basis for geologic structures such as fault lines in accordance to the evidence at the field. There were two geological features detected: linear and circular features. The linear features were interpreted as fault structures. A rose diagram showed four structural patterns, which were NW-SE, NE-SW, E-W, and N-S (Figure 3). Most structures were identified with NE-SW direction. These structures encompassed along hot spring, crater, and lake. Some manifestations also exhibited at NE-SW structures, such as the Walini area and Ciwidey Crater. It's inferred that manifestations in the study area were controlled by fault structures.

There were 25 circular features detected based on visual analysis in this study. These features were interpreted as volcanic morphology such as crater, caldera, or the eruption center. One of the circular features was found as a lake in the field. Based on large number of circular features found in the area, it could be interpreted that the study area had high intensity of volcanism.

3.2. Linear Features Density (LFD)

The Linear Features Density (LFD) map and geologic structures identification were derived from detected linear features by visual analysis on each polarization mode data (HH, HV, VH, and VV) of PolSAR image. Total total lenght of linear features was calculated within grid size 1×1 km^2. The total length each grid was used to generate density contour map. The density value was in range 0–3.6 km/km^2 and the blue – red scale bar was adopted to simplify the interpretation process (Figure 4).

A rose diagram of the detected linear features shows two dominant patterns, which are NW-SE and NE-SW. High LFD associated with surface linear features and located at NE part of the study area, SE-SW of Putih Crater, and around Tiis Crater. In general, high LFD value corresponded with high topographic of mountainous range.

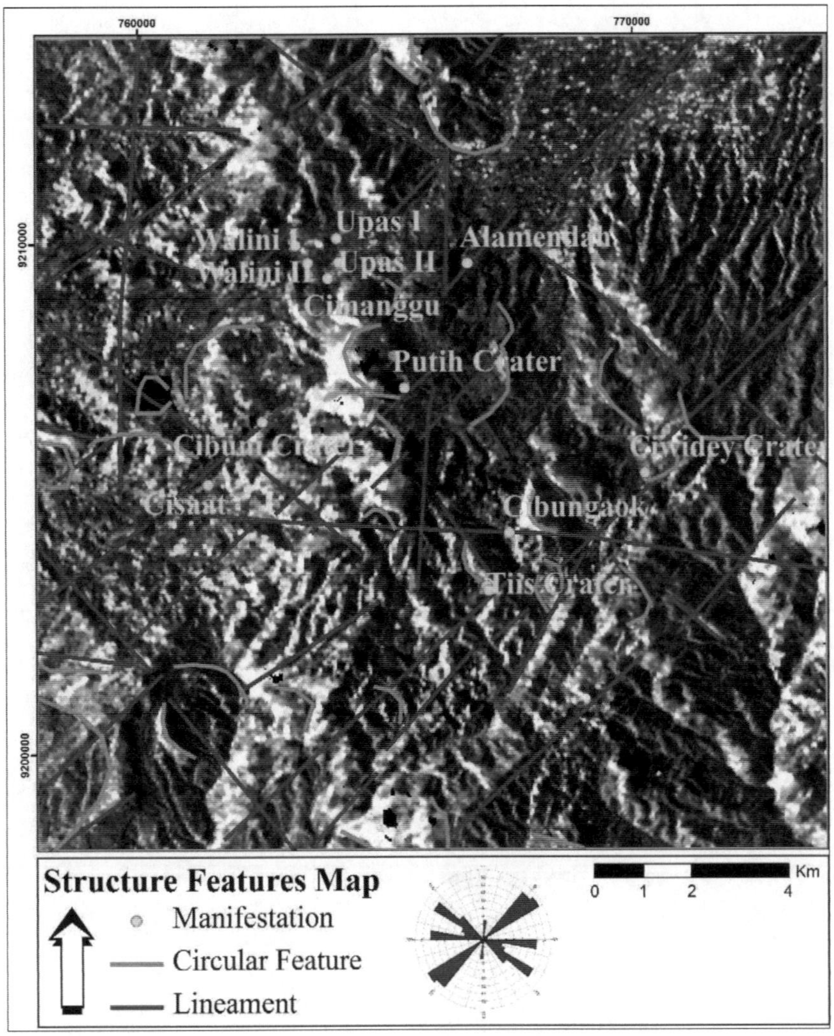

Figure 3. Visual analyses of study area in light blue box based on Polarimetric ALOS PALSAR with color composite for R=HV, G=HH, B=VV. Red lines and orange circulars are interpreted linear and circular features.

Low LFD value occurs at low topographic or valley area such as Walini and Upas area. Those low value forms an elongated shape from NE to SW in general and agrees with visual analysis in NE to SW trend. Based on manifestation occurences, those NE to SW lineament was interpreted as geologic structure.

This study shows that in general, surface manifestations associate with low LFD area such as Walini I-II, Upas I-II, Cimanggu hotsprings and medium LFD such as Cibuni Crater, Cisaat and Tiis Crater acid-springs, Cibungaok hotspring, and Ciwidey Crater. These phenomena are interpreted as structure trace dissipation due to high erosion, weathering, and alteration.

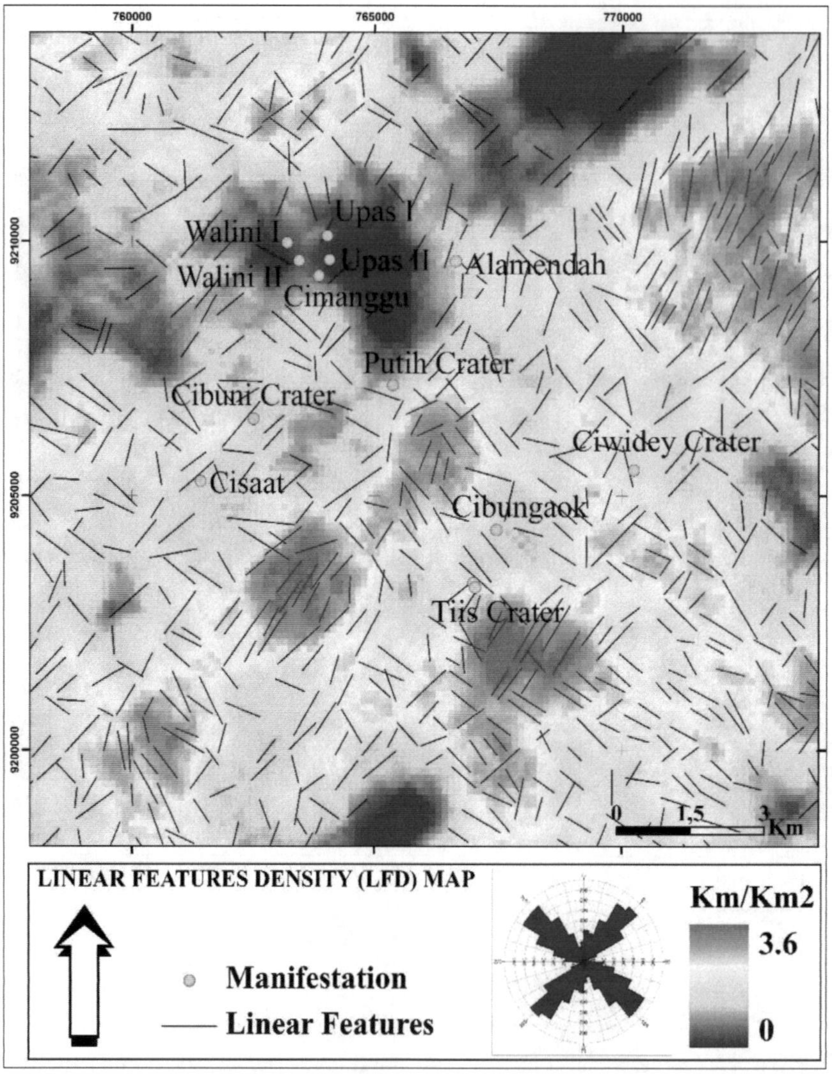

Figure 4. Linear Features Density (LFD) map shows surface manifestations from low to intermediate value with structural pattern NE-SW and NW-SW.

4. Geology Mapping

Geological mapping was performed and focused to geologic structure observation including measurements of shear, tension fracture, slicken slide, and brecciated zone. The measurements were conducted along detected faults based on SAR remote sensing as explained in the previous section. It is a clasical problem in tropical region that the geological structures at field is difficult to be found due to thick vegetation, intensive weathering, thick soil layer, and steep terrain.

The ten faults were identified in the study area based on SAR image analysis, field observation, and kinematics analysis (Figure 5). The faults type mainly are oblique-slip with strike directions are NE-SW and NW-SE, with N-S as the main stress formed the faults.

Manifestations are commonly located among the fault lines, as showed in Walini I, Walini II, Upas I, Upas II, and Cimanggu hotsprings (Figure 5). Those manifestations were located among Cimanggu Faults system with low LFD value. The Putih crater, Ciwidey crater, and Alamendah hotspring were also located among the fault lines with medium LFD value. The similar phenomena were also existed at Cibuni Crater, Tiis Crater, Cisaat acid-spring, and Cibungaok hotspring.

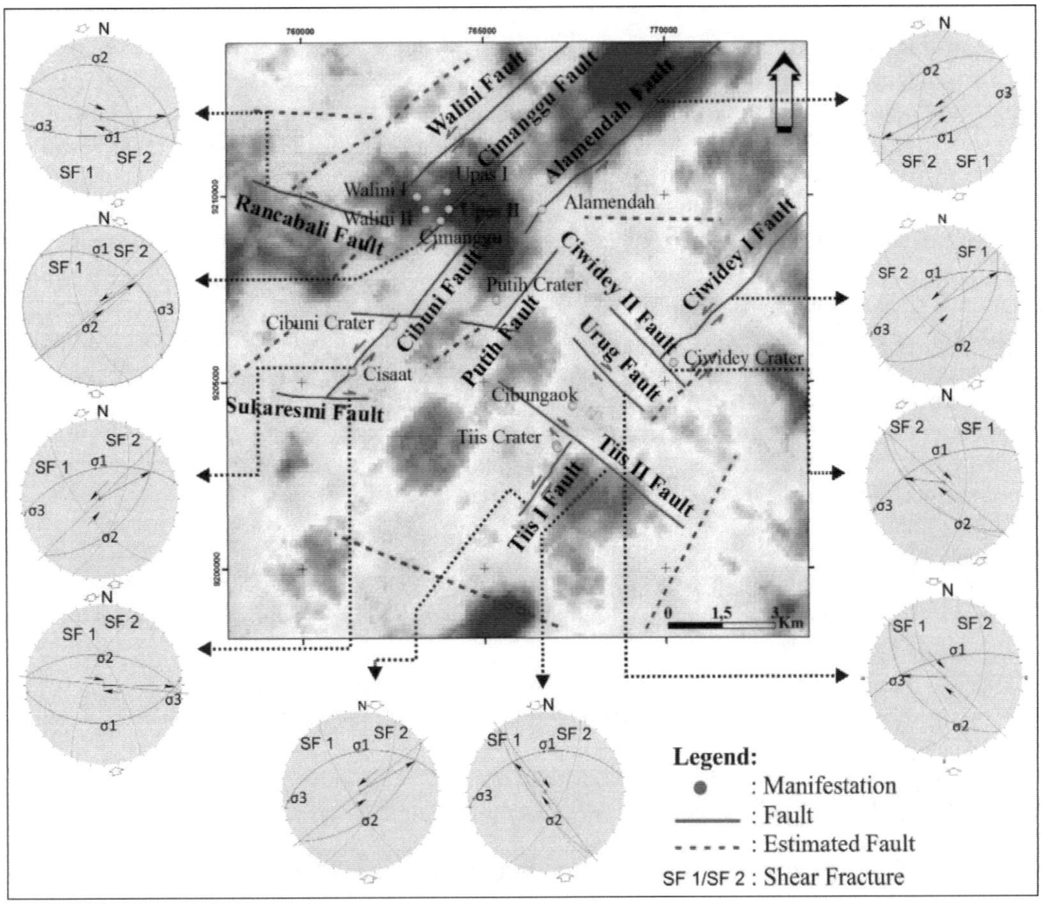

Figure 5. Geologic structures map and stereonet diagram analyses at study area overlaid on LFD Map. Geological structures (presented by red lines) were derived from PolSAR analyses confirmed by field observation as showed by black dots.

We interpreted that the surface manifestations located among the fault lines due to the faults system controlling the hydrothermal fluid paths. The segmentation and separation of the faults system caused offsets zones or extensional step-overs which increase the permeability of the rocks [11], [12] (Figure 6).

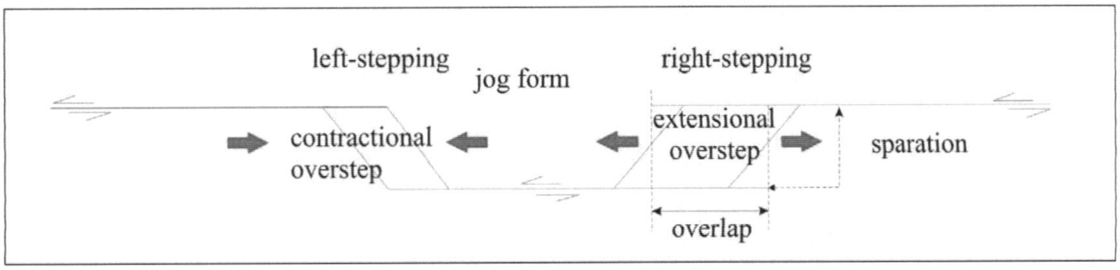

Figure 6. Conceptual model of restraining (contractional) and releasing (extentional) stepovers along a sinistral-slip fault for study area [11].

5. Conclusions

Visual analysis showed that there were four structural patterns, which were NW-SE, NE-SW, E-W, and N-S with dominant NE-SW trend. There were 25 circular features detected successfully and confirmed that the study area was located at volcanic terrain which plausible for the heat source occurrence in a geothermal system in Indonesia. Based on Polarimetric SAR image, the detected linear features are composed by two dominant patterns: NW-SE and NE-SW. Those patterns are justified as the existing faults direction and as discharge path of hydrothermal fluid.

Geothermal manifestations in the study area commonly associate with low-medium Linear Features Density (LFD). This phenomena caused structure trace dissipation due to high erosion and/or rock interaction with hydrothermal fluid.

According to PolSAR image analyses and geological mapping, the ten geological structures could be clarified. The faults type are mainly oblique-slip with strike directions NE-SW and NW-SE. The N-S is the main stress which formed the faults. Surface manifestations are commonly exhibit among the segmented fault lines. The segmented faults were separated by offsets or extensional stepover area and generated pulled apart system. The system provided subvertical conduits of high fracture density that enhance permeability and facilitate the rise of deep-seated thermal plumes.

References

[1] Smillie A, Satar S, Saptadji N, Aminzadeh F, and Setianingsih R 2015 *Capacity Building in the Geothermal Sector in Indonesia, a Unique Collaboration.* World Geothermal Congress. Melbourne.

[2] Suryantini and Wibowo H 2015 *Geologic Risks Assessment and Quantification in Geothermal Exploration Case Studies in Green Field and Developed Prospects.* World Geothermal Congress. Melbourne.

[3] West Japan Engineering Consultants (West JEC) 2007 *Feasibility study for Patuha geothermal power development: Final feasibility report.* Internal report: Japan Bank International Cooperation (JBIC).

[4] Suswati, Mulyana A R, and Sutawidjaja L S 2000 *Pemetaan Geologi Komplek Gunungapi Patuha Kabupaten Bandung, Jawa Barat. Bandung*: Pusat Vulkanologi dan Mitigasi Bencana Geologi.

[5] Saepuloh A., Koike K., and Omura M. 2012 *Applying Bayesian decision classification to Pi-SAR polarimetric data for detailed extraction of the geomorphologic and structural features of an active volcano*, IEEE Geoscience and Remote Sensing Letters (GRSL), Vol. 99, No. 4, pp. 554-558.

[6] Curnel I Y 2014 *Basic principles of remote sensing*. Gembloux: Walloon Agricultural research Centre (CRA-W).

[7] van Bemmelen R W 1949 *The Geology of Indonesia*. The Hague, Netherland: Government Printing Office, Nijhoff.

[8] Hamilton W 1979 *Tectonics of the Indonesian Region. Washington*: United States Goverment Printing Office.

[9] Koesmono M, Kusnama, and Suwarna N 1996. *Peta Geologi Lembar Sindangbarang dan Bandarlawu, Jawa. Bandung*: Pusat Peneliti dan Pengembangan Geologi.

[10] Layman E B 2003 *The Patuha Vavor-Dominated Resource West Java, Indonesia.* Twenty-Eighth Workshop on Geothermal Reservoir Engineering (hal. SGP-TR-173). Stanford: Stanford University.

[11] Burg J P 2015 Strike-Slip and Oblique-Slip Tectonic.

[12] Faulds J E, Coolbaugh M F, Vice G S, and Edwards M L 2006 *Characterizing Structural Controls of Geothermal Fields in the Northwestern Great Basin: A Progress Report.* GRC Transactions Vol. 30.

Acknowledgement
The authors wish to thank Geothermal Capacity Building (GeoCAP) Indonesia-Netherland for providing support in field survey and remote sensing data.

Geological, isothermal, and isobaric 3-D model construction in early stage of geothermal exploration

M P Saputra[1], Suryantini[1,2], D Catigtig[3], R Regandara[1], S N Asnin[1] and A B Pratama[1]

[1] Graduate Program in Geothermal Technology, Faculty of Mining and Petroleum Engineering, Institut Teknologi Bandung, Bandung 40132, Indonesia
[2] Engineering Geology Study Program, Faculty of Earth Sciences and Technology, Institut Teknologi Bandung, Bandung 40132, Indonesia
[3] Emerging Power Inc., Philippines

Email: mahesa.nawie@gmail.com

Abstract. Construction of geology, thermal anomaly and pressure distribution of a geothermal system in the early stage of exploration where data is limited is described using a 3-D software, Leapfrog Geothermal. The geological 3-D model was developed from a topographic map (derived from DEM data), geological map and literature studies reported in an early geological survey. The isothermal 3-D model was constructed using reservoir temperature estimation from geothermometry calculated from chemical analyses on surface manifestations, available shallow gradient temperature hole data and the normal gradient temperature (3°C/100m) for a non-thermal area. The isobaric 3-D model was built using hydrostatic pressure where the hydrostatic pressure is determined by the product of the fluid density, acceleration due to gravity, and depth. Fluid density is given by saturated liquid density as a function of temperature. There are some constraints on the modelling result such as (1) within the predicted reservoir, the geothermal gradient is not constant but continues to increase, thus, creating an anomalously high temperature at depth, and (2) the lithology model is made by interpolating and extrapolating cross-sections whereas usually only two to three geology sections were available for this study. Hence, the modeller must understand the geology. An additional cross section was developed by the modeller which may not be as suitable as the geologist constructed sections. The results of this study can be combined with geophysical data such as gravity, geomagnetic, micro-tremor and resistivity data. The combination of geological, geochemical, isothermal, isobaric and geophysical data could be used in (1) estimating the geometry and size of the geothermal reservoir, (2) predicting the depth of top reservoir, and (3) creating well prognosis for exploration and production wells.

1. Introduction

In the early stage of geothermal exploration, the surface and subsurface geological data are limited. Nevertheless, an interpretation of the subsurface condition and lateral extent of a geothermal area must be developed. This includes an interpretation of reservoir geometry, structural geology that may form permeability, lithology cover and reservoir depth. Such interpretation will encounter some difficulties due to the limited initial data and impact the accuracy of the interpreted model features such as the size of the geothermal system. Another problem encountered in developing a model is data integration of disparate data sets. Typically, the data to be integrated consists of a variety of data types (such as geology, geochemistry and geophysics) from different previous workers and different years with possibly different degrees of accuracy. Disparate datasets cause confusion and is time consuming. Moreover an interpretation of geoscience data to develop geothermal conceptual model is usually conducted in 2-D models such as maps or cross-section but to better visualize and infer the size of the geothermal system, the interpretation must be conducted in a 3-D model. Hence, the geoscientist needs a tool for developing a geothermal conceptual model recognizing the limitations of the data being used.

The case study presented herein is for the Montelago geothermal prospect (Figure 1) which for the first time a 3-D subsurface model is attempted. Exploration was conducted from 1978 to 2016. The exploration survey consists of surface geologic mapping, geochemical sampling (water and soil-soil gas), and geophysics - gravity, geomagnetic, CSMT and MT surveys. Furthermore, eight shallow temperature holes (up to 300 m depth) were drilled in 1979-1980.

This paper only discusses the construction of 3-D geological and isothermal model. The input for the geological model is a surface geologic map, information on the thickness and basement depth is from published literature on gravity and geomagnetic survey results in this area. The isothermal model constructed by using surface temperature of hot and warm springs, geothermometry, and temperature gradients from shallow holes up to 300 m depth. The isobaric model constructed by using pressure as a function of temperature derived from the isothermal model. Geologic, isothermal and isobaric 3-D models were interpolated using Leapfrog Geothermal software [1].

2. Montelago Geothermal Prospect

The Montelago geothermal prospect is located in Oriental Mindoro, Philippines, situated between Naujan Lake and Tablas Strait (Figure 1). The area has low-lying to moderate elevation but sometimes rugged terrains.

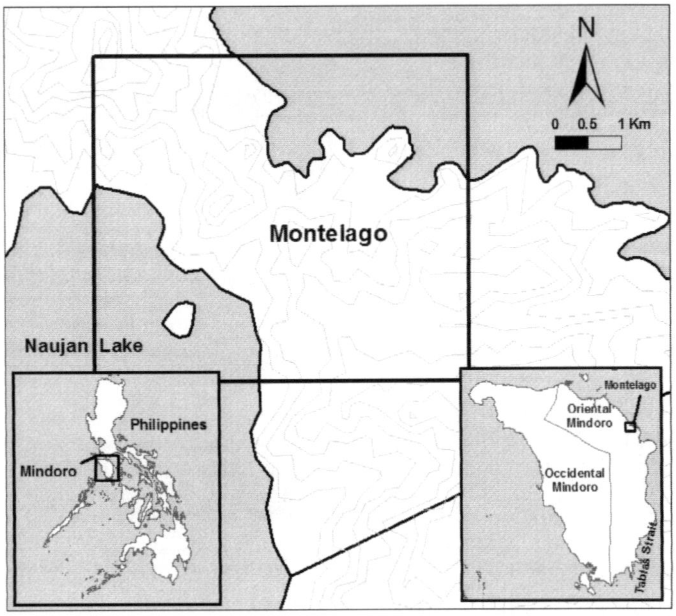

Figure 1. Location map for the Montelago geothermal prospect.

In general, lithology in Montelago consists of basement rock, sedimentary rocks, and Quaternary volcanic rocks which are dominated by andesite, tuff, basalt, and volcanic breccia. Paleozoic metamorphic such as schists, slates, and amphibolites; Mesozoic diorites and ultramafics that compose the core of Mindoro and here are collectively referred to as basement rocks. Although this rock is not exposed in the prospect area, their subsurface presence is inferred on the basis of geophysical surveys and the presence of accidental xenoliths in Quaternary lavas [2].

Delfin and Zaiden-Delfin [2] reported on a PNOC-EDC 1986 aeromagnetic survey that mapped the top of the basement rock at a depth greater than 2,500 or at least 3,000 m-b.s.l [2]. Possibly this basement rock overlain by a thick pile of Tertiary (?) clastic sedimentary rocks of East Mindoro Basin (in which Naujan Lake occurs). The thickness of sedimentary rocks was believed to exceed 4,000 m based on seismic reflection data [3]. Sedimentary rocks are exposed in southern part of Naujan Lake and they are comprised of semi-consolidated conglomerates, pyroclastic, sandstones, and siltstones. Reyes [4] mentioned that basement of the geothermal system in Luzon-Mindoro segment is sedimentary rocks, intruded by an igneous plutonic rock [4].

Volcanism in Montelago was reported as starting in the Pliocene [2]. Eruptive centers in the prospect area consist of four small stratovolcanoes and three exogenous domes. The deposits of these centers dominate the entire prospect area. The stratovolcanoes are Mt. Montelago, Mt. Pola, Mt. Naujan, and Mt. Dumali. The domes are Matabang Bundok, Usuyan, and Dome Hill. As a unit, these volcanics have a minimum thickness of 700 m.

Regandara [5] divided the stratigraphy of the Montelago area into eight stratigraphic units: Bagaygay Andesite, Bungao Vulcanic-Andesite Breccia, Lugta Breccia Volcanic, Matabang Bundok Basalt, Pungao Volcanic-Andesite Breccia, Montelago Tuff, Buloc Andesite, and Bambang Tuff-Andesite (Figure 2) [5]. A total of 19 faults were identified in the prospect area and these are identified as F1 to F19 in Figure 3. The four eruptive centers that suggest intrusions at the depth. Those are Mt Montelago, Mt Pungao, Mt Buloc and Mt Matabang Bundok.

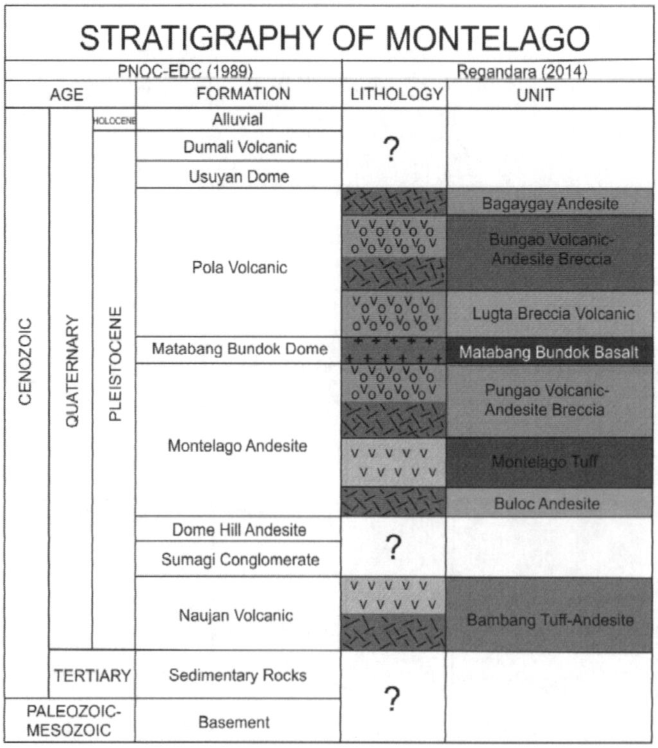

Figure 2. Stratigraphy of Montelago. The left side based on PNOC-EDC research 1989 and the other side is from Regandara.

Hydrothermal features in the Montelago geothermal prospect occur in three areas, Montelago-Pungao on the southern part (MP1 to MP9), Buloc-buloc on the northern part (BB), and Lugta (Montemayor) on the northeastern part (MM) (Figure 4). The Montelago-Pungao group of manifestations comprises the most impressive hot spring occurrences at Montelago. The spring area is near the shore of Naujan Lake. Spring temperatures range from 40-77°C with a neutral pH [6]. The Buloc-buloc warm spring is subaqueous occurring in Buloc Bay. The springs are only observable during low tide. Buloc-buloc manifestation has a temperature 36°C with a neutral pH. Montemayor hot spring is located near the sea shore in Barangay-Lugta. The temperature of this spring is 58°C and a neutral pH.

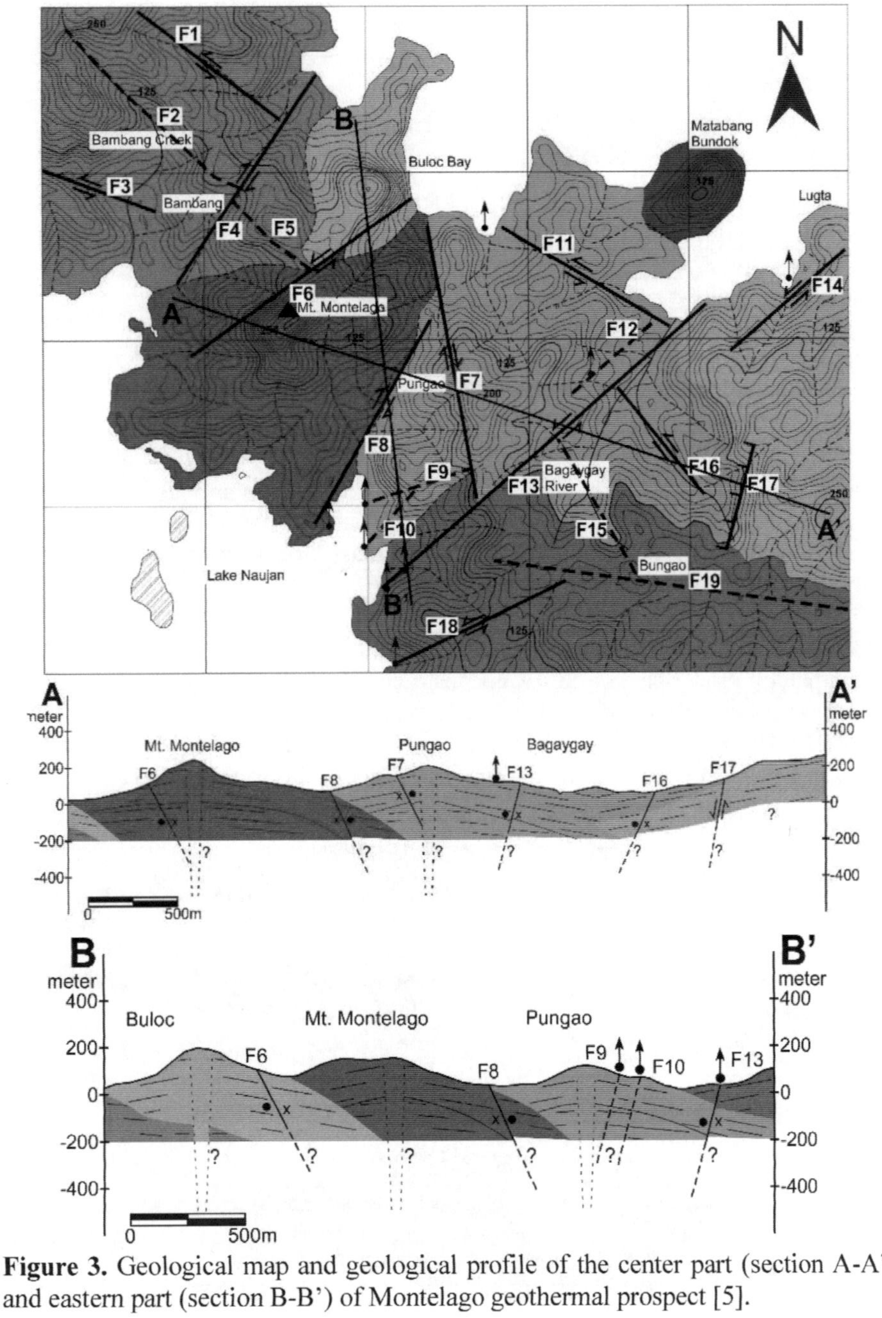

Figure 3. Geological map and geological profile of the center part (section A-A') and eastern part (section B-B') of Montelago geothermal prospect [5].

The estimation of reservoir temperature using the Na-K geothermometer is about 180-200°C. Hochstein and Browne classified geothermal reservoirs with a temperature of 125-225°C as an intermediate temperature system [7]. Therefore, Montelago geothermal prospect classified as an intermediate temperature system.

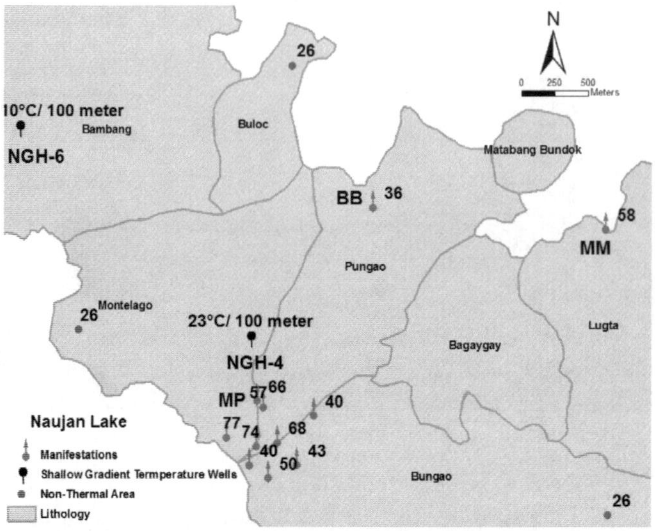

Figure 4. Map of temperature distribution in thermal springs, shallow gradient temperature holes, and the non-thermal area in Montelago. This temperature data is used to construct the 3-D isothermal model reported herein.

3. 3-D Geological Model

The subsurface lithology model up to 5 km was constructed from topography data, a geologic map, geologic sections, surface structural data, and regional stratigraphy. Topography was used with a digital elevation model (DEM) data from the NASA shuttle radar topography mission (SRTM). Figure 5a shows DEM data from SRTM which has a 3 arc second (approx. 90 m) resolution.

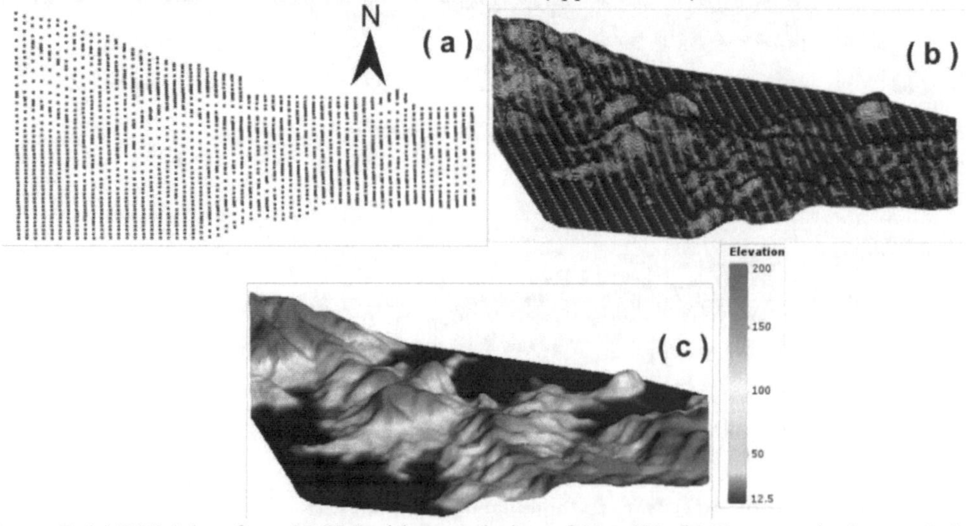

Figure 5. (a) DEM data from SRTM with a resolution of 90m [8], (b) Topography interpolation for the 3-D model, (c) Montelago 3-D topography model.

IOP Conf. Series: Earth and Environmental Science **42** (2016) 012009 doi:10.1088/1755-1315/42/1/012009

Leapfrog Geothermal provided the interpolation methods such as linear and spheroidal interpolation. This interpolation could be used to make topography points denser than the original data (Figure 5b). Based on these points, the mesh of the 3-D topography models could be created 3-D topography model in Figure 5c can be combined with a 2-D geological map on Figure 3 to generate a 3-D surface geology map. To construct subsurface lithology, geologic cross-sections are needed. The north-trending geologic sections are C-C', D-D', and E-E' (Figure 6). The structural model was constructed using strike and dip of a geologic structural measurement on the surface. The 3-D structural model illustrated in Figure 7.

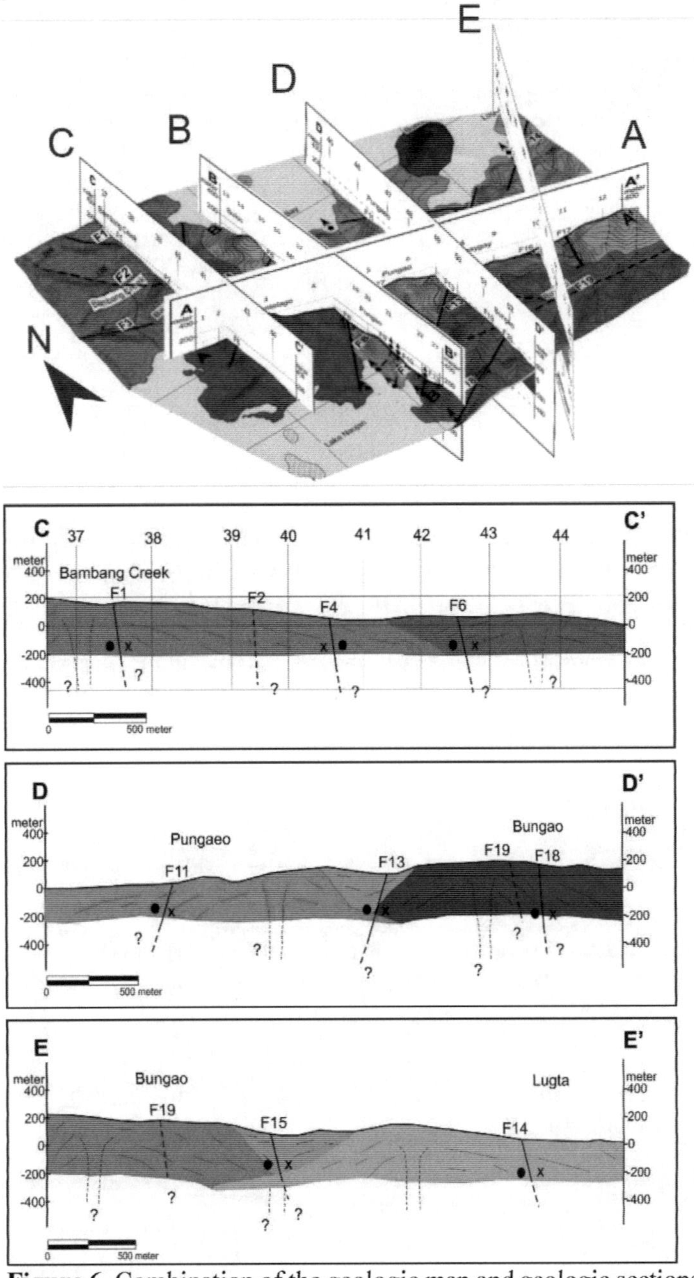

Figure 6. Combination of the geologic map and geologic sections in Leapfrog Geothermal. Section C-C', D-D', and E-E' have a north-trend.

Based on an interpretation of geologic sections A-A' to E-E', the geologic map, topography model, structural model, and stratigraphy of Montelago, a 3-D geological model was constructed by creating synthetic boreholes. There are 248 synthetic boreholes in this model. The vertical boundary for the geologic model is from an elevation of 312 m above sea level (m-a.s.l.) to 5,000 m-b.s.l.

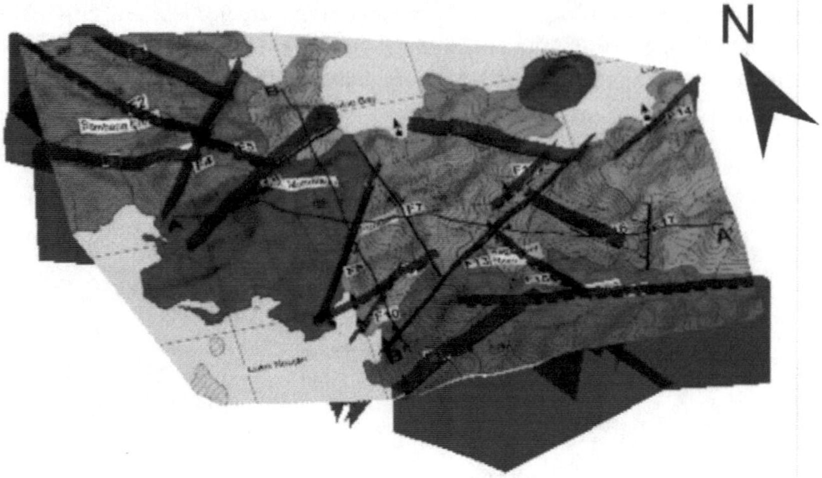

Figure 7. The structural model for 3-D geological model Montelago geothermal prospect.

The Basement rocks, as mentioned previously, at least start from 2,500 meter b.s.l. on the northwestern part of model. Sedimentary Rock have thickness at least 2,000 meter start from 1,000 or 1,500 meter b.s.l. to 3,000 or 3,500 meter b.s.l. (Figure 8a). The Sedimentary Rock was intruded by Montelago Tuff, Buloc Andesite, Pungao Volcanic-Andesite Breccia, and Matabang Bundok Basalt (Figure 8b). Bambang Tuff-Andesite on the northwestern part of Montelago have thickness around 700 – 1,000 meter spreading toward southern part. Montelago Tuff and Buloc Andesite superimposed with Pungao Volcanic-Andesite Breccia have thickness at least 500 meter. On the upper part Lugta Breccia Volcanic superimposed with Bungao Volcanic Andesite Breccia and Bagaygay Andesite (Figure 8a).

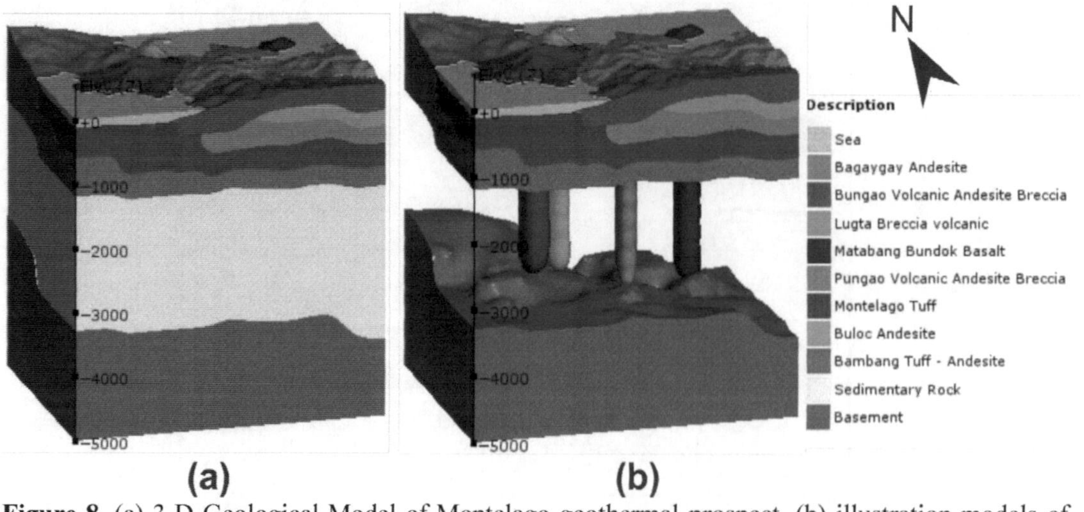

Figure 8. (a) 3-D Geological Model of Montelago geothermal prospect, (b) illustration models of volcanic rocks intruded sedimentary rocks.

4. 3-D Isothermal Model

The 3-D isothermal model was constructed using data from 11 surface manifestations, two shallow temperature gradient holes, and three non-thermal areas (Figure 4). The temperature gradient of the shallow holes was reported as approximately 23°C/100 meter (m) and 10°C/100 m (Figure 9). Where the shallow temperature gradient hole data is available, the reservoir temperature can be extrapolated to the depth by assuming conductive constant geothermal gradient (ΔG).

At -1000 m-b.s.l., the temperature of NGH-6 is 135.5°C while in NGH-4 is around 268°C. Based on ΔG, the NGH-4 temperature at -2000 m-b.s.l. is 498°C which is not reasonable. Therefore, the geothermal gradient from NGH-4 not applicable if temperature above the reservoir estimation (200°C).

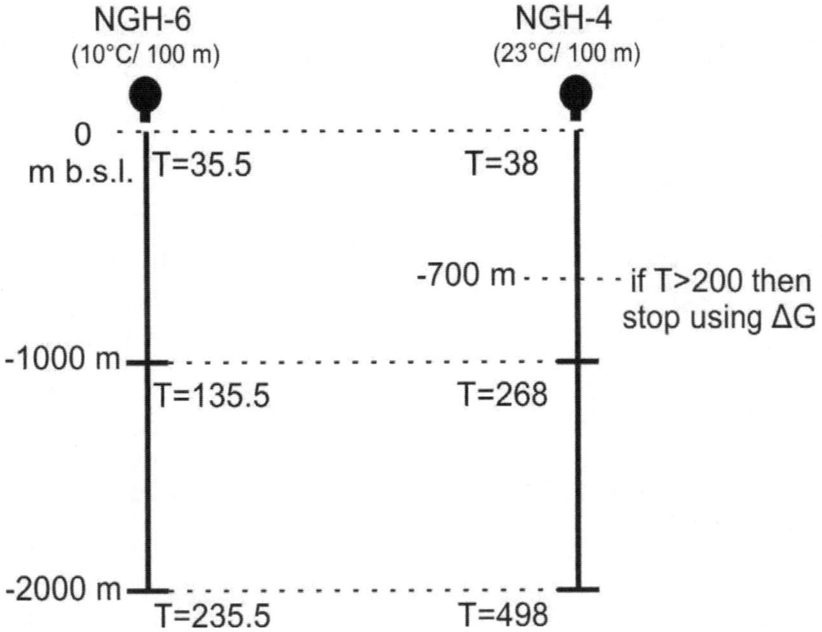

Figure 9. Illustration of gradient temperature vs depth in NGH-4 and NGH-6. ΔG is the geothermal gradient.

In the area where shallow temperature gradient hole data is absent, the surface temperature of the springs and its geothermometer derived temperature is used to estimate the geothermal gradient from the surface to the top of the reservoir. The first step in constructing a 3-D isothermal model using surface manifestation data is to determine the temperature (T_m) and the elevation (Z_m) of thermal spring. The second step is to estimate the reservoir temperature (T_{Res}) using the geothermometry of the thermal spring. In this model T_{Res} is about 180-200°C [6]. The next step is to determine the depth of top reservoir (H_{Res}). The depth of top reservoir is determined using linear interpolation from the graph of well depth (in meters) against bottom temperature (°C) developed by Hochstein and Sudarman, presented in Figure 10 [9].

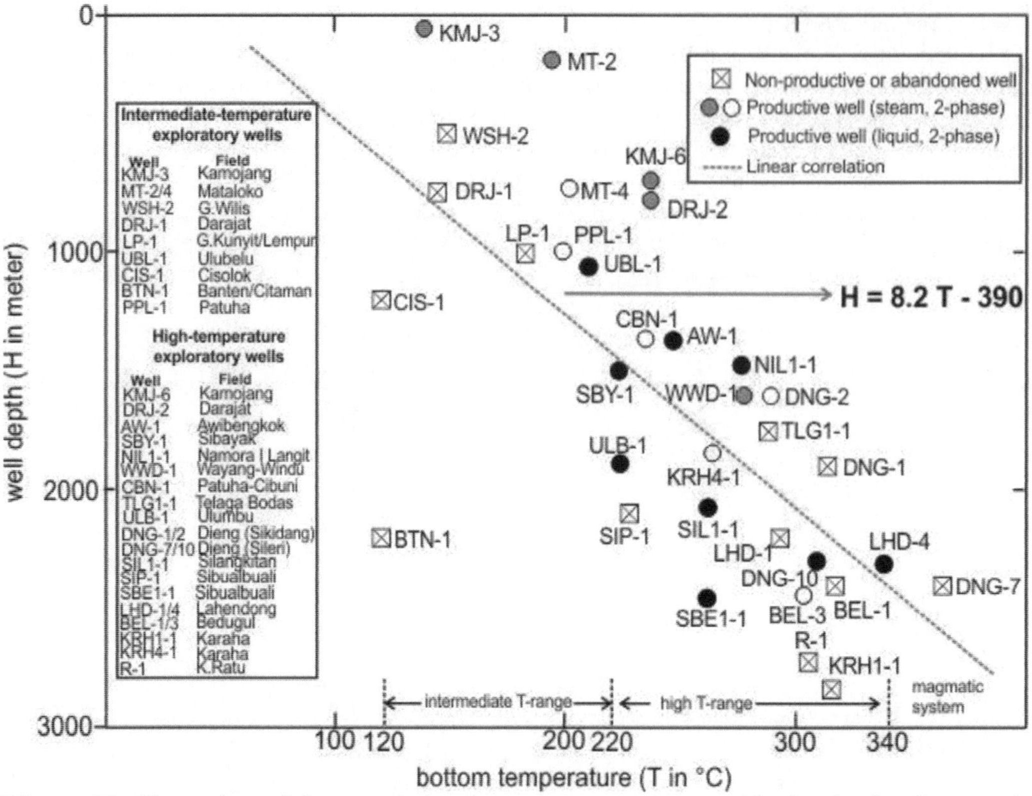

Figure 10. Illustration of the correlation between temperatures with the depth of reservoir modified from Hochstein and Sudarman.

This graphic was made by plotting bottom temperature (x-axis) and depth (y-axis) of geothermal wells (both low and high temperature as well as productive and non-productive) in many Indonesian fields and prospects. The linear correlation of productive wells (solid circles in Figure 10) have a function as follow,

$$H_{Res} = 8.2\, T_{Res} - 390 \quad (1)$$

Where H_{Res} is expected to be equal to the depth of top of the reservoir at any given reservoir temperature. Using this equation the top of the reservoir can be estimated.

Once the top of the geothermal reservoir has been determined, its depth must be converted to elevation to be able to modelled in the Leapfrog Geothermal software. Input of elevation b.s.l. in Leapfrog Geothermal have negative values thus, the elevation of top reservoir (Z_{Res}) is Z_m minus H_{Res},

$$Z_{Res} = Z_m - H_{Res} \quad (2)$$

Where, Z_{Res} is an elevation of the top reservoir in meter.
H_{Res} is the depth of top reservoir in meter.
Z_m is an elevation of surface manifestation in meter.

Once Z_{Res} has been determined, the next step is to calculate temperature (T) distribution for the depth interval from the top of the reservoir to the surface manifestation by using following linear interpolation:

$$T = T_m + \frac{(T_{Res} - T_m)}{(Z_{Res} - Z_m)}(Z - Z_m) \quad (3)$$

Where, T is temperatures in °C.
T_m is a temperature of the surface manifestations in °C.
T_{Res} is the estimate of top reservoir temperature from geothermometry in °C.
Z is elevation in meter.
Z_m is the elevation of surface manifestations in meter.
Z_{Res} is elevation of the top reservoir in meter.

This procedure is illustrated in Figure 11. Applying this procedure to the calculation of temperature distribution above the reservoir assumes that the temperature in this interval is conductive and is dominated by a clay cap, thus, the gradient is considered linear.

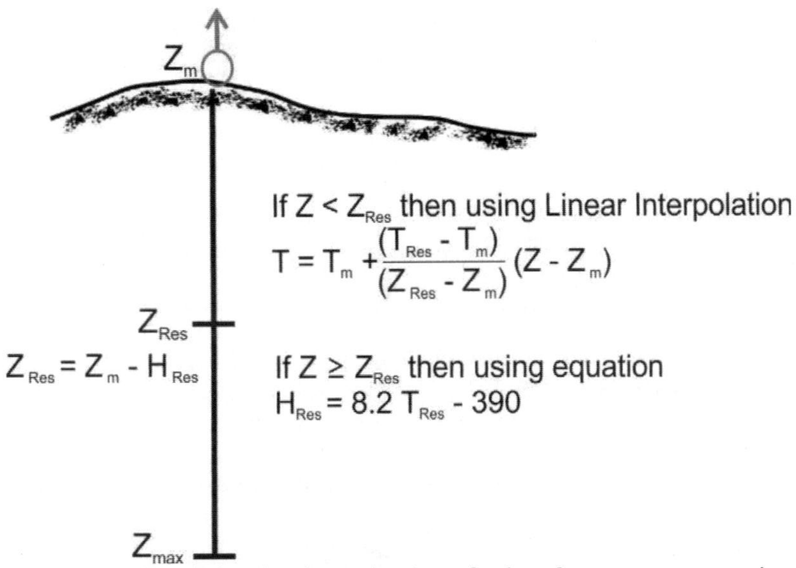

Figure 11. Illustration for determination of subsurface temperature using temperature from surface manifestation.

Because the temperature model is calculated using linear interpolation, consequently, within the reservoir there is still temperature gradient causing unrealistic high temperature at depth greater than the reservoir. However, within reservoir zone, the temperature is affected by convective heat transfer causing a low geothermal gradient or in a steam reservoir the temperature is almost constant about 240°C causing a temperature gradient close to zero. Therefore, the isothermal model created in Leapfrog Software must be interpreted with caution. The isothermal model must be constructed up to the reliable reservoir temperature according to geothermometry data or the type of geothermal system. For example, Montelago is predicted as medium to the high-temperature reservoir system, thus, the maximum reservoir isothermal model that can be estimated is 250°C. The temperature cut-off about 250°C is taken from common geothermal reservoir temperature with high temperature worldwide. The last step is to input all the data such as coordinate, elevation, and temperature up to the depth of boundary condition in the model. The summary of the calculation from surface manifestations is shown in Table 1.

In areas where surface thermal manifestations are absent or in non-thermal areas, the normal geothermal gradient (3°C/100m) and surface temperature 26°C are used to estimate isothermal temperature model to depth. The distance between the thermal area and non-thermal area is at least 1 km and must be located outside the mapped low resistivity anomaly.

Table 1. Calculated reservoir elevation and associated reservoir depth based on the linear interpolation from the graph by Hochstein and Sudarman [9].

No.	Manifestation	T_m	T_{Res}	Z_m	Z_{Res}	H_{Res}
		(in °C)		(in meter)		
1.	MP1	77	200	7	-1239	1246
2.	MP2	74	190	8	-1156	1164
3.	MP3	40	190	6	-1158	1164
4.	MP4	50	200	6	-1240	1246
5.	MP5	43	200	14	-1232	1246
6.	MP6	57	180	12	-1070	1082
7.	MP7	68	200	11	-1235	1246
8.	MP8	40	200	19	-1227	1246
9.	MP9	66	200	13	-1233	1246
10.	BB	36	200	19	-1227	1246
11.	MM	58	200	15	-1231	1246

The isothermal model from shallow gradient hole, surface manifestations and non-thermal areas are presented in Figure 12. Based on this model, 200°C is at 1,100–1,200 m-b.s.l. in the central part of the prospect. In northwestern part of the prospect a temperature of 200°C is at 1,700–1,900 m-b.s.l. which is significantly deeper than the center part. The anomaly temperature 200°C occurs in NGH-4 at 700 m-b.s.l. The thermal model of NGH-4 is overestimated because the depth of shallow gradient hole is only 300 m.

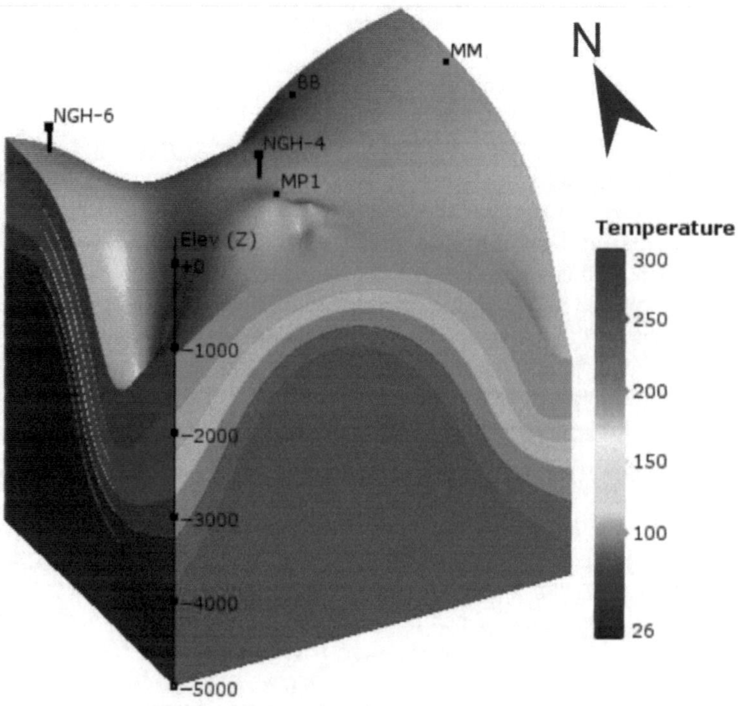

Figure 12. Isothermal 3-D model.

5. 3-D Isobaric Model

The 3-D isobar model was constructed using pressure as a function of temperature derived from the isothermal model (Figure 12). In the area where shallow holes and surface manifestations are available, the pressure can be interpolated using depth by assuming hydrostatic pressure (P_{hyd}) [10]. Hydrostatic pressure is determined as a product of density, the acceleration due to gravity, and depth. The reservoir in Montelago is assumed to be a liquid dominated system thus density is given by saturated liquid density as a function of temperature. Temperature data was obtained from the 3-D isothermal model section. The equation to calculate hydrostatic pressure is as follows:

$$P_{hyd} = \rho_{liquid\ (T)}\, g\, h + P_0 \quad (4)$$

Where, P_{hyd} is hydrostatic pressure (bar).
ρ_{liquid} (T) is saturated liquid density as a function of temperature (kg/m^3).
T is temperature (°C).
g is acceleration due to gravity (9.8 m^2/s).
h is depth (m).
P_0 is pressure at the surface (1.01325 bar)

In the area where any surface thermal manifestation is absent or in a non-thermal area, the hydrostatic pressure from equation (4) can be used. The non-thermal area is assumed as recharge area, hence, the density is still given by saturated liquid density. The distance between the thermal areas with non-thermal areas is at least 1 km and must be located outside the mapped low resistivity anomaly.

The isobaric model from shallow gradient hole, surface manifestation and non-thermal area are presented in Figure 13. Based on this model, a pressure 110 bar occurs at 1,100–1,300 m-b.s.l. in the central part of the prospect. In the northwestern part of the prospect, pressure 110 bar is at 1,000–1,100 m-b.s.l. which is shallower than the central part. In the southeastern part of the prospect, pressure 110 bar is at 900–1,100 m-b.s.l. The pressure 110 bar at NGH-4 occurs at 1,600 m-b.s.l.

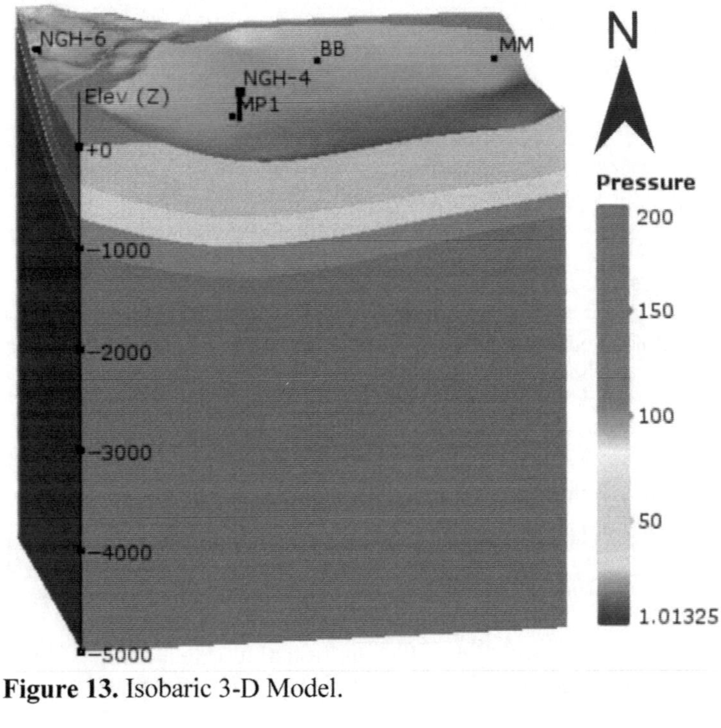

Figure 13. Isobaric 3-D Model.

6. Discussion

A 3-D geological model and an isothermal model could be used to formulate a well prognosis for exploration and production wells. The geometry and size of the reservoir could be estimated using the isothermal model. The boundary of the top of the reservoir would be used layers from the isothermal model where in this case used geothermometry 200°C while the thickness of the reservoir is assuming with 1 or 2 km [11]. Based on the isothermal model, the depth to the top of the reservoir is 1,100–1,200 m-b.s.l in the central portion of the Montelago prospect. Furthermore, the depth of the lithology units between the volcanic rocks and sedimentary rocks in the project area could be used as a well target because these intervening units are assumed to be a permeable zone; continuity of the faults could be used as well for targeting purposes. However, the result of the 3-D geologic model may not suitable as it was not constructed by the original project geologist rather the software modeler made additional geological sections for the model which may or may not be accurate.

Any model, especially in the early stages of an exploration project has its positive and negative features. For example, one difficulty in constructing a 3-D geological model in a geothermal system in a volcanic environment is identifying the boundaries between the various stratigraphic units and their thicknesses. The lithology in geothermal systems in this type of setting generally has volcanic rocks in the upper part, not sedimentary rocks. If the lithology in a project area is volcanic then the model would generate subsurface intrusions. The gap between intrusions will be blank or interpreted as unknown lithology. Usually, this gap will be filled by sedimentary rocks, but this doesn't happen if the models have all intrusion in their upper part. Additionally, the predicted temperature at depth using the geothermal gradient alone could generate anomalously high temperatures at depth. Therefore, the geothermal gradient in the model should be restricted based on the estimated temperature of the reservoir. The result of the 3-D isothermal model in the Montelago area could be used as an illustration of temperature distribution in a reservoir with a temperature of at least 200°C.

On the other hand, a 3-D geological model does begin to place constraints on the geothermal system and allow for a systematic determination of potential reservoir size among other factors. It provides an integrated, consistent summary of the relevant data and it can easily be updated with other or new data such as from geophysical model or data from temperature gradient holes and geothermal wells. It is a "living model" that allows planning for future exploration and development activities and updating as new data sets are generated.

References

[1] Leapfrog Geothermal 3.0.1 December 2015 ARANZ Geo (New Zealand: Christchurch)

[2] Delfin F G and Zaiden-Delfin M C 1989 Geology of the Montelago Geothermal Prospect, Mindoro, Southwestern Luzon *PNOC-Energy Development Coorporation Report* (Metro Manila: EDC) p 26

[3] Leynes R D C and Sanchez D R 1998 Geological and Geochemical Evaluation of the Montelago Geothermal Prospect, Orriental Mindoro, Philippines *Proc. Conf. GEOCON 98* (Metro Manila: Geology Society of the Philippines) pp 150-162

[4] Reyes A G 1990 Petrology of Philippine Geothermal Systems and the Application of Alteration Mineralogy to Their Assessment *Journal of Volcanology and Geothermal Research* **43** (Elsevier) pp 279–309

[5] Regandara R 2014 Struktur Geologi dan Potensi Zona Permeabel di Prospek Geotermal Montelago, Pulau Mindoro, Filipina Thesis Magister Teknik Panas Bumi (Bandung: Institut Teknologi Bandung) p 123

[6] Asnin S N 2014 Sistem Geotermal Montelago, Pulau Mindoro, Filipina, Berdasarkan Kajian Geokimia Air dan Tanah Thesis Magister Teknik Panas Bumi (Bandung: Institut Teknologi Bandung) p 99

[7] Hochstein M P and Browne P R L 2000 Surface Manifestations of Geothermal System with Volcanic Heat Sources *Encyclopedia of Volcanoes 1st edition*, ed H Sigurdsson et al (New York: Academic Press) pp 835-856

[8] SRTM 90m Digital Elevation Data 2008 CGIAR - Consortium for Spatial Information http://srtm.csi.cgiar.org/
[9] Hochstein, M.P. and Sudarman, S. 2008 History of Geothermal Exploration in Indonesia from 1970 to 2000 *Geothermics* **37** (Elsevier) pp 220-266
[10] Grant M A and Bixley P F 2011 *Geothermal Reservoir Engineering - 2nd edition* (Burlington: Elsevier) p 349
[11] Badan Standardisasi Nasional 1999 *Metode Estimasi Potensi Energi Panas Bumi – Standar Nasional Indonesia* SNI 13-6171-1999 (Jakarta) p 28

IOP Conf. Series: Earth and Environmental Science **42** (2016) 012010 doi:10.1088/1755-1315/42/1/012010

A Comprehensive Well Testing Implementation during Exploration Phase in Rantau Dedap, Indonesia

M T Humaedi, Alfiady, A P Putra, R Martikno, J Situmorang

PT Supreme Energy
Sentraya Tower 23rd Floor. Jl. Iskandarsyah Raya 1A, Kebayoran Baru. Jakarta 12160 - INDONESIA

E-mail: tamrin-humaedi@supreme-energy.com

Abstract. This paper describes the implementation of comprehensive well testing programs during the 2014-2015 exploration drilling in Rantau Dedap Geothermal Field. The well testing programs were designed to provide reliable data as foundation for resource assessment as well as useful information for decision making during drilling. A series of well testing survey consisting of SFTT, completion test, heating-up downhole logging, discharge test, chemistry sampling was conducted to understand individual wells characteristics such as thermodynamic state of the reservoir fluid, permeability distribution, well output and fluid chemistry. Furthermore, interference test was carried out to investigate the response of reservoir to exploitation.

1. Introduction

Rantau Dedap Geothermal Field is located in South Sumatera; Indonesia. The prospect is situated in the northern east of the Great Sumatra Fault. Abundant of fumaroles and chloride springs defined the presence of geothermal potential in the area. The later geologic, geochemical, and geophysical surveys delineated the potential extension of the geothermal system. Figure 1 presents the location of Rantau Dedap field and the surface features distribution.

In 2014-2015 PT Supreme Energy Rantau Dedap (SERD) drilled six full diameter exploration wells in order to confirm the existence of geothermal system and to assess the field production capacity. During the course of exploration drilling comprehensive well testing surveys were executed, consisting of static formation temperature test (SFTT), completion test, heating-up test, production test, and interference test. The well testing programs were designed with an aim to provide a reliable and inclusive data that will be used as a basis to conduct resource assessment. Key parameters that the well testing intended to confirm include initial thermodynamic state of reservoir fluid, permeability structures, reservoir size, well output, and fluid chemistry [1]. Moreover, some well testing surveys was also planned to provide technical justification on decision making during the sixth well drilling execution. In the subsequent paragraph the implementation of the well testing programs are presented.

IOP Conf. Series: Earth and Environmental Science **42** (2016) 012010 doi:10.1088/1755-1315/42/1/012010

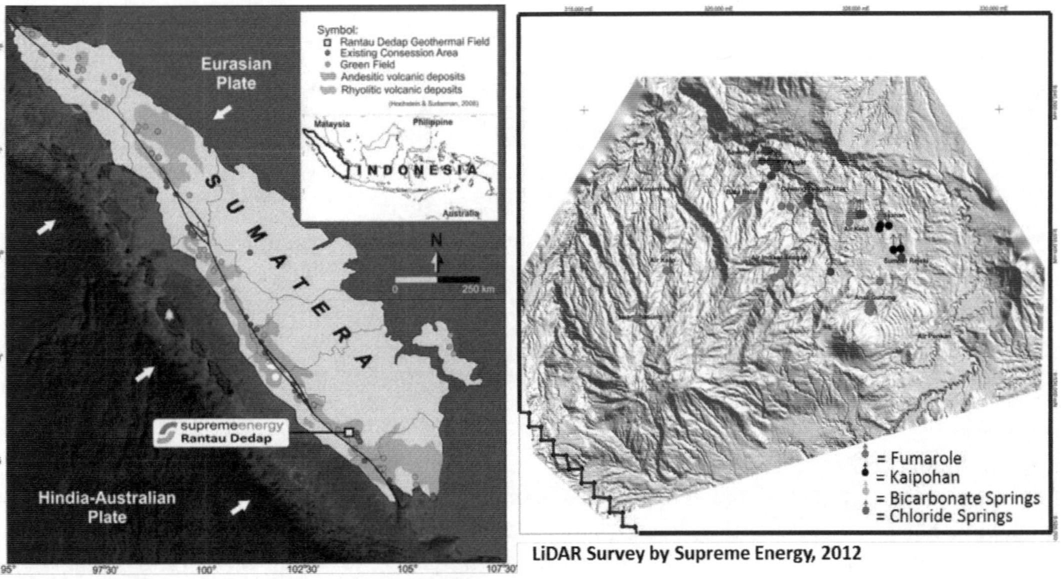

Figure 1. Rantau Dedap prospect location and surface features distribution.

2. Static Formation Temperature Test (SFTT)

The main objective of this test is to estimate the undisturbed temperature at certain depth during drilling that can assist the drilling program in setting production casing shoe (PCS) at appropriate depth where the temperature is expected to be 220-230°C. This is normally done at exploration stage since no downhole measurement data available [2].

To determine PCS setting depth, SFTT has been conducted in the first three wells, providing undisturbed temperature estimate which supplement other information gathered from drilling records such as Methylene Blue (MeB), the appearance of high temperature mineral (epidote), drilling breaks, and loss of circulation. Ideally the test should be conducted before loss of circulation to allow conductive heat flow only, so in practical when loss of circulation occurred close to the initially planned survey depth, drilling was stopped immediately and SFTT was performed 50 m above the bottom of the hole. The heating up duration was set to be longer than circulation time at specified survey depth.

Several methods were used to estimate undisturbed temperature such as Roux, Brennand, Ascencio and Horner Method. In RD-B2 well SFTT was conducted at a depth of 792 mMD at which the drilling records (i.e. MeB<10, appearance of epidote) indicated the temperature of rock being drilled is 200-220°C. Figure 2 (a) shows how the transient temperature recorded from SFTT being extrapolated using Ascencio Method, providing an estimate temperature of 205°C. Brennand Method provided 192°C while Horner Method provided more conservative number of 174°C. Eventually, pressure-temperature (PT) heating up measurements revealed the actual temperature of 207°C, concluding the Ascencio's as the closest estimate.

Figure 2 (b) displays complete set of SFTT calculation results using the above methods compared to the actual stabilized temperature in the first three wells. Good agreement can be found in the RD-B2 where no partial losses occurred prior the test, giving deviation of less than 10°C using Ascencio Method. In general, estimated temperature is less than the actual one due to limited heating-up time (i.e. related with operation rig cost) and non-ideal conductive heat flow due to partial losses.

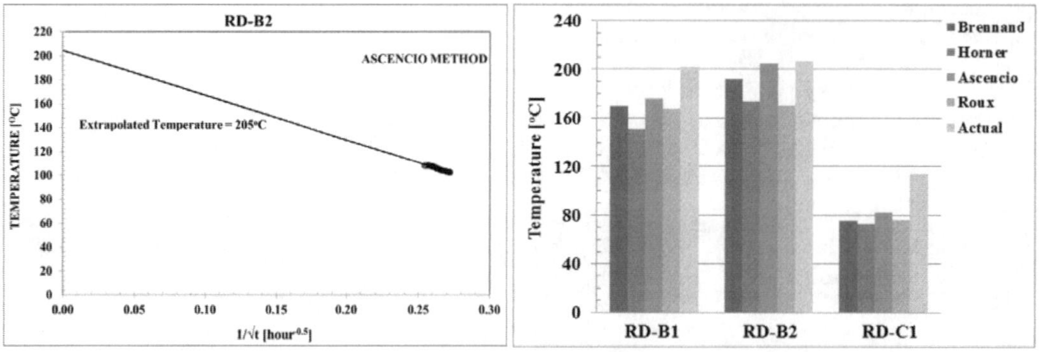

(a) (b)

Figure 2. Calculated undisturbed temperature using different methods.

3. Completion Test

Completion test was carried out immediately after drilling completed by utilizing the drilling rig equipment. The test consist of a series of surveys including dummy run, pressure-temperature-spinner (PTS) injection, multiple-rate injectivity test (II test) followed by fall-off test (FOT), run sequentially, spending approximately 27 hours rig time. The objective of these measurements is to acquire information regarding feed zone(s) depth, individual feed zone permeability, well injection capacity, permeability thickness, and near wellbore permeability [1]. Surface Read-out (SRO) measurements were done to observe the measured data in real-time, which allowed faster data acquisition, thus minimizing rig time. A completion test program has been designed to provide reliable data with minimum rig time. In general the completion test program is described in Figure 3. The test was started with dummy run to assess wellbore conditions and to investigate the well maximum clear depth. Such information consequently provides aid to design the subsequent surveys.

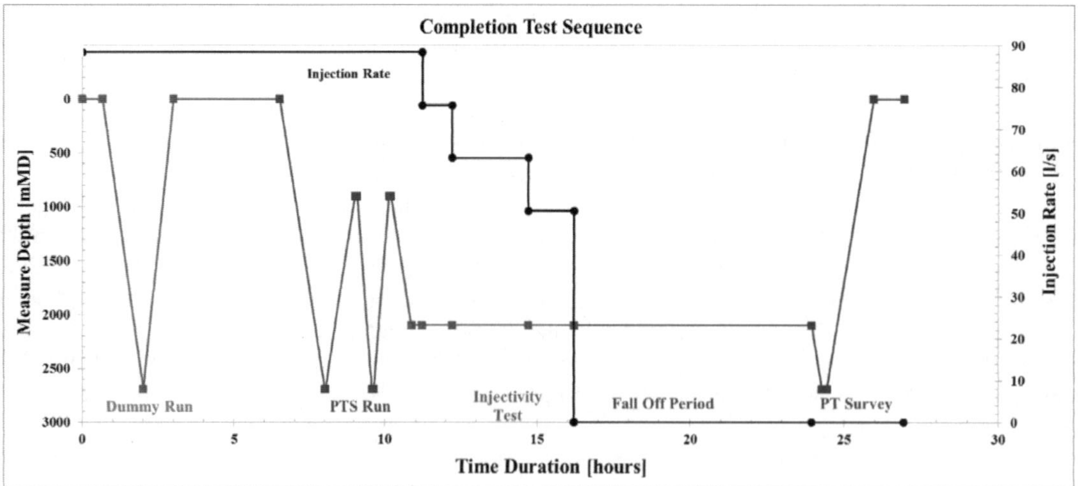

Figure 3. General completion test sequence.

3.1. *PTS Injection*

PTS injection test was conducted to identify location of the permeable zones (feed zones) within the tested well. The data was furthermore analyzed to quantify mass loss to each identified zone. In some cases reservoir fluid may flow into the wellbore (interzonal flow), requiring unconventional analysis to carry out. A comprehensive analysis involving spinner interpretation and mass and heat balance calculation was done to provide a detailed report of feed zones characteristics [3]. Figure 4 demonstrates calculation workflow in RD-B1 where downflow was observed during the test.

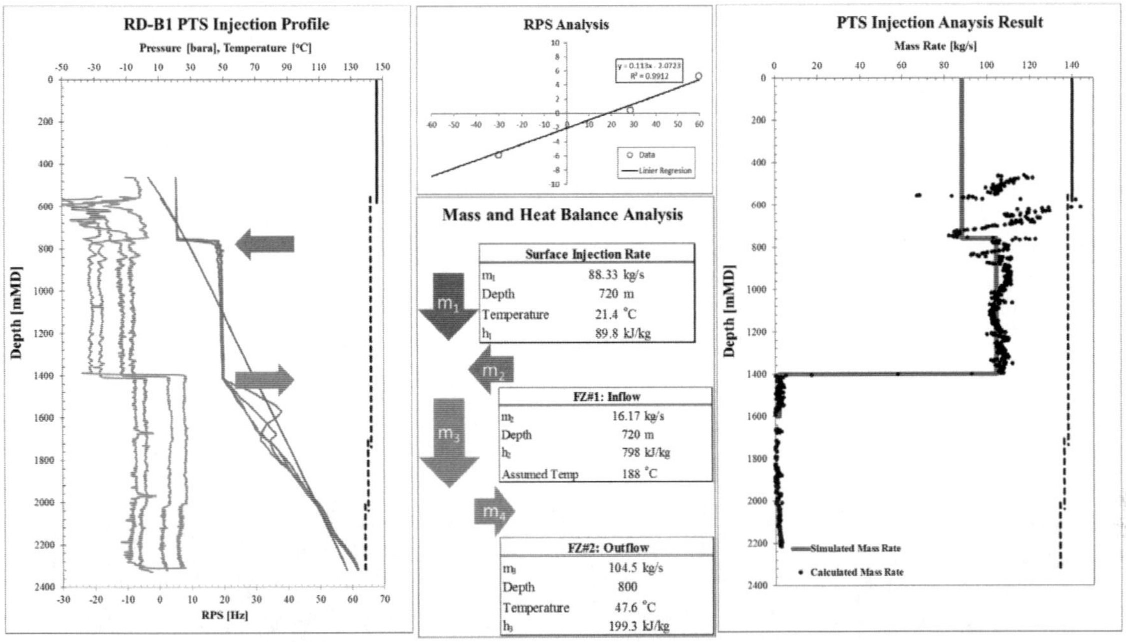

Figure 4. PTS injection calculations for a well with downflow.

3.2. Multi-rate Injectivity Test

Multi-rate injectivity test is conducted to obtain well Injectivity Index (II) value from which the well injection capacity can be estimated. The II value is essential in providing a measure of well permeability and when combine with pressure and temperature data could provide a preliminary indication of the well production capacity. Step down isochronal multi-rate injectivity test is selected to be performed in order to minimize rig time utilization. In this test, the well was injected with multiple injection rates while downhole pressure recorder was set close to pivot point which expected located adjacent to the identified major feed zone location from PTS Injection. SRO utilization during the test was very helpful in determining the corresponding pseudo steady state downhole pressure condition at each injection rate. Figure 5 provide an example of typical step down multi-rate injectivity test applied in each exploration well.

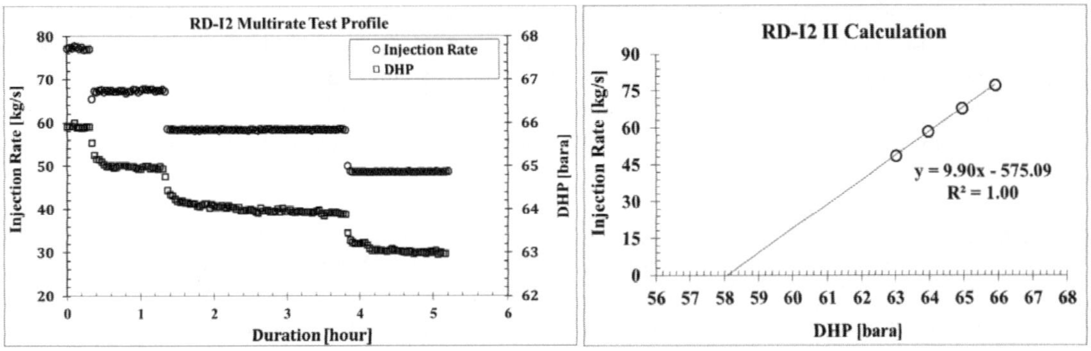

Figure 5. Injection test calculation.

3.3. Fall-off Test

Generally this test is performed right after II test completed by recording downhole pressure at particular depth after injection has stopped. The well pressure recovery data was analyzed using type curve matching to provide initial estimate of the rock transmissivity (kh) which translates well permeability, near wellbore condition and boundary condition.

Due to dynamic wellbore conditions such as interzonal flow, gas trap in the wellbore, steam rising, and temperature heat-up, the FOT test provided poor data quality, leading to noisy pressure derivative that would give unreliable interpretation. However, attempt to analysis fall-off data was made where unfavorable circumstances have less effect on data quality. Figure 6 illustrates fall-off data analysis in RD-B2. A good match of the pressure and its derivative was obtained by using a Dual Porosity model (Ω=0.05, λ=7E-8) with no boundaries. The reservoir properties inferred to obtain this match are:

- Initial reservoir pressure (Pi) = 52.4 Bara (consistent with the Multi-rate Injectivity Test observation)
- Transmissivity (kh) = 25.8 Dm
- Skin (S) = -1.5

Using 350 m of reservoir pay thickness (from 870 to 520 masl), the reservoir permeability at macroscopic scale is evaluated to be around 75 mD.

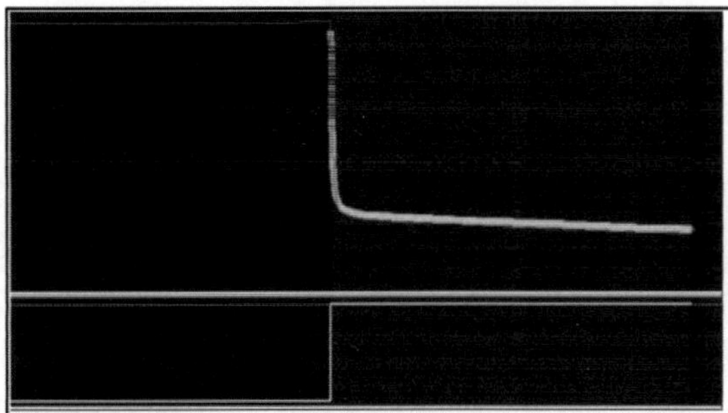

History plot (Pressure [bara], Liquid Rate [m3/D] vs Time [hr])
(a)

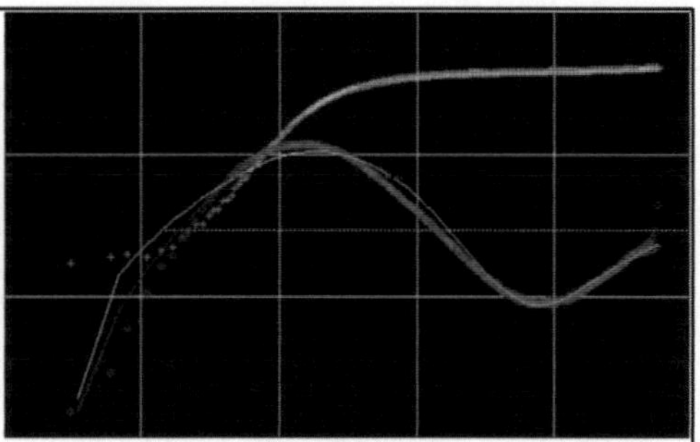

Log-Log plot: dp and dp' [bar] vs dt [hr]
(b)

Figure 6. Fall-off data analysis in RD-B2. (a) Pressure match (data in green, model in red) and rate history (in yellow); (b) Match of the pressure (data in green, model in red) and its derivative (data in red, model in white).

4. Heat-up Surveys

After completion test completed, the well was put in shut-in to allow the reservoir temperature heats up. During this period, pressure-temperature (PT) surveys were conducted to monitor natural pressure and temperature recovery. As general rule of thumb in geothermal industry, at initial, PT heat-up surveys was planned to be conducted on day 1, 3, 7, 14, 25 and 45 after completion test with flexibility to be extended depending on the temperature recovery required [2]. Such stabilized profiles consequently provide essential elements for initial state numerical model calibration. In addition the temperature profile from day 1, 3, and 7 were usually extrapolated to estimate undisturbed temperature profile, providing additional information for the next well drilling program.

4.1. Temperature

In the implementation most of the wells required longer time to fully heated-up, leading to more surveys conducted. The slow temperature recoveries observed in some wells is affected by huge amount of drilling fluid loses into the reservoir and natural state condition of the reservoir. Consequently air cap discharge stimulations were conducted in several wells in order to expedite temperature recovery.

Figure 7 illustrates heating-up process in RD-B1. Temperature extrapolation is drawn based on data on day 1, 7 and 14 which indicates good agreement with temperature after flowing (day 389) and initial temperature data. The extrapolation data was required to support decision on next drilling target.

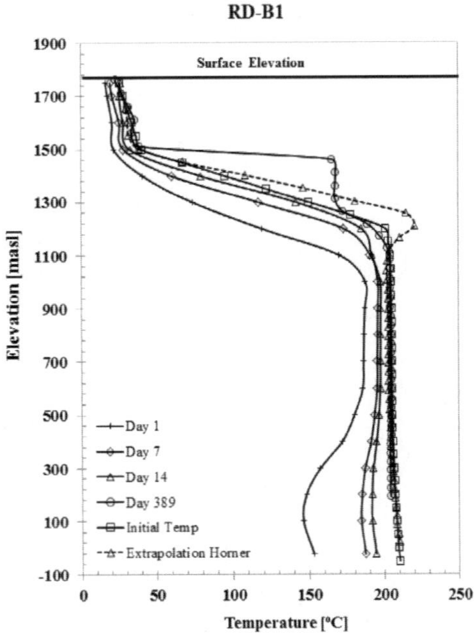

Figure 7. Heat-up temperature profiles in RD-B1.

4.2. Pressure

As the well heats-up, the pressure in the wellbore is dynamically shifting due to hydrostatic gradient changes on liquid column. There is a pivot point located at depth where pressure remains stable. Normally, the pivot point corresponds to feed zone depth at single feed zone well. At multiple feed zones well, the pivot point located between feed zone depths weighted by their injectivity index and fluid enthalpy [1].

Figure 8 shows the initial reservoir pressure correlation constructed from the pivot point of all wells. It seems that the wells have been drilled into the same pressure regime with the gradient 0.08255 bar/m, corresponding to 230°C reservoir temperature. Exception is RD-I2 where the pivot point is lower than the pressure correlation. This probably indicates that RD-I2 drilled into the upflow of the system, which is consistent with the fact of being the hottest well in the field.

4.3. Fluid State

Temperature and pressure in the wellbore will stabilize as heating-up period ends, providing initial temperature and pressure of the field. Such information provides fundamental elements for the construction of conceptual model and numerical model. Boiling curve with depth (BCWD) was also constructed to understand fluid state in the initial condition of reservoir, whether the fluid is compressed liquid, two-phase or dry steam.

As illustrated in Figure 9, initial temperature and pressure profile completed with BCWD suggested that the reservoir is liquid dominated with thin steam cap. It is also noticed that all wells have similar pressure near 1000 masl indicating good horizontal permeability on the field scale [4].

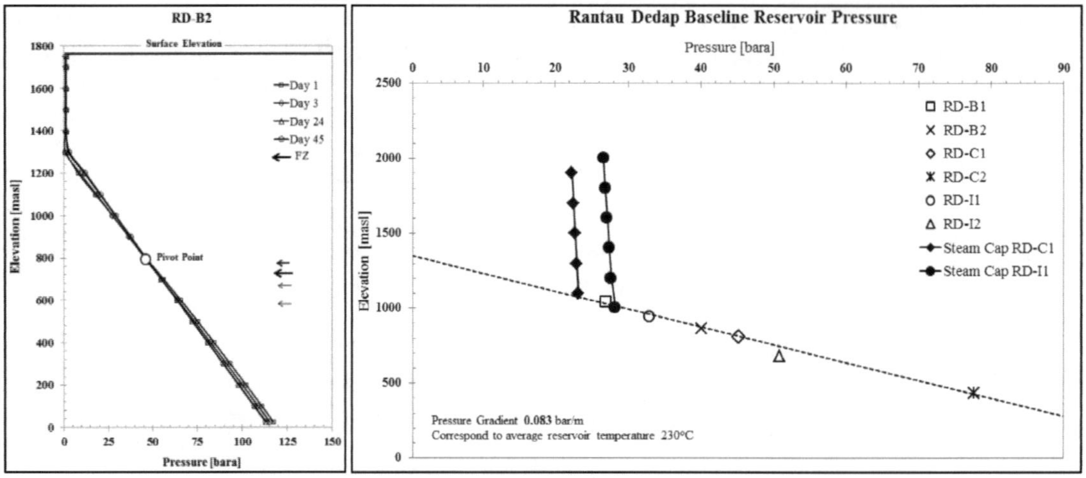

Figure 8. Pivot point identification and pressure distribution.

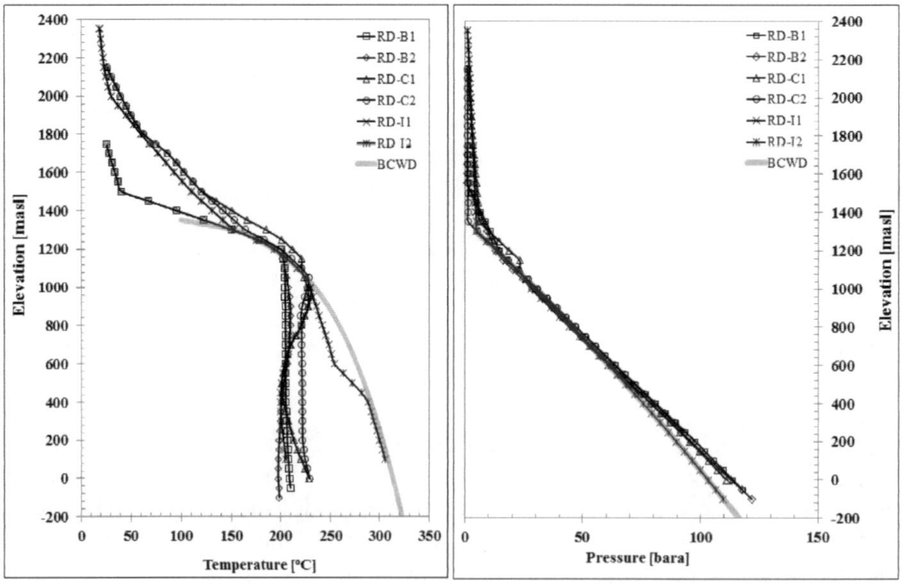

Figure 9. Initial reservoir temperature and pressure RD.

5. Production Testing

5.1. Discharge Initiation

Normally discharge attempt is made once heating-up reached conditions ready to flow. Due to slow heating-up process, however, some wells needed stimulation to initiate discharge. It was expected that by allowing the well to flow will expedite the heating up process, allowing more key information to be gathered earlier (i.e. well production capacity, actual temperature profile). This key information is then used as important input to possibly revise the subsequent wells drilling program. Air lift and air cap/air compression have been conducted to stimulate flow [5].

Air lift method requires drill pipe as medium for the high-pressure air injected from air compressor up to 2300 scfm. This method can target specific depth yet operationally limited since killing the well is required by re-injection cold fluid in order to pull the drill pipe out. In addition, the production parameters estimated such as total flow rate, steam and enthalpy did not reflect actual well performance. However, air lift method that had been applied in the first and the second well improved the chance of air cap stimulation by heating-up the upper cold section.

Air cap stimulation is the common stimulation method to initiate discharge by injecting high-pressure air through wing/side valves, compressing a liquid level down to production casing shoe/permeable formation. This allows the liquid to get heated up by higher downhole temperature. Once the liquid gets heated up to sufficient temperature and pressure decrease during opening the well, phase changing will occur and allowing lower density fluid travel up to the surface [2]. In most of the wells, wellhead pressure after air injection can be up to 42 barg, depending on liquid level and PCS shoe depth. All stimulated wells are successfully discharged by this method.

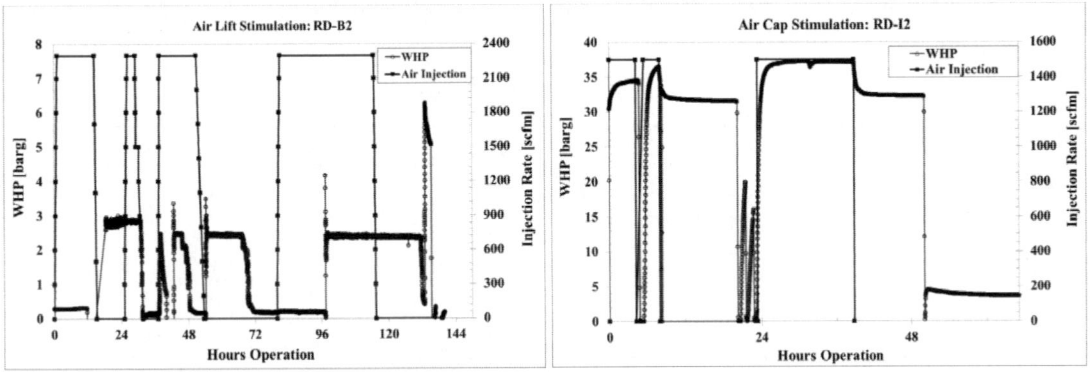

Figure 10. Typical airlift and air cap wellhead profiles and air injection rate.

5.2. Production Test

During production test, production parameters such as fluid enthalpy and total flow rate were continuously measured using James Lip Pressure method. In addition, Pressure-temperature-spinner (PTS) flowing is also taken to confirm productive feed zones location, measure their temperature, assess their individual contribution and calculate productivity index (PI). Furthermore, chemistry sampling and Tracer Flow Test (TFT) were conducted to identify fluid chemistry and validate James Lip Pressure measurements. Both short-term and long-term production test had been carried out for all exploratory well to allow evaluation of well production sustainability with time.

As initial plan for production wells, PTS flowing was planned to be conducted twice along with TFT survey. The PTS flowing is conducted at the early time of the well reaching stabilize flowing wellhead pressure condition and at the end of production test, which is usually parallel with Flow Performance Test (FPT). However, the actual number of the implemented PTS flowing surveys again is highly depend on well performance and downhole data requirement to support dynamic exploration program.

As an example, RD-I2 comprehensive surveys are provided here. The discharge was started on day 43. Calculation on production parameters with James lip method has been validated by PTS flowing and TFT result, leading to weirbox height calibration by adding +1.5 cm to the field data reading [6]. The resulting discrepancy on calculated enthalpy is 5% and 6% on total mass flow. Comparison on discharge test parameters between TFT and James Lip (weir box readings corrected by 1.5 cm) is presented in Table 1 while summary of discharge parameters over the time and various flow control valve (FCV) is presented in Table 2.

Figure 11 shows calculated production parameters and flowing wellhead pressure monitoring. As shown in the figure, steady increase on flowing wellhead pressure was observed from day 43 to day 111 as the well heated-up. Sharp increase in RD-I2 flowing WHP was observed on day 111 – day 116 and gradually decreased afterwards. These flowing behaviors were evaluated by running PTS during steady increase of WHP (day 59 and day 68) and gradual decrease of WHP (day 156).

Table 1. Comparison between TFT and James Lip Pressure method after calibration.

WHP	**3.8**		**5.6**		**10.7**	
Method	James Lip	TFT	James Lip	TFT	James Lip	TFT
Enthalpy (kJ/kg)	972	909	1072	1013	1393	1351
Total Flow Rate (kg/s)	47.4	47.6	68.1	64.5	104.8	93.4
Steam Rate (kg/s)	7.8	7.1	12.9	11.2	31.7	27.9
Brine Rate (kg/s)	39.7	40.5	55.2	53.3	73.0	65.5

Table 2. RD-I2 discharge parameters summary (after calibration on weir box head).

Day after Completion	**WHP**	**FCV**	**Enthalpy**	**Total Mass Flow Rate**	**Steam Rate**
	(barg)	**(%)**	**(kJ/kg)**	**(kg/s)**	**(kg/s)**
46	3.47	100	933	46.4	7.0
77	5.75	100	1,065	71.0	13.0
107	7.95	100	1,130	92.3	17.9
138	10.57	100	1,407	101.4	31.5
162	13.75	10	1,313	66.4	16.2

The analysis of PTS flowing involves spinner interpretation and wellbore simulation. The method had provided confirmation of feed zone characteristics such as location, mass rate contribution and in-situ fluid enthalpy. Drift flux correlation was applied in wellbore simulation to match pressure and temperature measurements and interpreted velocity [3].

An example of PTS flowing analysis from measurement during high flowing wellhead pressure (after sharp increase) is illustrated in Figure 12. A good match of wellbore model with observed data obtained by assigning three feed zones, suggesting that the deepest feed zone was at boiling condition. Such boiling could explain the sharp increase of wellhead pressure recorded at the surface.

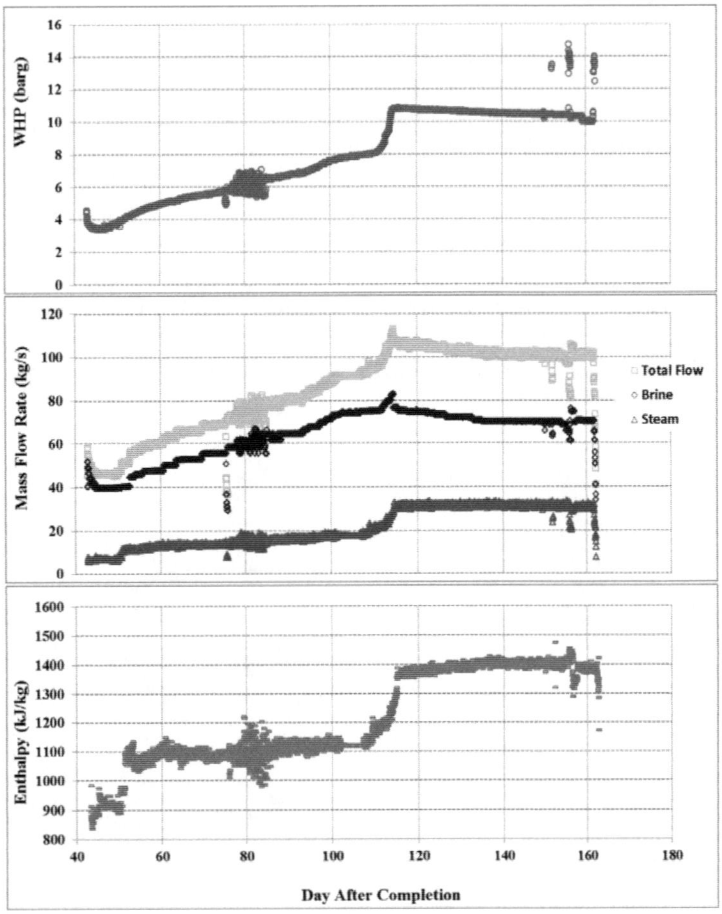

Figure 11. RD-I2 production parameters monitoring.

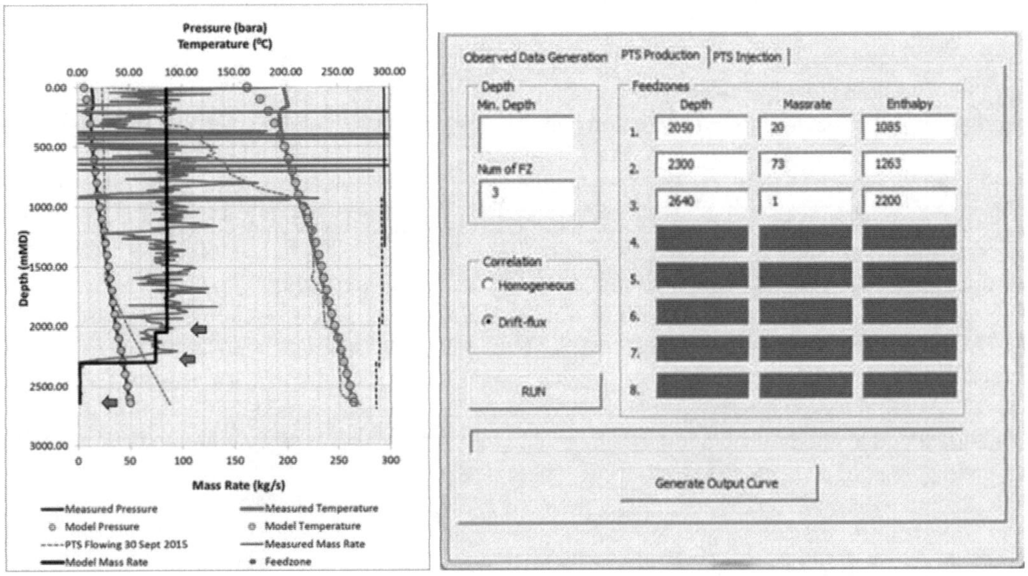

Figure 12. PTS flowing analysis using wellbore hydraulic model.

5.3. Flow Performance Test

Just before shutting-in the well at the end of the long-term discharge test, a flow performance test (FPT) was performed. The objectives of this test are to construct well output curve and determine maximum pressure discharge (MPD). This test was accomplished by throttling the well at different openings of the Flow Control Valve and monitoring the production parameters over these different FCVs.

In RD-I2, the test was performed at FCV openings, consecutively 100, 50, 20, 15 and 10% after four months of flowing period. The MPD was reported at 14.8 Bara at 10% opening. Production parameters were calculated by calibrated James Lip Pressure method. Construction of output curve was accomplished by plotting these parameters against correspondence flowing wellhead pressure. Measured production parameters computed by TFT are also reported (empty dots). The changes of enthalpy with the different openings quite clearly indicate that the respective contribution of each feed zone changes as well (the higher the WHP, the lower the enthalpy and thus the lower the contribution of the deep boiling feed zones).

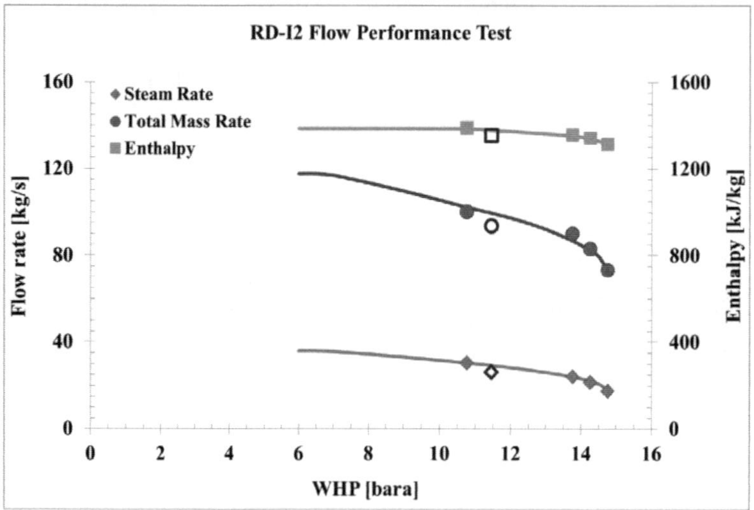

Figure 13. Output curve (James Lip Pressure method in solid dots, TFT in empty dots).

5.4. Chemistry Sampling

Comprehensive fluid chemistry sampling consisting of sampling analysis from brine, NCG, condensate and isotope was conducted during production test. The objectives are to characterize fluid origin and its features, reservoir process, scaling analysis and ultimately support conceptual and hydrology model.

Normally, sampling activity is performed along with TFT survey. In Rantau Dedap, chemistry sampling was planned to be conducted 4 (four) times during production test. This program was made based on assumption that the well needs sufficient time to reach stabilization, thus monitoring during developing phase and reach stabilization will allow the analysis thoroughly.

Throughout the extensive production test, fluid chemistry sampling was taken in 6 (six) exploration well. The fluid chemistry results suggest that the reservoir fluid is benign with neutral pH (range 7-8) and considered as chloride type fluid. The fluid have no tendency of corrosive and scaling also relatively low NCG content 0.07 – 0.8 wt% in steam. The geo-thermometer suggested that reservoir temperature in Rantau Dedap Geothermal field ranging from 220 to 280°C which is align with measured downhole reservoir temperature and calculated production parameters.

6. Interference Test

Interference test aims to reveal information regarding inter-well connectivity in a field scale, thus it involves minimum 3 wells acted as producer, injector and monitoring well. Well assignment is done after all data regarding estimate production or injection capacity; and stabilized pressure and temperature are obtained. The test was accomplished over 5 months duration by assigning all six wells; RD-B1, RD-C2, RD-I1 and RD-I2 as producer, RD-B2 as injector and RD-C1 and RD-I1 (later) as monitoring well. Figure 14 briefly demonstrates well location and its assignment.

Figure 15 shows the interference monitoring history during exploration stage. During the first month of monitoring, RD-C1 suggested no decline in pressures as only RD-I1 was being produced. As RD-I2 start to produce (gradually increasing in flow), RD-C1 began declining during the second month. Together with simultaneous production of RD-B1, RD-C2 and after the transition to high flow rates at RD-I2, RD-C1 reached a decline rate of about 0.5 bar/year. This decline is considered as relatively modest rate. After production, RD-I1 assigned as monitoring well. No measureable decline on the pressure trend over 2 months of monitoring. The low pressure decline rates in these wells could possibly explained by inducing boiling in upper permeable zones within the reservoir thus may buffer the pressure declines induced by production. Existence of steam cap zone in the natural state condition observed on RD-I1 and RD-C1 support the explanation of downhole pressure monitoring response.

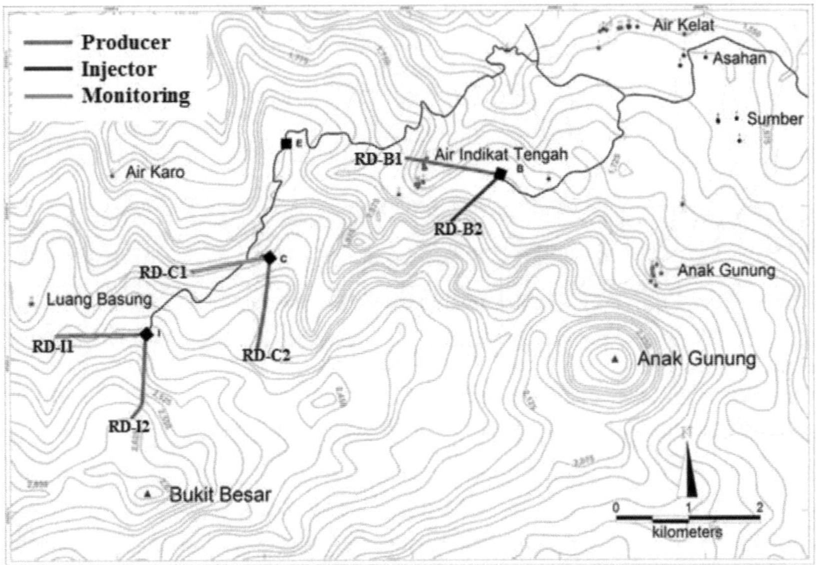

Figure 14. Well location and its assignment.

7. Conclusion

The intensive well testing had been performed in Rantau Dedap geothermal field during exploration stage from year period of 2014 - 2015. The data sets obtained have provided adequate reliable information to perform complete individual well characterization process and thorough resource assessment for supporting Rantau Dedap development stage decision. Early well data obtained from SFTT, completion test and heat-up surveys evaluation had allowed necessary adjustment in the exploration program to achieve a successful geothermal exploration. In addition, implementation of both short-term and long-term production test period as well as fluid chemistry sampling for all exploratory wells had allowed prudent evaluation in well output sustainability with time and confirmation of benign fluid chemistry.

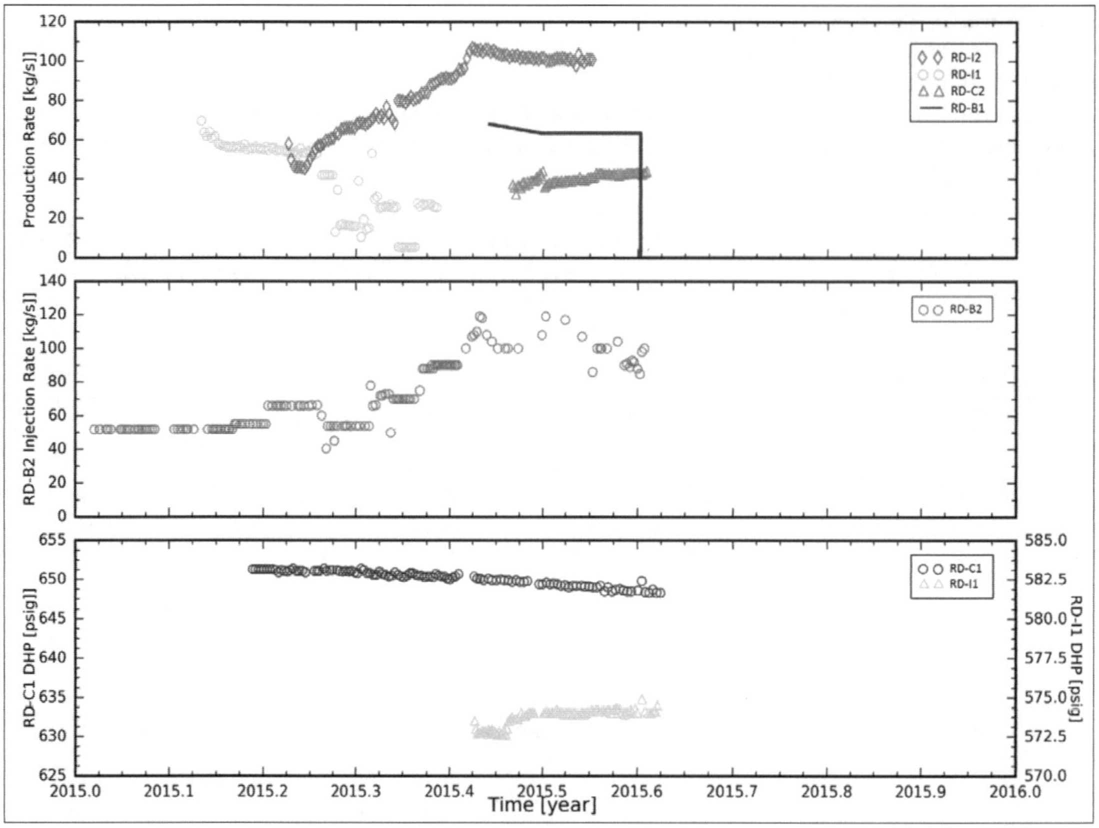

Figure 15. Rantau Dedap interference test history.

Acknowledgment
The authors wish to thank the management of SERD for the permission to publish this paper. Discussions with SERD Subsurface team members are very much appreciated.

References

[1] Grant, M A and Bixley P F 2011 *Geothermal Reservoir Engineering – Second Edition* (Oxford: Elsevier)

[2] Sarmiento Z F 2011 *Application of Well Testing in Assessing Geothermal Resources* (El Salvador: UNU-GTP and LaGeo)

[3] Situmorang, J 2012 *Pengembangan Program Komputer "PTS3" Untuk Karakterisasi Zona Permeabel dan Aliran Fluida dalam Sumur Panas Bumi Berdasarkan Survei PTS* (Institut Teknologi Bandung: Program Studi Magister Teknik Panas Bumi)

[4] Bacquet et al 2016 *Inverse Numerical Modeling of Rantau Dedap Geothermal Field after Six Exploration Wells* (California: Proceedings 41st Workshop on Geothermal Reservoir Engineering, Stanford University)

[5] Bacquet et al 2014 *Geothermal Wells Discharge Initiation Using Air Lift And Air Compression: Case Study With And Without Downflow In The Wellbore* (New Zealand: Proceedings 36th New Zealand Geothermal Workshop)

[6] Martikno et al 2013 *Evaluation of Weirs Calculation to Estimate Well Capacity: A Numerical Study* (Jakarta: Proceedings 13th Indonesia International Geothermal Convention & Exhibition)

Economic and Thermodynamic Analysis for Preliminary Design of Dry Steam Geothermal Power Plant (GPP) with Multifarious Gas Removal System (GRS) in Kamojang, West Java, Indonesia

Aloysius Damar Pranadi, Sihana, Kutut Suryopratomo, Fiki Rahmatika Salis

Nuclear Engineering and Engineering Physics Department, Faculty of Engineering, Universitas Gadjah Mada, Jl. Grafika 2, Sleman, Yogyakarta, Indonesia

E-mail: damarpranadi@gmail.com

Abstract. Indonesia has great number of geothermal potential separated by two kind of potential, 16.13 GW for high enthalpy and 7.88 GW for low enthalpy speculative resources [4]. In the end of 2013, Ministry of Energy and Mineral Resources stated that Geothermal Power Plant (GPP) in Indonesia have been built about 1.34 GW in capacity and wanted to seriously develop geothermal potential up to 6.64 GW by 2025 [8]. Cost is one of famous obstacle in Indonesia's GPP Development. To reduce grand total cost of GPP, this paper will present the relation between thermodynamic and economic analysis in purpose to find the most economical gas removal system in GPP. By gleaning data at Kamojang Steam Field on behalf of PT Pertamina Geothermal Energy, this study will thermodynamically analyze and calculate a GPP preliminary design with software, named as Cycle Tempo 5.0. In additional, writers create motive steam calculator (based on C++ language) to enhance thermodynamic analysis for gas removal system (GRS) and adapted the results in Cycle Tempo 5.0. After thermodynamic analysis has been done, economic study will be undertaken by Net Present Value Analysis to compare the utilization cost of three different GRS and find which kind of GRS is more economical for nearly 30 years operation. For the result, Dual LRVP has higher performance than the others, spend less utilization cost and more economical for nearly 30 years operation. Moreover, the economic analysis for replacement of gas removal system shown in this paper too. In conclusion, GPP with Dual LRVP is proper to be developed in the future Geothermal Power Plant or to replace the existing GRS in some existing GPP in Indonesia.

1. Introduction

Indonesia is the only archipelago country with over seventeen thousand islands. It is strategically located between two continents (Asia and Australia) and two oceans (Hindi and Pacific), so geographically Indonesia stands among the eastern end of Mediterranean volcanic belt and western side of Circum Pacific volcanic belt [13]. By this, there are more than 200 volcanoes located along Sumatra, Java, Bali and the islands of eastern Indonesia [12]. It is well known as part of "Pacific Ring of Fire", the terminology refers to the chain of volcanoes that was created by the upward intrusion of magma (molten rock) at the edge of the Pacific Plate. It makes Indonesia abundant of geothermal resources [16].

Based on calculation, Indonesia has great number of geothermal potential separated by two kind of potential, 16.13 GW for high enthalpy and 7.88 GW for low enthalpy speculative resources [4]. In the end of 2013, Ministry of Energy and Mineral Resources stated that Geothermal Power Plant (GPP) have been existed about 1.34 GW and wanted to seriously develop geothermal potential up to 6.64 GW by 2025 [8]. Cost is one of famous obstacle in Indonesia's GPP Development. To reduce grand total cost of GPP, this paper will present the relation between thermodynamic and economic analysis in purpose to find the most economical gas removal system in GPP. By gleaning data at Kamojang Steam Field on behalf of PT Pertamina Geohtermal Energy, this study will thermodynamically analyze and calculate a GPP preliminary design with software, named as Cycle Tempo 5.0.

2. Thermodynamic Analysis

Thermodynamic analysis are fundamentally based on energy balance and mass balance which is stated in First Thermodynamic Law. This analysis are undergone for each apparatuses in Geothermal Power Plant and assisted by software calculation, Cycle Tempo.

2.1. Thermodynamic Laws

In thermodynamic system, there are two fundamental flows occurred into/out from it. They are mass and energy flow. Mass flow in a thermodynamic system depends on the inlet and outlet of it, so it can be written in mass balance as shown below in equation (1).

$$\frac{dm_{system}}{dt} = \dot{m}_{in} - \dot{m}_{out} \tag{1}$$

Energy flow that occurs in a thermodynamic system obeys the first law of thermodynamics. Quantity of energy in thermodynamic system is conserved, so energy which saved in system depends on the energy flow into the system and from the system to the environment/other systems [2]. All of these should be written as energy balance in equation (2).

$$\frac{dE_{system}}{dt} = \dot{E}_{in} - \dot{E}_{out} \tag{2}$$

Energy flow is specifically occurred because of mass and energy transfers in form of heats and works. Total energy flow in a thermodynamic system can be written as equation (3).

$$\frac{dE_{system}}{dt} = \dot{E}_{m,in} - \dot{E}_{m,out} + \dot{Q}_{in} + \dot{Q}_{out} + \dot{W}_{in} - \dot{W}_{out} \tag{3}$$

2.2. Thermodynamic Laws Analysis for GPP

Further to the general scheme of GPP system, there are main apparatuses which is important to be analyzed using thermodynamic analysis:

1. Production well
2. Restriction Orifice and Master Control Valve
3. Separator/Demister
4. Turbine
5. Gas Removal System
6. Main Condenser
7. Auxiliary Condenser
8. Cooling Tower
9. Hot Well Pump

2.2.1. Production well. GPP process is begun with total energy flow from production well to GPP system. This flow can be written in equation (4).

$$\dot{m}\, h_{out,well} = f(P_{opt}, T_{opt}) \tag{4}$$

2.2.2. Restriction Orifice (RO) and Master Control Valve (MCV). In RO and MCV, steam is expanded into adiabatic isenthalpic condition. RO reduces pressure and temperature of geothermal fluid until desired value has been reached. Equation (5) is a function of pressure difference(ΔP) to find RO outlet mass flow.

$$\dot{m}_{out,RO} = f(\Delta P) \quad (5)$$

Inlet pressure of well head and mass flow of RO are known, so outlet pressure can be determined by equation (6).

$$(\Delta P) = P_{out} - P_{in} \quad (6)$$

2.2.3. Separator/Demister. Degree of dryness (vapor quality) in Kamojang steam is 99.9%, therefore Kamojang has dry steam production. Brine water from demister will be removed into brine pond and its ratio is only 0.1% from the inlet steam. Enthalpy from demister can be determined using vapor quality in the following equation [3].

$$x = \frac{h_{RO} - h_{brine}}{h_{turbine} - h_{brine}} \quad (7)$$

2.2.4. Turbine. Steam will be separated before entering the turbine through governour valve. Some steam flow into turbine and the rest of steam will flow into ejector which extract NCG from condenser (GRS function). This node's mass balance can be written in equation (8).

$$\dot{m}_{turbine} = \dot{m}_{out,demister} - \dot{m}_{motive,ejector} \quad (8)$$

Energy production by turbine in isentropic condition can be written in equation (9).

$$w_{turbine} = \left(h_{in,turbine} - h_{outlet,ideal}\right) \quad (9)$$

Equation (10) will be determined by the value of enthalpy that produced by actual turbine condition. The isentropic efficiency of turbine is determined by designer, generally in 80%.

$$\eta_i = \frac{(h_{in,turbine} - h_{out,turbine,actual})}{(h_{in,turbine} - h_{out,turbine,isentropic})} \quad (10)$$

The actual power output of turbine is determined using equations (11) as follow:

$$w_{actual} = \eta_i(h_{in,turbine} - h_{outlet,ideal}) \quad (11)$$

2.2.5. Gas Removal System. Actually, the most used gas removal system in GPP are ejector and liquid ring vacuum pump (LRVP). Mass balance in ejector can be written into equation (12).

$$\dot{m}_{SJE,out} = \dot{m}_{motive} + \dot{m}_{vapor} + \dot{m}_{NCG} \quad (12)$$

Mass of NCG that being extracted from condenser is equal to NCG ratio in the main steams can be written in equation (13).

$$\dot{m}_{NCG} = x_{NCG}\, \dot{m}_{vapor} \quad (13)$$

In this study, NCG and vapor are mixed under Dalton ideal gas mixture model can be written in equation (14).

$$P\dot{V}_{gas} = \dot{m}_{gas}RT \quad (14)$$

NCG and vapor are mixed with Dalton model will have constant volume and temperature condition. Total pressure from condenser suction is a sum of NCG partial pressure with partial pressure of vapor.

$$P_T = P_{NCG} + P_{wv} \tag{15}$$

Generally, total of suction pressure is approximated with 90% of condenser pressure.

Mixing process is under fixed volume, so that the volume rate of NCG and vapour are equal to total volume rate or can be written into equation (16).

$$\dot{V}_{total} = \dot{V}_{NCG} = \dot{V}_{wv} = \dot{V} = \dot{m}_{NCG}RT/(P_T - P_{wv}) \tag{16}$$

The mass flow rate of suction vapour in following equation.

$$\dot{m}_{wv} = \dot{V}_{wv}/v_{wv} \tag{17}$$

After determining mass flow rate of vapor, mass flow rate of motive steam can be known using practical approximate. Average of molecular weight in overall wells can be written in equation (18).

$$Mr_{average} = \sum_{n=0}^{\infty} \frac{m_{gas}}{m_{gas\,total}} \times Mr_{spesific\ gas} \tag{18}$$

After determining molecular weight, the next is determining the entertainment ratio (ER) of each gas with each molecular weight. ER_{ncg} and ER_{wv} can be found using equation (19) and (20). ER is used to find air rate that needed to draw a gas from condenser [14].

$$ER_{ncg} = \left[(5.737\ \ 10^{-4} \times 18.37) + \frac{\left(2.01 \times (Mr_{ncg})^{0.86}\right)}{\left(10.37 + (Mr_{ncg})^{0.86}\right)}\right] \tag{19}$$

$$ER_{wv} = \left[(5.737\ \ 10^{-4} \times 18.37) + \frac{(2.01 \times (Mr_{wv})^{0.86})}{(18.37 + (Mr_{wv})^{0.86})}\right] \tag{20}$$

After knowing ER of each gas, total air equivalent can be determined using the following equation [14].

$$TAE = \left[\frac{\dot{m}_{ncg}}{ER_{ncg}} + \frac{\dot{m}_{wv}}{ER_{wv}}\right] \tag{21}$$

Air-vapor ratio can be found by using equation of compression ratio and expansion ratio and put into air-steam ratio graph in Figure 1.

$$Er = \frac{P_{motive\ steam}}{P_{suction}} \tag{22}$$

$$Cr = \frac{P_{outlet/dicharge}}{P_{suction}} \tag{23}$$

Air-vapor ratio can be obtained by the intersection two lines (expansion ratio line and also compression ratio line) in Figure 1. Motive steam mass can be determined using equation as follow:

$$\dot{m}_{motive} = \frac{TAE}{AS} \tag{24}$$

In LRVP, the mass balance of LRVP can be written in equation (25).

$$\dot{m}_{water,out} = \dot{m}_{water,in} + \dot{m}_{vapor,in} \tag{25}$$

The presence of motor and LRVP efficiency can be considered as additional power need in power plant which can be written in equation (26).

$$\dot{W}_{LRVP} = \frac{k}{k-1} \frac{\dot{m}RT}{\eta_m \eta_{LRVP} M_r} \left[\left(\frac{P_o}{P}\right)^{\frac{k-1}{k}} - 1\right] \tag{26}$$

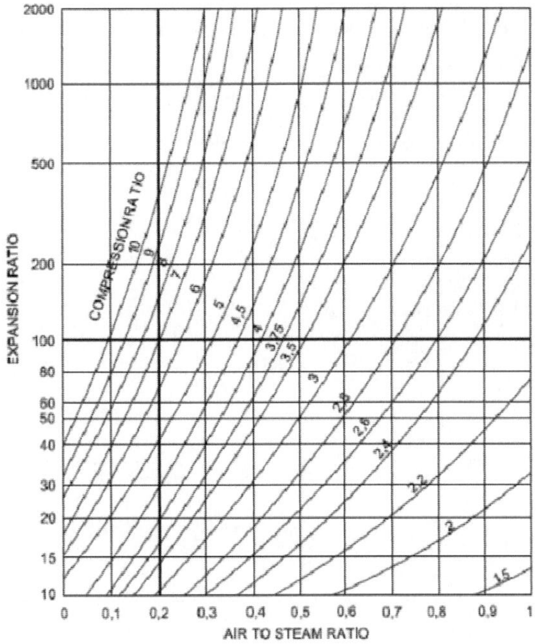

Figure 1. Air steam ratio graph [7].

2.2.6. Main condenser and Intercondenser. Direct contact condenser is the most effective condenser in GPP because the vapor and water are mixed directly [1]. The direct heat exchange can minimize the design of condenser and also reduce the needs of cooling water. Through this latent heat transfer, heat transfer will occur more rapidly and require smaller heat transfer area. In direct contact condenser, mass balance can be known using equation (27).

$$\dot{m}_{out} = \dot{m}_{cw} + \dot{m}_{vapor} \tag{27}$$

Energy balance in condenser can be written into equation (28).

$$\dot{m}_{out}h_{out} = \dot{m}_{cw}h_{cw} + \dot{m}_{vapor}h_{vapor} \tag{28}$$

Cooling water rate needs in condenser can be known by using mass and energy balance combination in the following equation.

$$\dot{m}_{cw} = \dot{m}_{vapor}\frac{h_{vapor} - h_{out}}{h_{out} - h_{cw}} \tag{29}$$

2.2.7. Cooling tower. There are two mass balances in cooling tower, they are water mass balance and air mass balance. In cooling tower, an amount of inlet air mass flow is equal to outlet air mass flow, so that can be written in the following equation.

$$\dot{m}_{air,in} = \dot{m}_{air,out} = \dot{m}_{air} \tag{30}$$

In water mass balance there is a change of amount water content because of evaporation process, so mass of inlet water in cooling tower is not equal anymore with mass of outlet water. Drift loss is negligible because it isn't significant to others, water mass balance in cooling water can be written in equation (31).

$$\dot{m}_{water,in} + \omega_{air,in}\dot{m}_{air,in} = \dot{m}_{water,out} + \omega_{air,out}\dot{m}_{air,out} \tag{31}$$

In cooling process, water is sprayed into small water droplet and mixed with an air. The air flows into cooling tower by fan inducing. The fan power can be determined by the followong equation.

$$P_{fan} = \frac{\dot{m}_{air}\Delta p}{\eta_{is}\ \eta_{motor}\ \rho_{air}} \tag{32}$$

2.2.8. Hot well pump. Power of hot well pump is written into general form as following equation [1]:

$$\dot{W}_{HWP} = \frac{\dot{m}H}{\eta_m \eta_{Pump}\ 3.67\ 10^5} \tag{33}$$

2.3. System Performance Parameter
In this study, there are some performance parameters, which is by these parameters the system will be detected whether the system is in good condition or not. These are exergetic efficiency, gross spesific steam consumption and net spesific gross consumption which their equations has shown below, respectively [1].

$$\eta_{Ex} = \frac{\dot{W}_{nett}}{\sum \dot{m}_{well}\,(h_{out} - h_{amb})} \tag{34}$$

$$\mathrm{SSC_{gross}} = \frac{\dot{m}_{steam}}{P_{gen}}\ 3600 \tag{35}$$

$$\mathrm{SSC_{net}} = \frac{\dot{m}_{steam}}{P_{gen} - P_{aux}}\ 3600 \tag{36}$$

For GPP, there is another parameter called as parasitic load, which shown as total electricity consumption for auxilliary components in system, such as pump, fan and others. Parasitic load will be accounted totally to these simulations to get the net electricity produced and sold by GPP.

3. Economic & Performance Analysis

To calculate Gas Removal System (GRS) cost in Geothermal Power Plant (GPP), Performance Analysis for Preliminary Design must be described from the actual data of the other existed GPP in Kamojang. The most important performance properties in Economic Calculation is GPP's availability factor. Availibility factor is defined as the ratio of GPP's operation time while really producing the electricity ($t_{operation}$) and the total operation time of plant (t_{actual}).

$$\mathrm{Av_{factor}} = \frac{t_{operation}}{t_{actual}} \times 100\ \% \tag{37}$$

In Preliminary Design, GPP is designed to achieve plant performance as equal as other existed plants. Based on data, other plants have 3 kinds of overhaul time, they are First Inspection Day, Minor Overhaul and Major Overhaul. First inspection day is the earliest overhaul time with estimated taking 24 x 26 days off-plant after the new power plant commissioned and firstly operated around 2 years. Minor Overhaul is the four-year maintenance which is firstly done 2 years after First Inspection Day. This overhaul only checks the small apparatuses (such as piping, sensor, separator) and just takes briefer time around 24 x 14 days. In the other hand, Major Overhaul is the four-year maintenance and overhaul for all plant apparatuses include turbines, condenser, cooling tower and other main apparatuses. This overhaul takes 24 x 26 days every 2 years after minor overhaul taken.

Throughout 30 years existing (t_{actual}), this plant will be able to have 1 times First Inspection Day, 7 times Minor Overhaul and 7 times Major Overhaul. By this, the operation time will be known by equation (38).

$$t_{operation} = t_{actual} - t_{overhaul} \tag{38}$$

This availability factor is the basic of economic calculation of GRS cost comparison. By this, GRS will be operated as plant available time ($t_{operation}$). For LRVP, the cost will be calculated based on the electricity consumption in 30 years and for ejector, the cost will be calculated based on steam potential loss in 30 years. Steam potential loss is steam which has no potential to produce electricity because it is used by the ejector. Actually it has potential to produce electricity if it is not used by the ejector. So, this steam potential loss can be calculated as electricity because of potential to be processed and changed to electricity on turbin. By this, the equation (39) will explain about electricity equivalent from those steams.

$$\mathrm{P_{equi}} = \dot{m}_{motive} \times 3600/\mathrm{SSC_{gross}} \tag{39}$$

For double LRVP, the total power calculation will be written on equation (40).

$$\mathrm{P_{30\,years}} = (\,\mathrm{P_{LRVP1}} + \mathrm{P_{LRVP2}})\ \mathrm{Av_{factor}}\ t_{actual} \tag{40}$$

For hybrid GRS, the total power calculation will be written on equation (41).

$$\mathrm{P_{30\,years}} = (\,\mathrm{P_{equi1}} + \mathrm{P_{LRVP2}})\ \mathrm{Av_{factor}}\ t_{actual} \tag{41}$$

For double ejector, the total power calculation will be written on equation (42).

$$\mathrm{P_{30\,years}} = (\,\mathrm{P_{equi1}} + \mathrm{P_{equi2}})\ \mathrm{Av_{factor}}\ t_{actual} \tag{42}$$

After that, basic GRS cost calculation will be done by equation (43).

$$C_{GRS} = \mathrm{P_{30\,years}} \times US\ Dollar/kWh \tag{43}$$

Not only from the calculation of GRS utilization cost, but also other costs is included as initial cost on Net Present Value (NPV) method. NPV method is one of economic method with most effective and accurate technique based on most modern books and courses [5]. Generally, NPV method can be defined in equation (44) and (45).

$$NPV = FV - I \tag{44}$$

$$NPV = \sum_0^{n-1} PV \times (1+r)^n - I \tag{45}$$

On those equations, *FV* is future value of revenue (USD), *I* is for investment cost (USD)., *PV* is for annual present value of revenue in this case is electricity sale revenue (USD), *r* is the number of interest rate in Bank of Indonesia averagely at 7.5% in last quarter of 2013 – December 2015 (BI Rate Data, 2016) and *NPV* is net present value of GPP revenue (USD).

Invesment cost consists of operation and maintenance (O&M) cost, capital cost and GRS utilization cost. Capital cost comprises of two basic cost, aparatuse initial cost (like ejector unit price) and installation cost. The installation cost will be predicted with Uppal's Installation Factor provided on Table 1.

Table 1. Installation factor for process plant equipment [5].

Name	IF	Name	IF
Air Cooler	1.7	Agitator	1.1
Blower	1.1	Boiler (package)	0.9
Bullets (pressured)	0.9	Centrifuge	1.5
Compressor	1.2	Cooling tower	0.6
Cyclone	1.5	Drums (horizontal)	1.6
Drum (vertical)	2.4	Dryer	1.6
Dust Collector	1.9	Evaporator	2.2
Exchanger	1.85	Filter (rotary)	1.6
Furnace (package)	1.2	Hopper (bolted)	1.5

Pump (API w/motor	1.4	Pump (chem. w/motor)	1.85
Refigeration Unit	1.4	Reactor	2.1
Screw Conveyor	1.4	Sphere	0.9
Stack	0.8	Strainer	1.5
Tank	0.9	Tower	2
Transformer	0.6	Weigh Scale	1.1

For initial cost, data on the site is prohibited to be published, so the writers try to create the approximation of cost by the data which is taken from Geremew's research on Alto Geothermal Power Plant with complete price from the manufacturer. This will assume that Kamojang GPP have the same product's brand and specification for ejector & LRVP as listed in Table 2. The reference equipment condition was defined by 100 lb/hr air plus 20 lb/hr water vapor, 10 mmHg vacuum, 100 PSIG steam and 85 Fahrenheit Degree of water, while required equipment defined for minimum 1.539 kg/s plus 0.4178 kg/s water vapor, Maximum 2.64 kg/s plus 0.497 kg/s water vapor, up to 12.47 bar steam and 0.09 bar suction [5].

Table 2. Initial/equipment cost for GRS [5].

No	Ref equipment specification	Ref Price USD	Required Equipment Specification	Scale Up Factor (SF)	Required Price USD
1	Steam Ejector(G0.04 kg/s)	$19,793	Steam Ejector (1.97 kg/s)	1.97	$38,992
			Steam Ejector Hybrid (2.622 kg/s)	2.622	$51,897
2	LRVP Vectra XL 35 6.3 HP	$27,413	LRVP (900 HP or 670 kW)	14.29	$391,614
			LRVP Hybrid (352 HP or 262 kW)	5.59	$153,165
3	LRVP Vectra XL 45 8.6 HP	$30,685	LRVP (900 HP or 670 kW)	10.47	$321,122
			LRVP Hybrid (352 HP or 262 kW)	4.09	$125,594
4	LRVP Vectra XL 60 17.5 HP	$47,967	LRVP (900 HP or 670 kW)	5.14	$246,687
			LRVP Hybrid (352 HP or 262 kW)	2.01	$96,482

From those, the capital cost for each GRS will defined by equation (46).

$$C_I = C_E + IF \times C_E \tag{46}$$

In 2011, Teke has defined Maintenance Cost Factor (MCF) for plant's simple economic calculation. MCF for ejector is 2% of Equipment/Initial Price (C_I) and LRVP's MCF is 5% of Equipment/Initial Price (C_I) [5]. This simplification is paralel with actual maintenance, where ejector just need low additional spare parts (low prices) and simple actions, otherwise the LRVP needs higher price spareparts and more complicated works for maintenance. Operation & Maintenance Cost is determined as the cost to pay the laborer, parts retrofitting and replacement and also total major/minor inspection during 30 years operation. To simplify the calculation, Teke just create MCF to summarize the all cost to be annual cost [5]. So the annual cost of operation and maintenance can be written in equation (47).

$$C_{O\&M} = MCF \times C_E \tag{47}$$

Operation and Maintenance (O&M) cost ($C_{O\&M}$) is annual cost that must be calculated with NPV calculation. So equation (47) must be rewrite as equation (48) for 30 years calculation (n = 30).

$$C_{O\&M,30} = \sum_0^{n-1} PV \times (1+r)^n \tag{48}$$

In conclusion, investment (I) can be rewritten in equation (49).

$$I = C_{O\&M,30} + C_I + C_{GRS} \tag{49}$$

After GRS calculation, the amount of electricity sale revenue must be defined as equation (50). So by this, the NPV after GRS investment will be obtained.

$$PV = w_{gen} \times US\ Dollar/kWh \tag{50}$$

To know, the replacement effect in economic calculation (based on NPV method), the cost benefit analysis must be adapted in NPV method. Cost benefit will be described in 3 cases, such as Dual Ejector to Hybrid, Dual Ejector to Dual LRVP and Hybrid to Dual LRVP. So general equation for those replacements will defined in equation (51).

$$NPV_{replacement} = NPV_{after} - NPV_{before} \tag{51}$$

4. Data and Basic Assumptions

Geothermal Power Plant Development Unit in Kamojang utilizes 4 main production wells, they are A, B, C and D in Table 3, and some process apparatuses listed in Table 4 [10].

Table 3. Steam quality in each well (A, B, C, D).

Parameter	A	B	C	D
Enthalpy (kJ/kg)	2781.29	2782.05	2781.78	2780.87
Steam Flow (kg/s)	6.614	22.157	37.728	11.87
Well Head Pressure (bar-a)	15.55	17.35	18.81	19.61
RO Outlet Pressure (bar-a)	11.19	11.43	11.34	11.06

Table 4. Components specification in GPP dry steam unit at Kamojang, West Java [9, 10, 14].

No	Component/Spesification	Amounts/Type
1	Demister	
	Pressure Loss (bar -a)	0.3
	Pressure Inlet (bar -a)	9.7
2	Turbine	
	Turbine Inlet Pressure (bar -a)	6.5
	Turbine Isentropic Efficiency	80%
	Turbine Inlet Temp. (Celcius)	166
3	Generator	
	Generator Efficiency	97%
4	Condenser	
	Type	Direct contact
	Inlet Pressure (bar -a)	0.1
	Outlet Water Temp. (Celcius)	42.7
	Cooling Water Temp. (Celcius)	26
5	Cooling Tower	
	Type	Wet mechanical induced draft
	Input Air Temperature	20
	Output Air Temperature	36.7
	Relative Humidity	91%
	Inlet Pressure of Fan (bar-a)	0.85

	Outlet Pressure of Fan (bar-a)	0.8512
	Water Output Temp. (Celcius)	26
	Fan Isentropic Efficiency	70%
	Fan Motor Efficiency	80%
6	Gas Removal System	
	Input Pressure Stage I (bar-a)	0.09
	Input Pressure Stage II (bar-a)	0.35
	Final Output Pressure (bar-a)	0.9
	Input Temp. Stage I (Celcius)	30
	Input Temp. Stage II (Celcius)	32
	LRVP Eff. (Isentropic & Motor)	60%

5. Models and Simulations

Thermodynamics analysis which included mass balance and energy balance in GRS variations will be performed in software, Cycle Tempo 5.0. Input value in simulation refers to data as shown in table 3 and table 4. Simulation will be performed into 3 variations; dual LRVP, hybrid and dual ejector. One of those models is shown in Figure 2. In actual condition, the weight of NCG molecule in Kamojang is about 1% averagely, but to present the thermodynamic and economic effect of NCG content, it will be varied from 1% content until 5% content (the range of possible NCG contents in Indonesia). In this simulations, the average molecular weight of NCG is 39.5 kg/mol, based on its chemical properties.

For economic analysis, this research will use NPV method. The calculation boundaries, this research just provide GRS utilization cost depends on kWh power for LRVP and kWh equivalent to steam for ejector, because of the limitation of maintenance cost data for LRVP and ejector and the confidential data about both initial cost. The detail assumption have been explained at Economic and Performance Analysis.

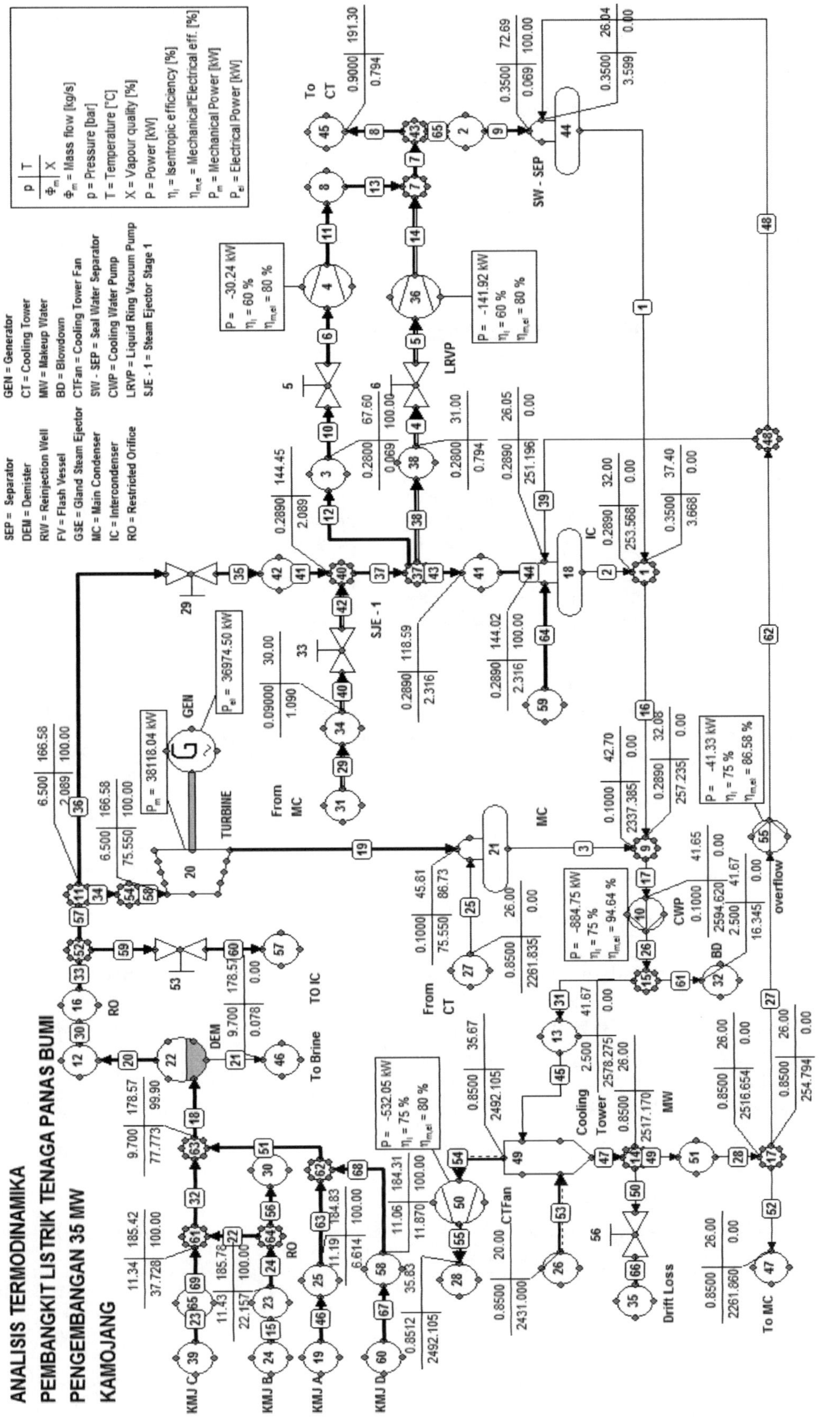

Figure 2. Calculation of geothermal power plant by cycle tempo 5.0.

6. Results

Based on thermodynamic analysis, the increase of parasitic load was experienced in all GRS as NCG increased. The highest parasitic load is shown by Dual LRVP in the same NCG, such as in 1% NCG Dual LRVP has parasitic load close to 1.81 MW, Hybrid has 1.73 MW and Dual Ejector has 1.61 MW. The growth of 1% NCG is multifarious for each GRS. In Dual LRVP, each 1% NCG growth, parasitic load will increase nearly 409 kW, in the other hand the lowest growth is experienced by Dual Ejector with 206 kW per 1% NCG increased. In hybrid GRS, the parasitic load growth is in 323 kW per 1% NCG growth (Figure 3).

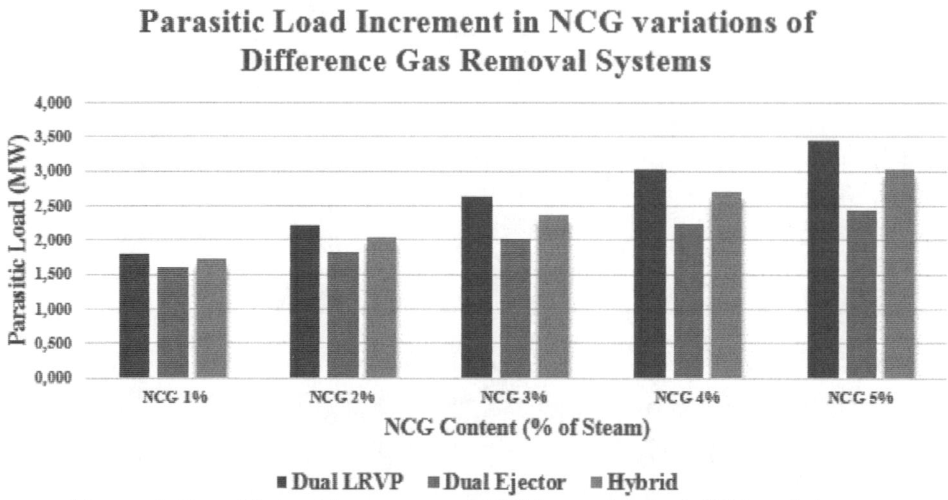

Figure 3. Parasitic load increment as NCG increasing in 3 GRS variations.

In this study, NCG contents are assumed to relatively affect the enthalpy change for its inlet turbine so there is no change for gross power result in dual LRVP. It is stagnated in 38 MW in any percentage of NCG content. In the meantime, hybrid and dual ejector diminish their gross power from the lower NCG content because of the motive steam needs increased. The gross power average reduction (per 1% NCG increased) is nearly 1.02 MW for hybrid and 2.18 MW for dual ejector system. As the results, in 1% NCG, gross power for hybrid GPP is 37.00 MW while for dual ejector GPP is 35.82 MW.

Because of parasitic load growth is undergone as long as NCG content increased, there is net production alteration in each GRS. In 1% NCG, Net Power for dual LRVP reached 36.19 MW, while hybrid producing 35.24 MW net power and dual ejector producing 34.21 MW net power. In the higher NCG contents, hybrid produces less 1.35 MW for each 1% growth. Dual LRVP losses its net power by 0.41 MW per 1% NCG growth and dual ejector, which is the highest one, decreasing its net power by 2.38 MW per 1% NCG growth. In 5% NCG, Net Power for dual LRVP reaches 34.55 MW, while hybrid producing net power for 29.86 MW and dual ejector producing for 24.69 MW. Gross and Net Power can be shown in Figure 4.

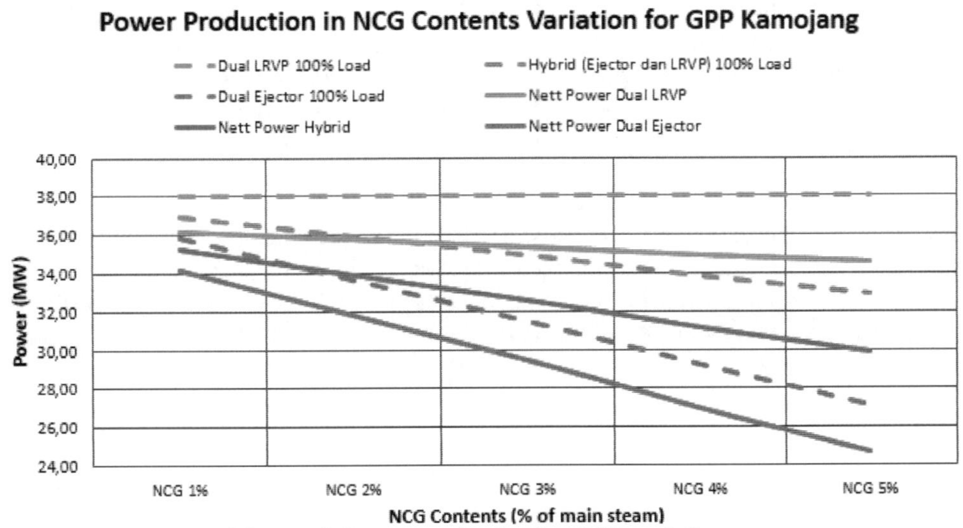

Figure 4. Power production in NCG variations.

Specific steam consumption (SSC) in this study has also resulted from the simulation. The net and gross SSC in each GRS has decreased as NCG content increasing. In 1% NCG, net SSC of dual LRVP is 7.80 kg/kWh, hybrid is 8.01 kg/kWh and dual ejector is 8.25 kg/kWh. The net SSC reduction in each NCG content increasing is approximately 0.092 kg/kWh, 0.361 kg/kWh and 0.795 kg/kWh respectively. For gross SSC in 1% NCG, dual LRVP has shown the result in 7.43 kg/kWh, while hybrid in 7.63 kg/kWh and dual ejector in 7.88 kg/kWh. By the assumption (no effect for enthalpy change in turbine), gross SSC reduction stand in stagnant number, but hybrid and dual ejector has shown the reduction (per 1% NCG increasing) are close to 0.237 kg/kWh and 0.632 kg/kWh respectively.

The most important factor for GPP performance is exergetic efficiency, which by the NCG increasing efficiency exergetic will be decreased because net product has also been decreasing. Exergetic efficiency for GPP with dual LRVP change from 49.91% (1% NCG) to 47.65% (5% NCG). In the other results, for GPP with hybrid GRS change its exergetic efficiency from 48.61% (1% NCG) to 41.19%(5% NCG) and for GPP with dual ejector change its from 47.19% (1% NCG) to 34.05% (5% NCG). Exergetic Efficiency in GPP with GRS variations can be shown in Figure 5.

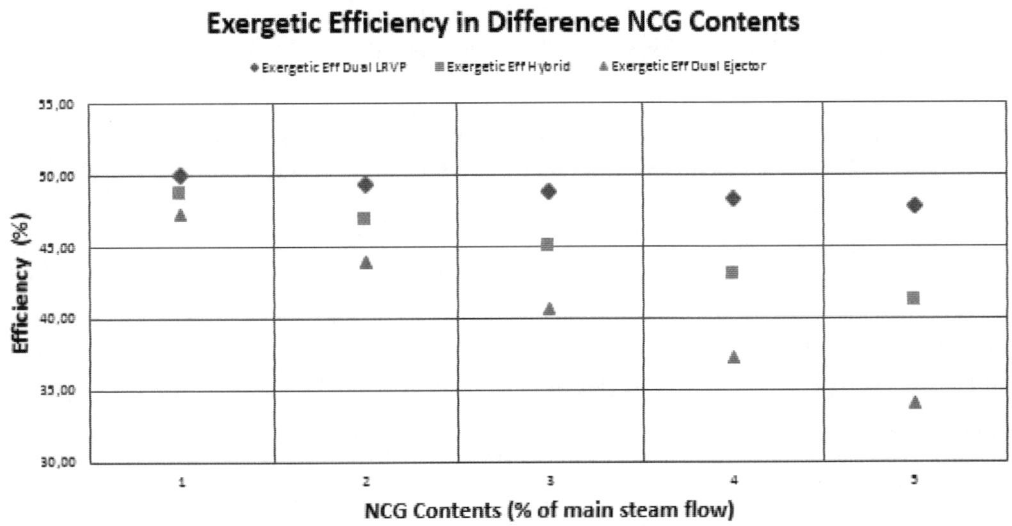

Figure 5. Exergetic efficiency of GPP in NCG and GRS types variations.

In another scenario, thermodynamic analysis are able to look for the effect of GRS selection to the steam need flows in system. Further to these analysis, the steam need is multifarious among the GRS variations. In dual LRVP, NCG growth has slightly rose the steam needs to produce the same net product. Steam needs has no additional steam flow while NCG is still in 2% and 3%, but in 4% NCG contents there is additional steam flow about 0.37 kg/s and in 5% NCG the additional steam flow is about 1.4 kg/s to produce 35 MW in any NCG contents. Another results show that dual ejector need additional steam about 1.68 kg/s to produce 35 MW in 1% NCG. In the higher NCG, the additional steam flow averagely increase about 4.14 kg/s. For hybrid, in 1% NCG there is no additional steam need, but in 2% NCG the additional steam need is around 2.3 kg/s, in 3% NCG the additional steam need is around 5.10 kg/, in 4% NCG it will be 7.93 kg/s and 5% NCG the additional steam need will increase to 10.72 kg/s. The final results of this scene can be shown in Figure 6.

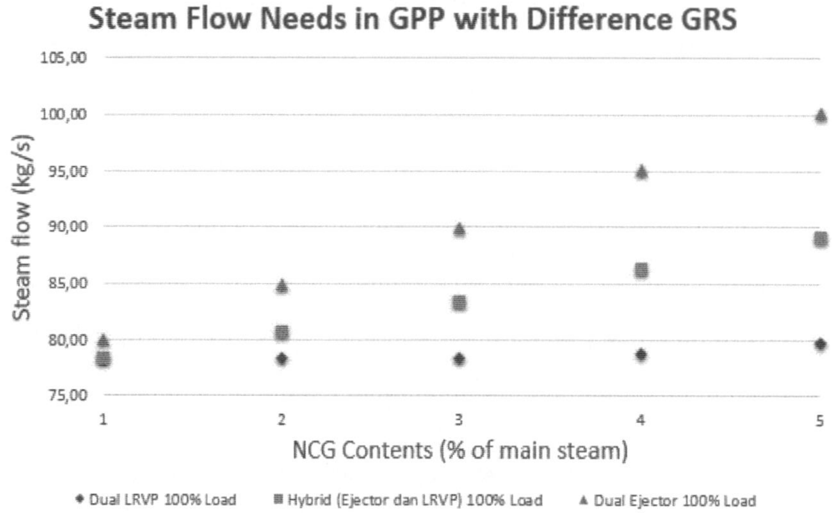

Figure 6. Steam needs increment due to NCG increment.

Under economic calculation, the GRS cost with 1% NCG is calculated for all of GRSs throughout 30 years (availability factor 97.205%), where hybrid GRS needs 28,122,129 US $ (24 million US$ for steam cost, 4 million US$ for electricity cost), dual LRVP needs 9,901,668 US$ electricity cost and dual ejector needs 51,074,343 US$ steam cost). The simple results are shown on Table 5. From the calculation, Dual LRVP has the highest capital cost and operational cost from other GRSs. In comparison, Hybrid help the designer of GPP to reduce the initial cost, but this action will reduce the electricity production as the ransom. So by selling electricity 30 years, the NPV gross profit for the company is highest with Dual LRVP as its GRS. The NPV gross profit for Dual LRVP, Hybrid and Dual Ejector are USD 3,099,755,769; USD 3,003,092,570; and USD 2,892,174,218 respectively.

Table 5. Thermodynamic & economic analysis for GPP dry steam Kamojang

Parameter	Dual LRVP	Hybrid	Dual Ejector
Gross Production (MW)	38.00	36.98	35.82
Parasitic Load (MW)	1.81	1.73	1.61
Net Production (MW)	36.19	35.24	34.21
Exergetic Efficiency (%)	51.13	49.81	48.34
Gross SSC (%)	7.43	7.63	7.88
LRVP Power Utilization 30 years (GWh)	104.23	44.30	0.00

Electricity Loss caused by Ejector 30 years (GWh)	0.00	252.78	537.62
Total Electricity Loss by GRS (GWh)	104.23	297.08	537.62
Electricity Selling 30 years (USD)	USD 3,113,781,398.11	USD 3,032,628,602.45	USD 2,943,609,208.79
Capital Cost 30 years (USD)	USD 1,307,469.00	USD 441,410.00	USD 172,689.10
O&M Cost 30 years (USD)	USD 2,816,491.16	USD 872,493.46	USD 187,957.37
Utilization Cost 30 Years (USD)	USD 9,901,668.40	USD 28,222,128.77	USD 51,074,343.44
NPV Gross Profit after 30 years (USD)	USD 3,099,755,769.55	USD 3,003,092,570.22	USD 2,892,174,218.88

To present the economic effect from NCG content, Figure 7 shows that the cost will have higher differences in higher NCG proportion on the steam. If NCG is about 5%, the cost of GRS with dual LRVP just spend approximately 50 millions US Dollar. The cost of GRS with Dual Ejector will spend greatest money about a quarter of billions US Dollar, while hybrid GRS is spending just under 150 million US Dollar.

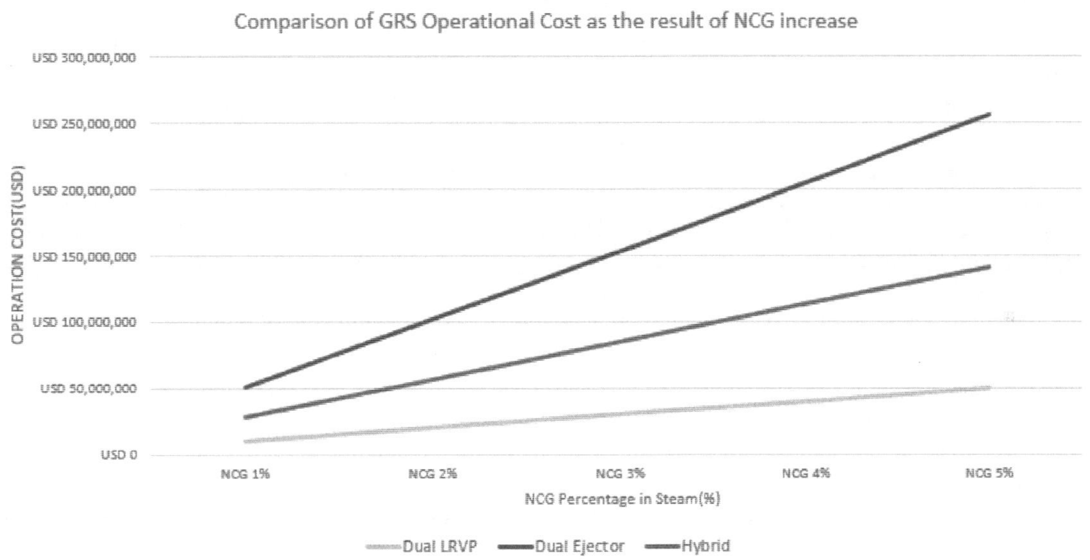

Figure 7. Utilization cost on 3 GRS with different proportion of NCG in steam

For maximum replacement benefit analysis based on NPV method, the results will be represented on Table 5. In the table, NPV replacement in GPP with NCG 1% for Hybrid to Dual LRVP will take benefit for company around 96 million US Dollar for 30 years operation, for Dual Ejector to Hybrid is around 110 million US Dollar, and for Dual Ejector to Dual LRVP is around 207 million US Dollar. More NCG containtments on the steam will boost NPV replacement and replacement will be strongly recommended.

Table 6. Optimization Replacement Profit for GPP dry steam Kamojang

NPV Profit for Replacement	NCG 1%	NCG 2%	NCG 3%	NCG 4%	NCG 5%
Hybrid to Dual LRVP	USD 78,342,738.97	USD 156,125,839.65	USD 236,600,400.95	USD 322,902,080.93	USD 392,463,791.20
Dual Ejector to Dual LRVP	USD 166,408,875.63	USD 331,915,956.72	USD 501,986,709.37	USD 676,304,938.09	USD 833,758,694.98
Dual Ejector to Hybrid	USD 88,066,136.66	USD 175,790,117.07	USD 265,386,308.42	USD 353,402,857.16	USD 441,294,903.77

7. Conclusions

In conclusion, Net Power (NP) for dual LRVP reaches 36.19 MW, while hybrid GRS producing net power in 35.24 MW and dual ejector producing for 34.21 MW. Exergetic efficiency for GPP with dual LRVP is 49.91%, hybrid GRS is 48.61% and dual ejector is about 47.19%. Under economic calculation, the GRS cost is calculated for all of them throughout 30 years (availability factor 97.205%, NCG 1%), where hybrid GRS needs 28,122,129 US $ (24 million US$ for steam cost, 4 million US$ for electricity cost), dual LRVP needs 9,901,668 US$ electricity cost and dual ejector needs 51,074,343 US$ steam cost). The NPV gross profit for Dual LRVP, Hybrid and Dual Ejector are USD 3,099,755,769; USD 3,003,092,570; and USD 2,892,174,218 respectively.

The maximum replacement benefit in Kamojang GPP (1% NCG) for Hybrid to Dual LRVP will take benefit for company around 96 million US Dollar after 30 years operation, for Dual Ejector to Hybrid is around 110 million US Dollar, and for Dual Ejector to Dual LRVP is around 207 million US Dollar. For conclusion, based on economic and thermodynamic analysis, Dual LRVP is the best GRS implemented in the future Dry Steam GPP although its NCG is just about 1% and it will be better implemented for more than 1% NCG Geothermal field. At last, Dual LRVP implementation will improve GPP production in the future, not only for Dry Steam GPP but also for Wet Steam GPP in Indonesia. So the research on Wet Steam GPP is strongly recommended to be implemented in the future.

8. Acknowledgments

We would like to show our gretatest gratitude to Mr. Aris Kurniawan (Thesis Supervisor/ Sr. Prod. Staff), Mr. Roy Bandoro Swandaru (Prod. Manager) dan Mr. Riyanto T.P. (Prod. As. Manager) for their opportunity, supervisory time, advices and insight sharings in geothermal process and power plant which is greatly assisted by this research and Aloysius Damar Pranadi's undergraduate thesis. This research was supported by PT Pertamina Geothermal Energy, Kamojang. Thank you so much for a home of newborn researcher, Nuclear Engineering and Engineering Physics Department, University of Gadjah Mada. We thank to Mr. Alexander Agung and Mr. Singgih H. for their pearl of wisdom and advices to revised the reseach results. We would like to thank to Delft University of Technology, Asimptote, and TNO which has developed Cycle Tempo 5.0 and very useful for international students learn more about power plant analysis [6, 11.15].

References

[1] Bagus, H. S., 2010. Basic Design of Lumut Balai 2 x 55 MW Geothermal Power Plant, Indonesia. *Report 2010 Number 29* (Reykjavik, Iceland: United Nations University Geothermal Training Programme)

[2] Cengel Y A and Boles M A 2006 *Thermodynamics: An Engineering Approach.* 5th edn (New York: McGraw-Hill)

[3] DiPippo, R., 2012. Geothermal Power Plants: Priciple, Application, and Case Study. *Elsevier Science.* pp. 450.

[4] Fauzi, Amir, 2015, Revision of Geothermal Resource Counting Presentation. *Presentation and Course Material.* (Bandung: ITB International Geothermal Workshop)
[5] Habtamu, G., 2012. A Study of Thermodynamic Modelling and Gas Extraction System Design for Aluto Langano Geotheral Power Plant II in Ethiopia. *Report 2012 Number 10.* (Reykjavik, Iceland: United Nation University Geothermal Training Program)
[6] Intoduction Cycle Tempo 5.0. *Technical Document* (TU Delft: Faculty of Mechanical, Maritime, and Materials Engineering Delft Institute of Technology)
[7] Millachine M A T 2011 *Guidelines for Optimum Gas Extraction System Selection* (University of Iceland: Faculty of Industrial Engineering, Mechanical Engineering and Computer Science)
[8] Nenny, S., 2015. Geothermal for Everyone. *Presentation in ITB International Geothermal Workshop* (Bandung, Indonesia: ITB International Geothermal Workshop)
[9] Perry B H and Green D W 2008 *Perry's Chemical Engineers' Handbook* (New York: McGraw-Hill)
[10] PT. Pertamina, 2015, Pertamina Geothermal Energy's Data. *Technical Document.* (Kamojang: PT. Pertamina)
[11] Reference Guide. *Technical Document.* (TU Delft: Faculty of Mechanical, Maritime, and Materials Engineering Delft Institute of Technology)
[12] Suryadarma, Ibrahim, R.F., and Fauzi, A.(2005) The Progress of Geothermal Energy Resources Activities in Indonesia, *Proceedings of World Geothermal Congress 2005*, 24-29 April 2005, (Turkey: World Geothermal Congress 24-29 April 2005)
[13] Suryantoro, S., Dwipa, S., Ariati, R., and Darma S. (2005) Geothermal Deregulation and Energy Policy in Indonesia. *Proceedings of World Geothermal Congress 2005* (Turkey: World Geothermal Congress 24-29 April 2005)
[14] Swandaru, Roy B., 2006, Thermodynamic Analysis of Preliminary Design of Power Plant Unit I Patuha, West Java, Indonesia. *Report 2006 Number 7.* (Reykjavik, Iceland: United Nation University Geothermal Training Program)
[15] Technical Notes. *Technical Document* (TU Delft: Faculty of Mechanical, Maritime, and Materials Engineering Delft Institute of Technology)
[16] Wijaya, Muhammad E. and Limmeechokchai B. Optimization of Indonesian Geothermal Energy Resources for Future Clean Electricity Supply: A Case of Java-Madura-Bali System. (Thailand: King Mongkut's University of Technology Thonbur Paper)

Identifying the Distribution of Alteration Zone Using Very Low Frequency Method in Candi Gedong Songo, Ungaran, Semarang, Central Java

A D Alfianto[1,2], D P Sari[1], S A Almas[1], W S Kurniawan[1], S G Ambariani[1] and I Suyanto[1]

[1]Geophysics Sub-Department, Faculty of Mathematics and Sciences, Universitas Gadjah Mada, Po BOX Bls 21, Indonesia

[2] Geology and Geophysics Engineer PT. Pertamina EP Asset 3, Third Party Contract PT. Pranalika Energi Nusa, Indonesia

E-mail: agungdwi.alfianto@yahoo.co.id

Abstract. The alteration zone could be a key link and a proof to know where the paleo heat source was previously located. An electromagnetic survey to identify and to map the lateral and vertical distribution of alteration zone had been done at geothermal area Candi Gedong Songo, Ungaran, Central Java, from 9th – 19th of June 2014, using VLF method. The survey consisted of 6 profiles, with NW – SE direction, which were located nearby the fumaroles spots and then went down to the observed alteration zone. Each profile was 600 m long and the distance between each profile was 20 m. The space between each measurement point of a profile was 20 m. In this study, tilt and ellipticity data with frequency of 19.8 kHz (Japan) and 24 kHz (Panama) were used. First, the data was processed to get the cross features anomaly between tilt and ellipticity data on the chart. Then, the derivative fraser and the relative current density pseudosection were also made to support the cross features anomaly. The interpretation of this data was done qualitatively using fraser and relative current density pseudosection. The result shows that the alteration zone gives high response of conductivity compared to its surrounding area. This is supported by the anomaly cross features between tilt and ellipticity data on the chart, also by high value of fraser and relative current density. Thus, the alteration zone are located in meter 150 – 250 in V1 and V2 profiles, also in meter 180 – 250 in V5 and V6 profiles. This result indicates that the ancient heat source was previously located nearby the fumaroles area and it is physically shown by the presence of sulphuric clay mineral content at the alteration surface area.

1. Introduction

Many techniques and efforts are being developed to fulfill the needs of energy these days. One of the many efforts is through the exploitation of geothermal resource. Geothermal has proven to be a fulfilling energy resources that has many advantages compared to fossil energy resources. It is renewable, produces almost zero pollution, and of course environmental-friendly.

Geothermal potencies spread from West to East of Indonesia. Java Island has many geothermal potencies spread along the mountains, one of them is located in Gedong Songo, Ungaran, Semarang, Jawa Tengah (Middle Java). The geothermal type of Gedong Songo is volcanic which has many visible manifestation such as fumarole, steaming ground, hot spring, and alteration zone. In this work, the VLF method was used in order to identify the alteration zone as it allows fast fieldwork measurements, low cost, and can detect conductive zones easily.

This kind of work is a final-year field work activity of Geophysics sub-Department, Universitas Gadjah Mada. The acquisition process took up to 8 days which has been done by all of student of Geophysics sub-Department, Universitas Gadjah Mada (Class of 2010). The other 2 days were needed to complete the processing and interpretation step to finalize the result. So, as the total, this work has been done in 10 days (from 10th to 19th June 2014).

2. Regional geology

Gedong Songo is a part of Ungaran Mt. Which located on East of Serayu Highlands. Ungaran Mountain bordered by alluvial land on North, quarter volcanic route (Sindoro, Sumbing, Telomoyo, Merbabu Mt.) on South, and Kendeng Highlands on East. North of Java is consist of geosyncline with West-East orientation (Van Bemmelen, 1949) [1].

According to Bemmelen, the measurement area is occupied by alluvial and young Ungaran deposition. This formation consisted of volcanic breccia which lies over Damar Formation series (tuffaceous marls and tuff limestone) and developed on Upper Pleistocene.

Java Island is an area of basin subsidence, fault system, folding system and volcanism under the influence of different stress regime on different time. There are three structural pattern namely Meratus pattern (North East – South West), Sunda pattern (North – South) and East – West Pattern.

Tectonic system in Ungaran is dominated by stress regime of Sunda pattern. This pattern created mainly extensional geology structure. On this area, it was found many collapse structure with West – South East orientation. This situation leads to graben which exist on South and horst on North of Ungaran Mountain

3. VLF method

The VLF method is one of many geophysics method that used electromagnetic wave on its acquisition. The aim is to measure electrical conductivity of the target by capturing secondary electromagnetic wave. Secondary electromagnetic wave is generated from induction of primary electromagnetic wave to a medium. The induction of a conductor by the primary electromagnetic wave resulting in Eddy current and causing secondary magnetic field to occur which is captured by VLF receiver. This general concept on how VLF method works is shown on **Figure 1**.

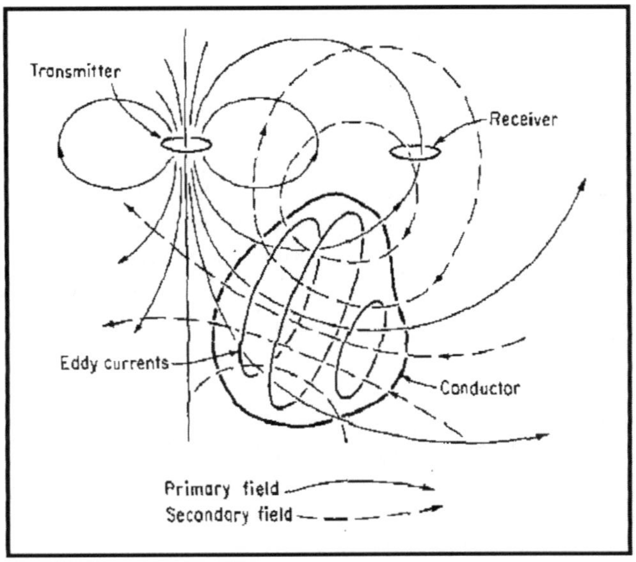

Figure 1. Principle of electromagnetic method induction (Grant, F.S. and West, G.F., 1965 [2])

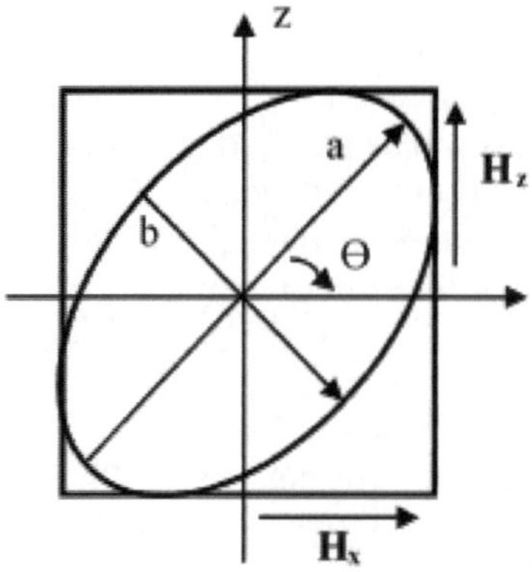

Figure 2. Tilt angle and ellipticity

VLF method studies the target using a certain designed survey in a shape of line by line. The parameters as a result of VLF survey along the line are the values of tilt angle (θ) and ellipticity (ε) (or known as I/P and Q) at each measurement point(Karous and Hjelt, 1983) [3]. The ilustration and formulas of tilt angle (θ) and ellipticity (ε) are shown in **Figure 2** and equation (1) and equation (2).

$$\varepsilon = \frac{b}{a} = \frac{H_2 H_x \sin \emptyset}{[\,H_2 e^{i\emptyset} \sin\theta + H_x \cos\theta]} \tag{1}$$

$$\tan(2\theta) = \frac{2\left(\frac{H_z}{H_x}\right)\cos\emptyset}{1 - (\frac{H_z}{H_x})^2} \tag{2}$$

4. Data

Data used on this study are tilt and ellipticity (%) with frequency of 19800 KHz (Japan) and 24000 KHz (Panama) based on measurement using T-VLF Iris Instrument.

5. Methods

VLF is an effective method to map some targets of geothermal area, it is cheap and fast. This fieldwork aims to map the alteration zone distribution of geothermal area Candi Gedong Songo. VLF data acquisition which covering the whole survey area was done in 10 days (from 10th to 19th June 2014). The complete workflow of this fieldwork is shown in **Figure 3**.

Survey design is the initial step that should be done before starting acquisition. Survey design was made based on the geological information of the geothermal manifestations location and the topography map of the survey area. Data acquisition was performed with the final survey design which comprises of 6 profiles (each profile was 600m long), the distance between each measurement point and between each profile were 20m. Each profile was measured in N 165° E direction. Two transmitters at 19.8 kHz (Japan) and 24 kHz (Panama) were used in this survey. These profiles layout can be found in **Figure 4**.

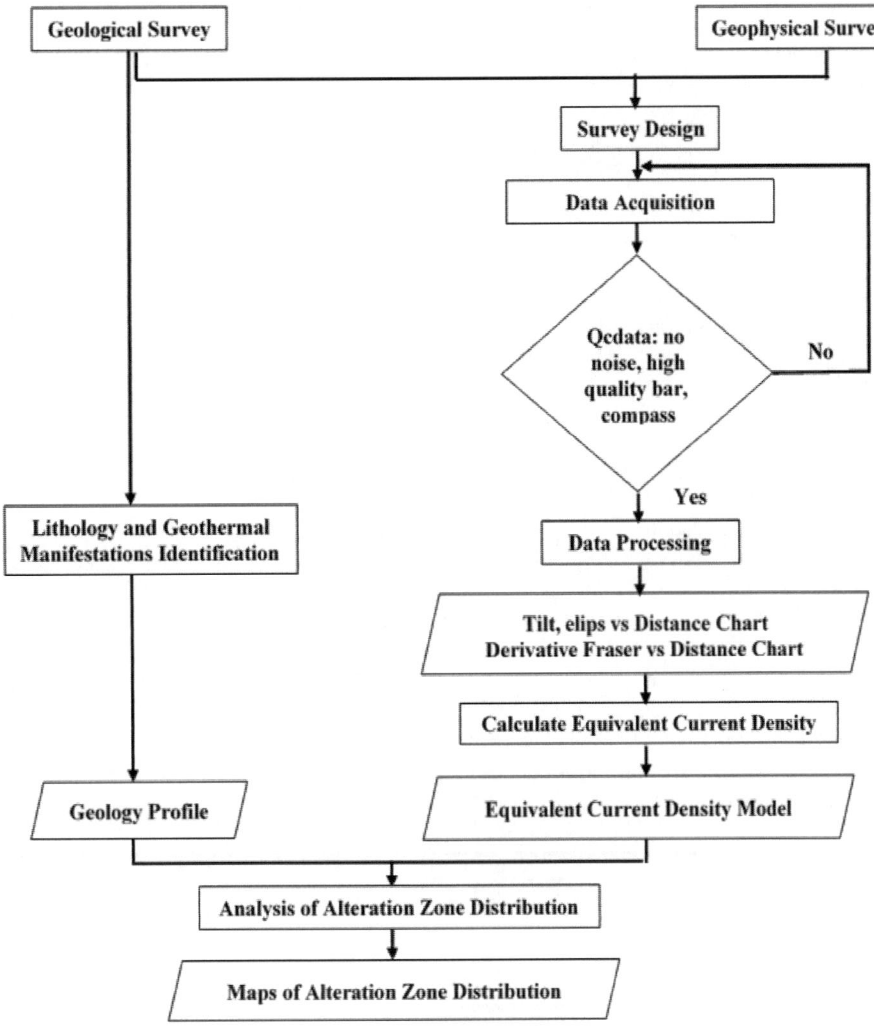

Figure 3. Workflow

The profiles were designed perpendicular to the regional structure and to the transmitters.VLF survey was performed using T-VLF from *Iris Instrument* with three receiver components (*Hx*, *Hy* and *Hz*). The measurement recorded tilt and ellips respond. During the acquisition, we also needed to do simple geological sampling and mapping which associate with the existence of geothermal manifestations. Furthermore, the elevation of each measurement point should be noted. That will give benefit in creating geological profiling along the measurement profile as shown in **Figure 5**. This geology profiling along the line measurement helps us to interpret and analyze the processed data.

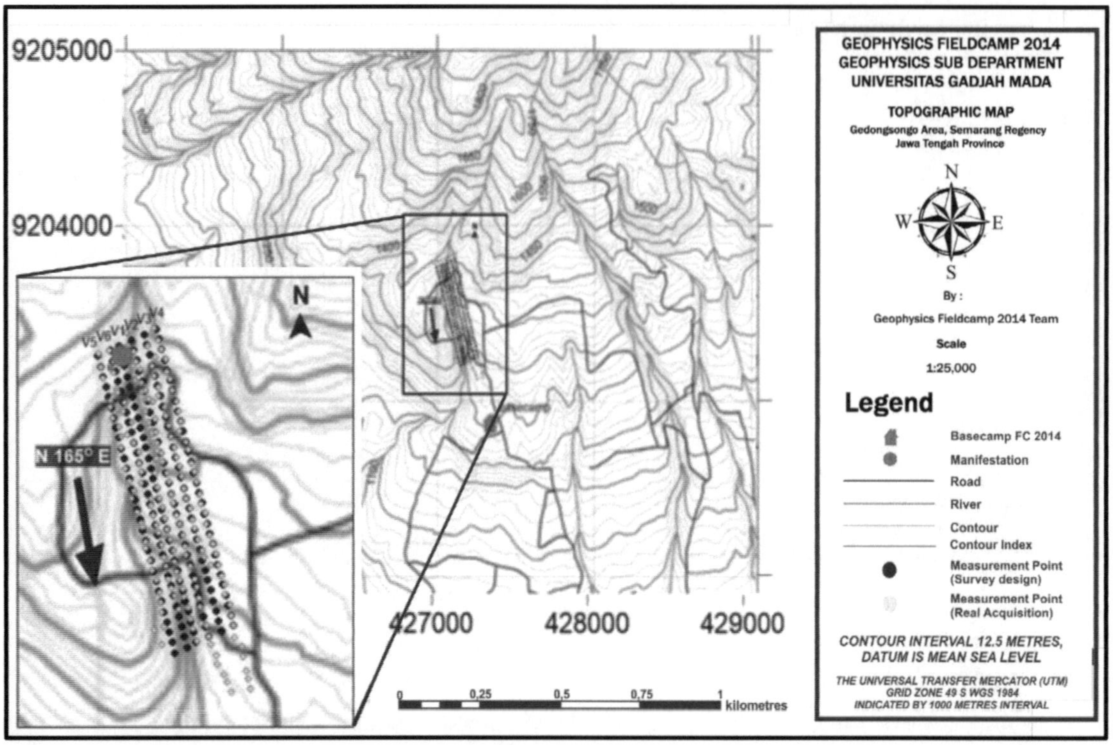

Figure 4. Design survey map (black dotted) and survey map (yellow dotted) VLF measurement with azimuth direction N1650°E

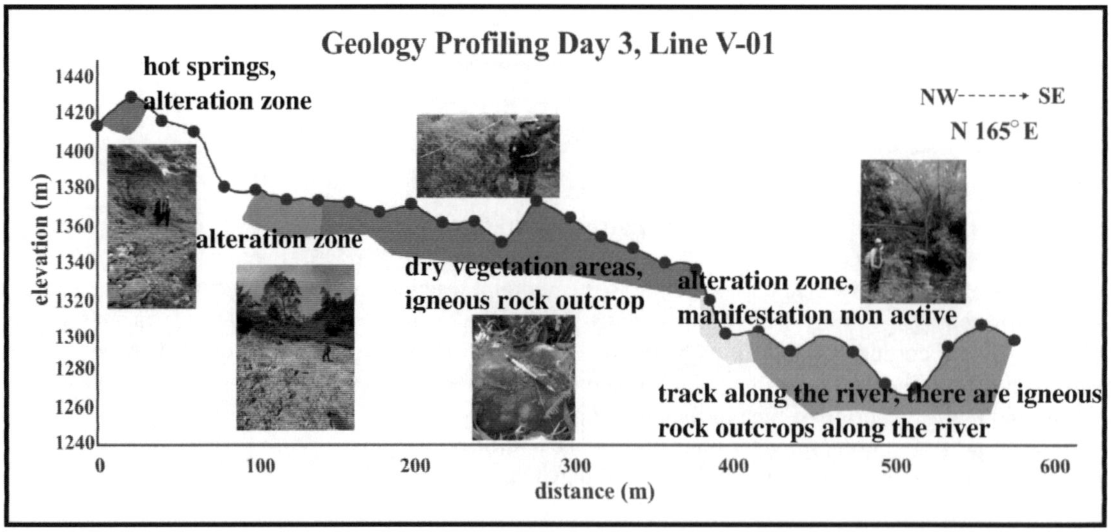

Figure 5. Geology profiling of line V1 in Geothermal area, Gedong Songo. Alteration zones found around in meter 20, 150 and 400 along the track measurement line V-01

The data obtained from acquisition was processed to get anomaly chart by plotting tilt & ellips versus distance along profile measurement. This anomaly chart was used to identify conductive zones. Derivative fraser calculation and equivalent current density model were made to support the anomaly chart. The result shows alteration zone gives high response on conductivity than its surrounding area.

This is supported by cross feature between tilt and ellipticity data on anomaly chart, also by high value of fraser and equivalent current density. The result of these plots and the equivalent current density model can be seen in **Figure 6**.

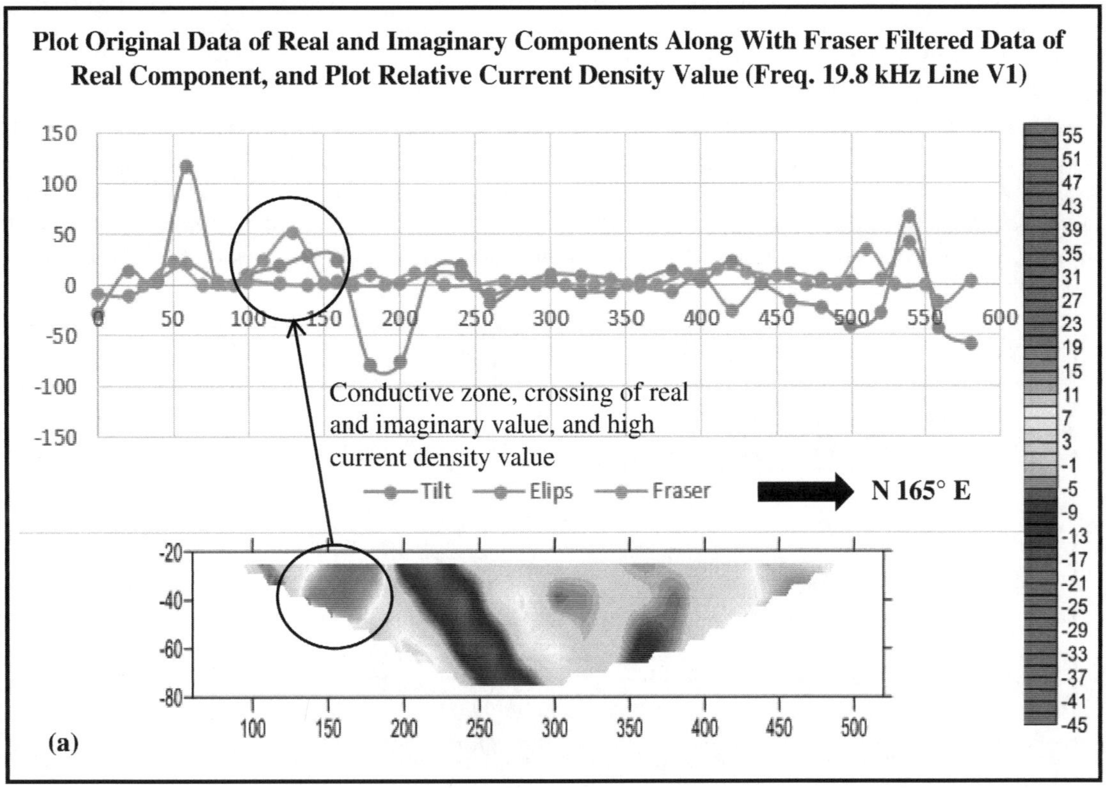

Figure 6. An example of Line V1 processing which resulting in plot of real & imaginary component, also fraser filtered data. This line graph plot was combined with the 2D pseudodepth relative current density value.

As can be seen on **Figure 6**, the original data of real (blue dotted line) and imaginary (orange dotted line) components were plotted along with the value of fraser filtered data (grey dotted line) using the real component along with the model of relative current density of six profiles. The relative conductive zones are shown on the original plots as crossing value of real and imaginary part, with high value of fraser filtered data, and high relative current density value.

The equivalent current density was created by calculating the tilt value as real component in electromagnetic field. In its calculation, there was distance correction every six measurement points. The maximum pseudodepth of equivalent current density distribution was 4n, where n is the distance between each measurement point. This means that with 20m measurement distance, the pseudodepth penetration was 80m, as shown in **Figure 6**.

The result of relative current density was then used to make interpretation of the measurement area. Furthermore, the slicing model of each profile in pseudodepth of 20m and 40m was made to get lateral equivalent current density. The slicing result was overlaid on contour maps and were plotted on each measurement point. The alteration zone distribution was interpreted using the lateral current density model, plots of tilt, ellips, and fraser (which shown anomaly area), with the aid of geological profile over the measurement area.

6. Discussion

Plots of the original and filtered of real and imaginary components (using Japan transmitter 19.8 kHz) was produced for each profile (V1, V2, V3, V4, V5, and V6) and they were interpreted in view of the existence of conductive zones that could be related to the alteration zones. As a first step, the original data of real (blue dotted line) and imaginary (orange dotted line) components are plotted along with the value of fraser filtered data (grey dotted line) using the real component. The crossing value of real and imaginary component and high value of fraser was interpreted as conductive area.

In **Figure 6**, plots of the original and filtered of real and imaginary part along with the model of relative current density of six profiles are presented. As we can see from this figure, the relative conductive zones are shown on the original plots as crossing value of real and imaginary part, with high value of fraser filtered data, and high relative current density value. In this picture, there is correlation between the plots of data and high relative density current that related to the conductive zone in meter 150 in line V1 as alteration zone.

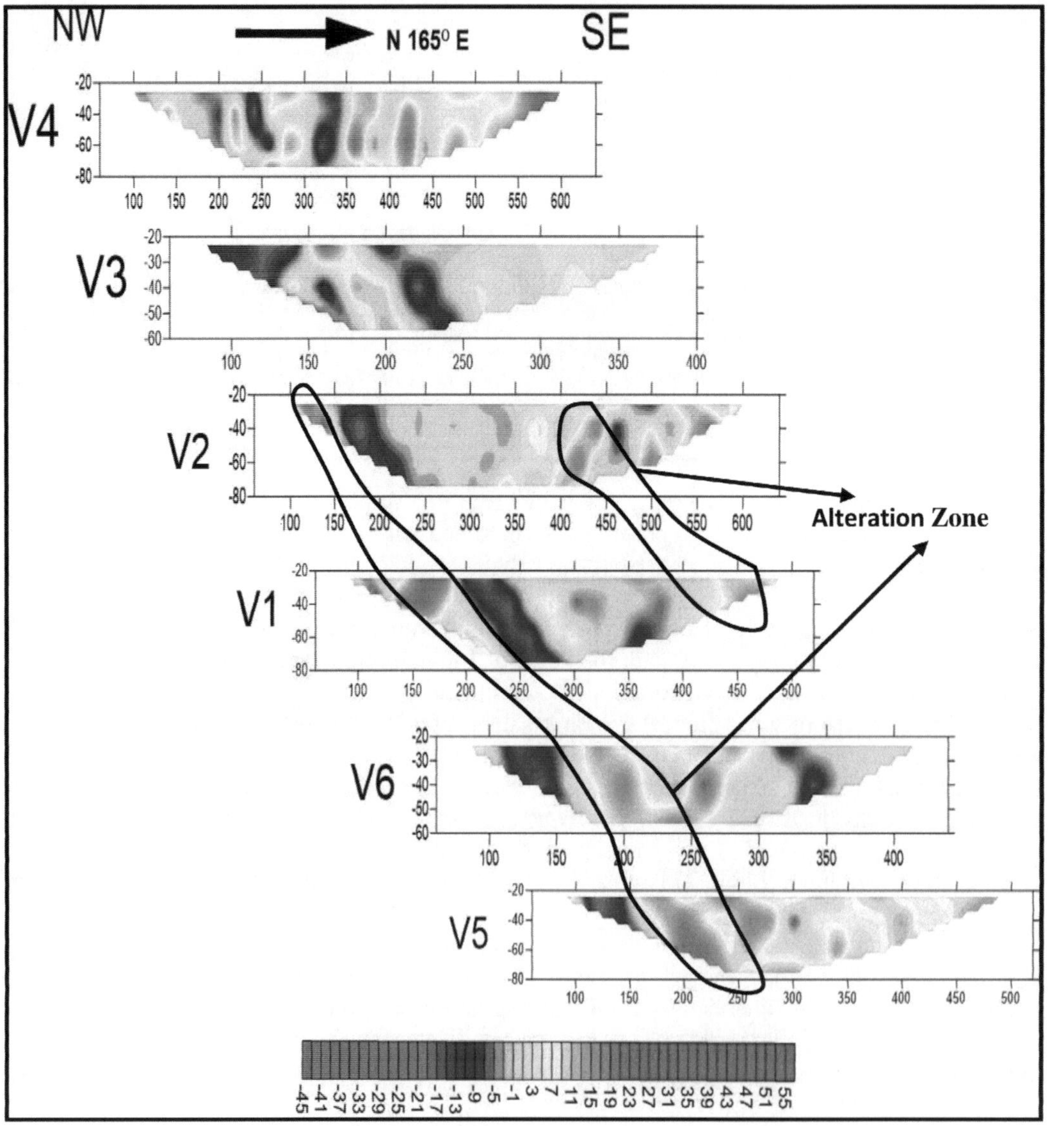

Figure 7. Correlation of relative current density of each profile.

As a final step, the models of relative current density of each profile are produced then we make a correlation. **Figure 7** shows a continuity conductive zone along line V2, V1, V6, and V5. It can be interpreted that the conductive zone shown in **figure 7** are related to the alteration zones. There is a continuity conductive zone in meter 150 – 250 along line V-02, V-01, V-06, and V-05 also in meter 450 in along line V-01 and V-02. The conductive zone are related to the alteration zone.

The slicing of relative density current of all profiles is also produced. The slicing is made in order to show the lateral view of the measurement area. The red area shown in **figure 8** has higher density current value rather than the surrounding area. This correlates with red zone in line V1 until V6. This can be interpreted as an area of alteration zone.

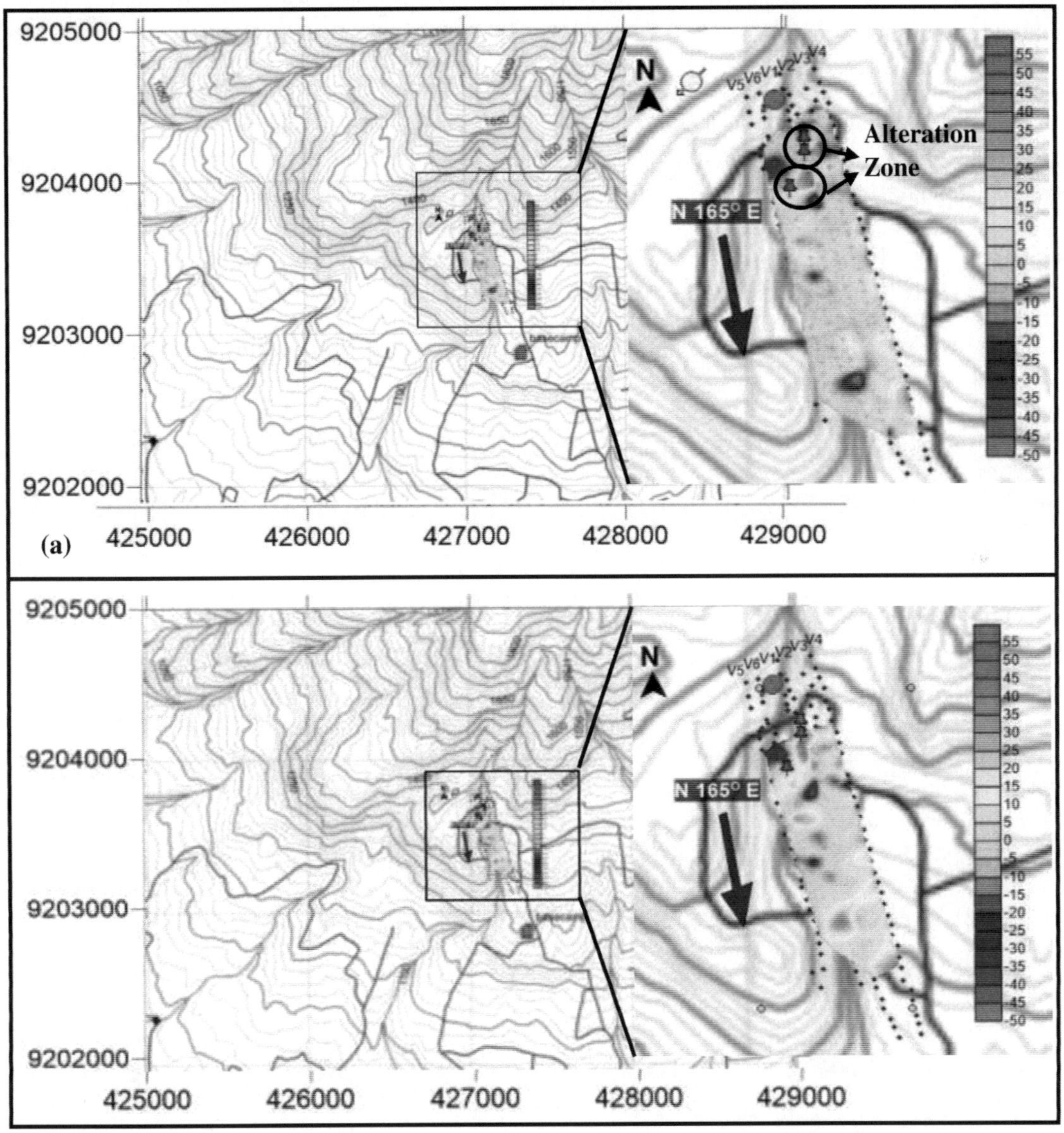

Figure 8. Slicing of line V1, V2, V3, V4, V5 and V6. (a) Slicing in apparent depth 20; (b) Slicing in apparent depth 40. The red area shown in this figure has higher current density value rather than the surrounding area. So that we interpreted this area as alteration zone

7. Conclusions

The alteration zones as surface manifestation of geothermal field in Gedongsongo, Ungaran, Central Java are shown in plots in original data as crossing value of the real and imaginary parts, with high value of fraser filtered data, and high relative current density value.

Those alteration zones are located in meter 150 – 250 in V1 and V2 profiles. Also in meter 180 – 250 in V5 and V6 profiles.There is another relative conductive zone in meter 450 in V1 and V2 profiles that could be related to the flow of sulphuric water around the measurement area.

This alteration zone distribution provides us an information about the paleo heat source of this geothermal field. Thus, by interpreting and analyzing the lateral and vertical pseudodepth of equivalent density model with the aid of geological and geochemical information found on the surface area (steaming ground and relic manifestation), we can assume that the paleo heat source was located nearby the fumarole and hot spring area.

Acknowledgements

We would like to thank to all of participants of Field Camp 2014 (Geophysics Study Program Universitas Gadjah Mada - Class of 2010), some participants from other universities, assistants, and lecturers who have been contributed and made great effort during the data acquisition and processing.

References

[1] Van Bemmelen RW 1949 *The Geology of Indonesia* Vol.1A (Den Haag, Netherland: Martinus Nijhoff) pp 28-29
[2] Grant, F S and West G F 1965 *Interpretation Theory in Applied Geophysics* (New York: McGraw-Hill) p 361
[3] Karous, M and Hjelt, S E 1983 *Linear Filtering of VLF Dip-Angle Measurements* Geophysical Prospecting 31 pp 782-794

Mathematical Modelling of Silica Scaling Deposition in Geothermal Wells

M Nizami[1], Sutopo[1,2]

[1]Petroleum Engineering Study Program, Institut Teknologi Bandung, Indonesia
[2]Geothermal Engineering Study Program, Institut Teknologi Bandung, Indonesia

Email: muhammad_nizami@students.itb.ac.id/muhammad_nizami@hotmail.com

Abstract. Silica scaling is widely encountered in geothermal wells in which produce two-phase geothermal fluid. Silica scaling could be formed due to chemical reacting by mixing a geothermal fluid with other geothermal fluid in different compositions, or also can be caused by changes in fluid properties due to changes pressure and temperature. One of method to overcome silica scaling which is occurred around geothermal well is by workover operation. Modelling of silica deposition in porous medium has been modeled in previously. However, the growth of silica scaling deposition in geothermal wells has never been modeled. Modelling of silica deposition through geothermal is important aspects to determine depth of silica scaling growth and best placing for workover device to clean silica scaling. This study is attempted to develop mathematical models for predicting silica scaling through geothermal wells. The mathematical model is developed by integrating the solubility-temperature correlation and two-phase pressure drop coupled wellbore fluid temperature correlation in a production well. The coupled model of two-phase pressure drop and wellbore fluid temperature correlation which is used in this paper is Hasan-Kabir correlation. This modelling is divided into two categories: single and two phase fluid model. Modelling of silica deposition is constrained in temperature distribution effect through geothermal wells by solubility correlation for silica. The results of this study are visualizing the growth of silica scaling thickness through geothermal wells in each segment of depth. Sensitivity analysis is applied in several parameters, such as: bottom-hole pressure, temperature, and silica concentrations. Temperature is most impact factor for silica scaling through geothermal wellbore and depth of flash point. In flash point, silica scaling thickness has reached maximum because reducing of mole in liquid portion.

Keywords: *mathematical modelling, silica scaling, silica solubility, well-bore modelling*

1. Introduction

Silica scaling is widely encountered in geothermal wells which produce two-phase geothermal fluid. Silica scaling could be occurred in production wells, reinjection wells, turbines and surface pipeline. This phenomenon is caused by transport of chemical in hot water from geothermal reservoir which allows deposition of chemical during exploitation process. In exploitation processes, heat loss and pressure drop through geothermal wells are occurred and can causes changes in enthalpy, the vapor fraction, and drop temperature. These phenomena will cause changes solubility of chemical content dissolved in the geothermal fluid. Changing solubility will cause precipitation of solid content through the geothermal wells. Deposition of silica will be deposited on the inside of the pipe radially [1]. This precipitation will change the size of the pipe diameter becomes smaller as a function of time and will also affect the ability of brine flow.

The primary objective of this study is to develop mathematical model to determine silica scaling deposition through geothermal wells. This mathematical model is developed by integrating the solubility-temperature correlation and two phase pressure drop coupled wellbore fluid temperature correlation in geothermal production well.

Solubility-temperature correlation is used because temperature is most impact factor compared to other factor such as enthalpy, salinity and pressure. Two-phase pressure drop coupled heat loss calculation which is used for model development is Hasan-Kabir correlation [2]. This modelling is divided into two main categories: single and two phase fluid model.

Silica deposition can be split into two sections: thermodynamics and kinetics of silica deposition. The thermodynamics of silica solutions allows the prediction of what must eventually happen when equilibrium is reached. The kinetics try to explain how fast equilibrium is achieved.

2. Fluid Flow and Heat Transfer Through Geothermal Wells

Mechanistic Model of Multiphase Flow

This mechanistic model consisting to gas-liquid fraction and pressure gradient. In predicting of pressure in the geothermal well as function of depth, the mechanistic model which were developed by Hasan-Kabir are used.

2.1.1 Gas Volume Fraction. Gas-volume fraction is calculated from the known flow rates in gas or liquid phase. Normally, flow regime of fluid flow is modelled because gas-volume fraction depends on the flow pattern; that is, bubbly, slug, churn, and annular. However, all of flow regimes show that gas phase moves faster than the liquid phase because of its buoyancy and its tendency to flow close to the channel center. Therefore, the generalized form of gas-volume fraction relationship with measured superficial velocity variable of gas and liquid phases can be written as [2]:

$$f_g = \frac{v_{sg}}{C_o v_m \pm v_\infty}, \tag{1}$$

where v_{sg} is superficial velocity of gas, v_∞ is average rise of velocity and $C_o v_m$ is the channel-center mixture velocity. In this equation, the values of flow parameter C_o and rise velocity v_∞ dependent on the type of flow and flow pattern. Parameters for flow type and flow pattern are presented in Table 1.

Table 1: Flow Pattern Criteria

Flow Pattern	Flow Parameter C_o	Rise Velocity v_∞
Bubbly	1.2	$v_{\infty b}$
Slug	1.2	$\bar{v}_\infty$
Churn	1.15	$\bar{v}_\infty$
Dispersed Bubbly	1.15	$v_{\infty b}$
Annular	1	0

In addition, we can define gas mass fraction, x, (called quality in steam/water flow) and can be expressed by the following equation:

$$x = \frac{h-h_L}{h_g-h_L}, \quad (2)$$

where h is enthalpy of fluid, h_L is enthalpy of saturated liquid and h_G is enthalpy of saturated vapor.
Moreover, superficial velocity of liquid and gas can be calculated by the following equation:

$$v_{sg} = \frac{m_{fluid}x}{\rho_g A}\ ;\ v_{sL} = \frac{m_{fluid}(1-x)}{\rho_L A} \quad (3)$$

Flow Pattern (Transition Criteria)

An alternative and more flexible approach to empirical flow-pattern maps is to examine each pattern transition and develop criteria. Later, will be explained how to determine transition criteria such as bubbly-slug flow transition, slug churn, flow transition and transition to annular flow by using several equations which were developed by some researcher.

v_m is mixture velocity of two phase flow and can be obtained by the following equation:

$$v_m = v_{sg} + v_{sL} \quad (4)$$

v_∞ is terminal rise-velocity of single bubble which can be calculated by using Harmathy equation:

$$v_\infty = 1.53\left[g(\rho_L - \rho_g)\frac{\sigma}{\rho_L{}^2}\right]^{\frac{1}{4}} \quad (5)$$

where surface tension (σ) is calculated by the following equation: [3]

$$\sigma = B\left(\frac{T_c - T}{T_c}\right)^{1.256}\left[1 + b\left(\frac{T_c - T}{T_c}\right)\right] \quad (6)$$

Where $B = 235.8 \times 10^{-3}$, $b = -0.625$, $T_c = 647.15$

And $v_{\infty T}$ is terminal rise velocity of small bubble of Taylor bubble which strongly dependent on the pipe diameter, proposed by Nicklin *et al*

$$v_{\infty T} = 0.35\sqrt{\frac{gd(\rho_L-\rho_g)}{\rho_L}} \quad (7)$$

where $v_{\infty T}$ is greater than v_∞ the flow would be transition from bubbly to slug flow.

Flow regime would be bubbly flow if $v_{sg} < v_t$. v_t value can be obtained by the following equation:

$$v_t = 0.429v_{sL} + 0.357v_\infty \quad (8)$$

At high gas flow rates, transition from churn or slug flow to annular flow takes place. The transition between churn (or slug) and annular flow can be understood by visualizing a long vertical tube with porous section at the middle. If the liquid is fed through the porous section at the middle and gas from the bottom, then at low gas flow rates, the liquid will flow down the tube as a film. However, at high gas flow rates, the shear force of the gas on the liquid will pull it upward, just as in the case of annular flow. The minimum of gas velocity need for reversal in the direction of the liquid flow can be expressed by the following equation

$$v_{sg} > 3.1\left[g(\rho_L - \rho_g)\frac{\sigma}{{\rho_g}^2}\right]^{\frac{1}{4}} \tag{9}$$

Bubbly Flow

In bubbly flow, gas bubbles tend to flow mostly through the center of the channel. At the channel center, the mixture velocity is significantly higher than the cross-sectional average velocity, which means that gas bubbles are moving faster than the average velocity of the mixture.

The simple relation of gas-volume fraction can be expressed by the following equation:

$$f_g = \frac{v_{sg}}{C_o v_m + \frac{v_d}{f_g}} \tag{10}$$

where v_d is the expression of drift flux which were proposed by Findlay and Wallis. The expression is semi theoretical in terms of the volume fraction and terminal rise velocity.

$$v_d = v_\infty f_g (1 - f_g)^n \tag{11}$$

For gas/liquid flow in vertical pipes, exponent n has a zero value. Therefore, the expression of drift flux could be simplified to:

$$v_d = v_\infty f_g \tag{12}$$

And the expression of simple relation of gas-volume fraction is expressed by the following equation:

$$f_g = \frac{v_{sg}}{C_o v_m + v_\infty} \tag{13}$$

Slug Flow

Fluid configuration in slug flow is quite different from that in bubbly flow. During slug flow, there are two distinct zones; one is dominated by the large Taylor bubble and the other consisting of small bubble dispersed in the liquid slug.

The average rise velocity is expressed in terms of the Taylor bubble-rise velocity, small bubble rise velocity, and bubbly slug-transition velocity as

$$\bar{v}_\infty = v_\infty\left(1 - e^{-\frac{v_l}{v_{sg}}}\right) + v_{\infty T} e^{-\frac{v_l}{v_{sg}}} \tag{14}$$

And f_g can be calculated by the following equation:

$$f_g = \frac{v_{sg}}{C_o v_m + \bar{v}_\infty} \tag{15}$$

Annular Flow

In annular flow, gas flow through the central of the pipeline, while liquid flows along the wall as a film. Therefore, the system can be viewed as single phase flow of gas through a pipe, in which its wall is formed by the liquid film. For such an ideal annular flow, it is not necessary to estimate holdup, as we are modeling a true single phase flow, except that diameter must be reduced somewhat. Unfortunately, annular flow is rarely ideal. Usually, a certain fraction of the liquid is carried as droplets in the gas

stream. Often, the amount of entrained liquid is a substantial portion of the total liquid flow. In annular flow, average rise velocity should be neglected.
The simple relation of gas-volume fraction can be expressed by the following equation:

$$f_g = \frac{v_{sg}}{C_o v_m} \tag{16}$$

2.1.2 Pressure Gradient. Total pressure gradient during fluid flow is the sum of static, friction, and kinetic gradient should be written as

$$-\frac{dp}{dz} = \frac{f_m {v_m}^2 \rho_m}{2 g_c d} + \frac{g \sin\alpha}{g_c}\rho_m + \frac{\rho_m v_m}{g_c}\frac{dv_m}{dz} \tag{17}$$

ρ_m is in-situ density of the two-phase mixture which can be expressed by:

$$\rho_m = f_L \rho_L + (1 - f_L)\rho_g \tag{18}$$

μ_m is mixture viscosity of two-phase mixture which can be expressed by [4]:

$$\mu_m = x\mu_g + (1 - x)\mu_L \tag{19}$$

Friction factor is usually expressed as a function of Reynolds number (Re). Friction factor can be estimated reliably from the Blassius equation by the following condition: smooth pipes such as plastic pipes and tubulars coated with PVC lining.

$$f_m = 0.32 Re^{-0.25} \tag{20}$$

where:

$$Re = \frac{\rho_m v_m D}{\mu_m} \tag{21}$$

Energy Balance for Wellbore Fluid
Temperature difference between the wellbore fluid and the surrounding formation results in energy exchange. An energy balance for the fluid may be performed following any standard text on thermodynamics. Wellbore heat transfer to surrounding formation gives impact for fluid properties and, in turn, the dynamics of fluid flow. Consequently, pressure drop coupled heat loss correlation require simultaneous solutions for both process. A general energy balance for the wellbore fluid which have developed by Hasan and Kabir are

$$\frac{dH}{dz} + \frac{g \sin\alpha}{J g_c} + \frac{v}{J g_c}\frac{dv}{dz} = \mp\frac{Q}{w} \tag{22}$$

where Q is rate of heat transfer between the fluid and the surrounding formation per unit depth of the wellbore; the negative sign applies to fluid production and the positive sign applies to fluid injection. Heat loss to the formation can be expressed by the following equation [2]:

$$Q \equiv -L_R w c_p (T_f - T_{ei}) \tag{23}$$

where L_R is the relaxation length parameter defined as

$$L_R \equiv \frac{2\pi}{c_p w}\left[\frac{r_{to}U_{to}k_e}{k_e+(r_{to}U_{to}T_D)}\right] \quad (24)$$

2.3 Wellbore Fluid Temperature

By substituting the expression for heat loss to the formation, *Q*, into the energy balance equation, fluid temperature to well depth can be obtained by the following equation

$$\frac{dT_f}{dz} = \pm(T_f - T_{ei})L_R - \frac{g\sin\alpha}{c_p J g_c} - \frac{v}{c_p J g_c}\frac{dv}{dz}C_j\frac{dp}{dz} \quad (25)$$

where the + sign applies to production, and the – sign applies to injection. We assume that the undisturbed formation temperature, T_{ei}, varies linearly with depth and can be expressed by the following equation:

$$T_{ei} = T_{eibh} - (L-z)g_G\sin\alpha \quad (26)$$

where g_G is the geothermal gradient in terms of vertical depth and T_{eibh} is the static earth temperature at the bottom hole. Note that *z* is positive in the downward direction.

2.3.1 Produced Fluid Temperature (Single Phase Liquid). Liquids, being essentially incompressible and allow a number of simplifications of wellbore fluid temperature equation. In addition, fluid velocity is changing with depth become negligible and liquid density variation with pressure is usually very small. For single phase liquid, wellbore fluid temperature equation become a simple first-order linear differential equation, which can be solved by the integrating factor method. The solution is

$$T_f = T_{eibh} - g_G\sin\alpha\left[(L-z) - \frac{\left(1-e^{(z-1)L_R}\right)}{L_R}\right] \quad (27)$$

Equation above shows that the fluid temperature decreases exponentially from the bottom hole ($z = L$) to the wellhead.

2.3.2 Produced Fluid Temperature (Two-Phase Flow). For two-phase fluid, the equation is also solved by the integrating factor methods and resulting first order linear differential equation. The solution is

$$T_f = T_{ei} + \frac{1-e^{(z-1)L_R}}{L_R}\left(g_G\sin\alpha + \emptyset - \frac{g_G\sin\alpha}{c_p J g_c}\right) \quad (28)$$

where $\emptyset$ is empirical expression which depend on many variables, such as flow rate, gas/liquid ratio, and wellbore pressure. Hasan and Kabir used the empirical expression for $\emptyset$ which were developed by Sagar *et al*, which is valid for flow rates less than 5 lbm/sec:

$$\emptyset = -0.002978 + 1.006\times10^{-6}P_{wh} + 1.906\times10^{-4}w - 1.047\times10^{-6}GLR + 3.229\times10^{-5}API + 0.004009\gamma_g - 0.3551g_G \quad (29)$$

For flow rates higher than 5 lbm/sec, Sagar *et all.* set $\emptyset = 0$. An alternative to this empirical approach is the use of Joule-Thompson coefficient for two-phase flow, which is derived from first principles and is given by

$$C_j \equiv \frac{1}{c_p}\left[\frac{x}{\rho_g}\left(-\frac{T}{z}\left(\frac{\partial z}{\partial T}\right)_p\right) + \frac{1-x}{\rho_L}\right] \tag{30}$$

where x is the mass fraction of the vapor phase.

2.4 Calculation Procedure

To obtain pressure, temperature, dryness and flow regime at each section of pipeline, there are several steps have to be undertaken based on the modelling of fluid flow and heat transfer through geothermal wells. The flow chart of calculation is presented in Figure 1:

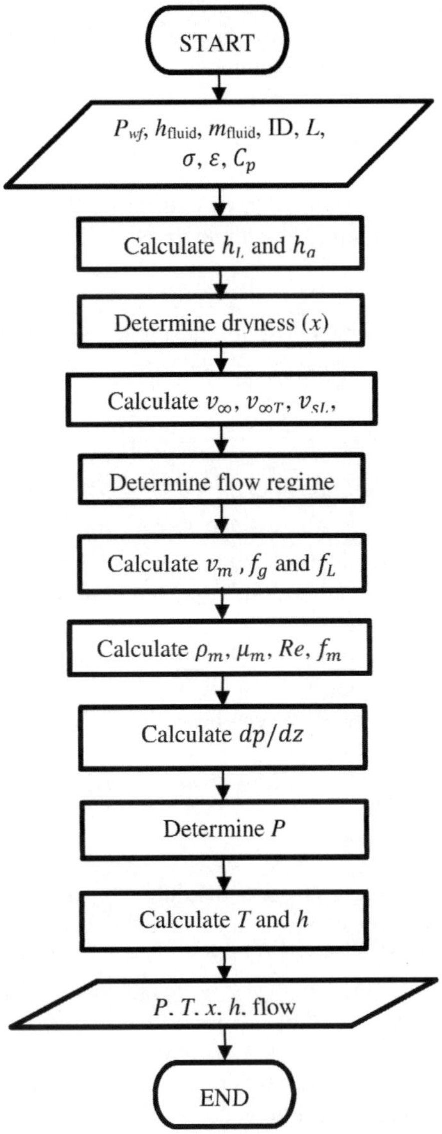

Figure 1: Pressure Drop and Heat Transfer Modelling Flow Chart

3.1 Silica Solubility

Silica is one of the chemical components found in many earth faults. At the regional manifestations, silica is commonly found in soil and water output surface of the earth. Silica can be formed in a variety of forms called polymorphs such as quartz, chalcedony, cristobalite, and amorphous [1]. All of these various polymorphs are known to occur in the nature. The surrounding rocks of most geothermal reservoirs contain quartz, and this will dissolve in hot water. In study of hydrothermal system, quartz and amorphous silica are the two polymorphs most commonly encountered.

Temperature is most impact factor that affects mineral solubility of geothermal fluid and cause scaling deposition not only in geothermal wells, but also in surface pipeline, turbines and geothermal wellbore. Other factor such as enthalpy, salinity, and pressure also give impact in changing of solubility of geothermal fluid, but not dominant.

In a geothermal system, there are 3 properties of the effect of temperature on solubility [1]:

1. Solubility increases with increasing temperature
2. Solubility increases with decreasing temperature
3. Solubility increases with increasing temperature but increased solubility only up to its maximum value and then decreases with increasing temperature

Silica has the third properties because it has a maximum value. There are two aspects which influence to silica scaling deposition: thermodynamic aspect and kinetic aspect

3.2 Thermodynamic Aspect

In geothermal reservoir (T > 230°C), the equilibrium of quartz between the solid and dissolved species can be expressed by the following reaction [4,5]:

$$SiO_{2(s)} + 2H_2O_{(l)} \Leftrightarrow Si(OH)_{4(aq)}$$

or

$$SiO_{2(s)} + 2H_2O_{(l)} \Leftrightarrow H_4SiO_{4(aq)}$$

The reaction is dependent to temperature, pH and salinity. However, temperature is most dependent parameter for silica scaling deposition. In this study, only temperature dependent is used to obtain silica scaling deposition

Normally, hot water is in equilibrium with quartz in subsurface. But in the surface, form of silica will be precipitated is amorphous silica. Amorphous silica has no crystalline structure and is more soluble than quartz. The solubility of amorphous silica can be obtained by the following equation at saturated vapor condition [1,5]:

$$\log C = -\frac{731}{T} + 4.52 \tag{31}$$

where C = silica concentration in mg/L and T = absolute temperature (K)

The solubility equations for quartz and amorphous silica have been calculated at the saturated vapor pressure of pure water. As the concentration of other dissolved species is increased (e.g. in NaCl solutions) the solubility of both quartz and amorphous silica is decreased.

The important parameter which is related to silica scaling deposition is silica saturation index (SSI). SSI is ratio between silica concentration in solution between silica solution at same condition. SSI parameter could be used to estimate possibility of silica scaling occurring, with this criteria [6]:

1. If SSI > 1, fluid is in supersaturated condition and possible to deposit silica
2. If SSI = 1, fluid is in saturated condition
3. If SSI < 1, fluid is in under saturated condition and impossible to deposit silica

3.3 Kinetic Aspect

Kinetic aspect in silica deposition is related to mechanism and silica deposition reaction rate. Silica deposition kinetic are influenced by several factor: Supersaturation level, pH, temperature, flow rate, aeration, and other ion in solution. Kinetic aspect is hardly estimated. Two possible mechanism to deposit silica are [6]:

1. Initial deposition of colloid is followed by in phase deposition of colloid
2. Direct deposition at solid surface

3.4 Mass Silica Deposition

Normally, heat loss which is occurred through geothermal well causes decreasing in temperature. Changing temperature can cause changing of silica solubility. Mass rate of silica deposition can be obtained by calculating the difference ΔC silica concentrations due to changes in temperature.

3.5 Changes of Inner Diameter of Pipelines

Deposition of silica scaling through the geothermal pipeline is assumed radially, then the radius of the well will be reduced at certain interval time. Silica scaling thickness h_s can be obtained by the following equation: [1]

$$h_s = r_w - \sqrt[2]{r_w^2 - \frac{m_{dep}t}{\rho_{silica}\pi L}} \tag{32}$$

where m_{dep} is a mass deposition rate of silica on the inner well, t is interval time, L is lengh of the pipeline and ρ_{silica} is density of silica deposition. From precipitated silica thickness radially on the pipe wall, it can be obtained new inside diameter (ID) of pipeline at any given time

$$ID_{new} = ID_{old} - 2h_s \tag{33}$$

3.6 Calculation Procedure

To thickness and change of diameter of pipeline at each section of pipeline, there are several steps have to be undertaken based on the modelling of silica dissolution and precipitation. The step by step which also could say as the flow chart is presented in Figure 2.

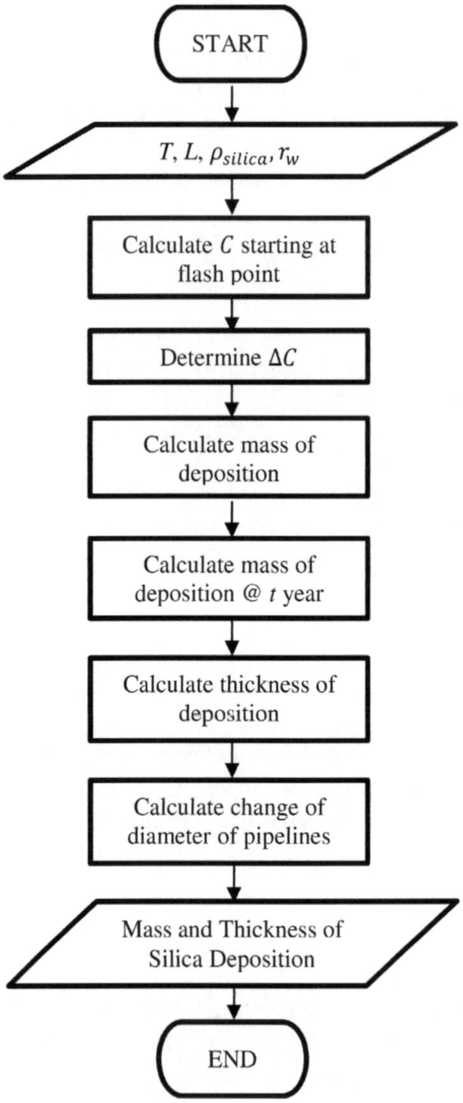

Figure 2: Silica Solubility Modelling Flow Chart

4. Data

The proposed modelling of silica precipitation, fluid flow as well as heat transfer through geothermal wells has to be validated with valid data from published paper [7]. The data is presented in Table 2 and Table 3.

Table 2: Completion of Well X

Casing 13 3/8" OD	12.415" ID	0-500	mVD
Liner 9 5/8" OD	8.618" ID	500-2000	mVD

Table 3: Production Data of Well X

Static Pressure @ 2000 mVD	150	Bara
Static Temperature @ 2000 mVD	300	Celsius
Bottom Hole Pressure	140	Bara
Reservoir Fluid Enthalpy	1338	kJ/kg
Mass Flow Rate	100	kg/s
Relaxation Length Parameter	9.51×10^{-4}	m^{-1}

5. Result and Analysis

5.1 Pressure Drop Modelling

From the pressure drop by using Hasan-Kabir Modelling shows that wellhead pressure about 32 bar absolute with dryness at the wellhead is 11%, Flow regime occurred at well X is presented in Table 4. From the modelling result, the single-phase liquid (compressed liquid) flow at bottom-hole, flashing occurs at a depth of approximately 1030 meters from the surface, and annular flow condition occurred at near the surface.

Table 4: Flow Regime of Well A

Depth	Flow Regime
2000 – 1020	Single Phase Flow
1010 – 870	Bubbly Flow
860 – 320	Slug Flow
310 – 0	Annular Flow

There are four flow regimes occurred in the well which are resulted by pressure drop modelling. Single phase flow occurs at depth 1040-2000 meter; bubbly flow occurs at depth of 870 - 1010 meter; slug flow occurs at depth of 320 - 860 meter; and annular flow occurs at depth 0 – 310 meter.

The profile of pressure, temperature and dryness versus depth is presented in Figure 3, 4 and 5, respectively:

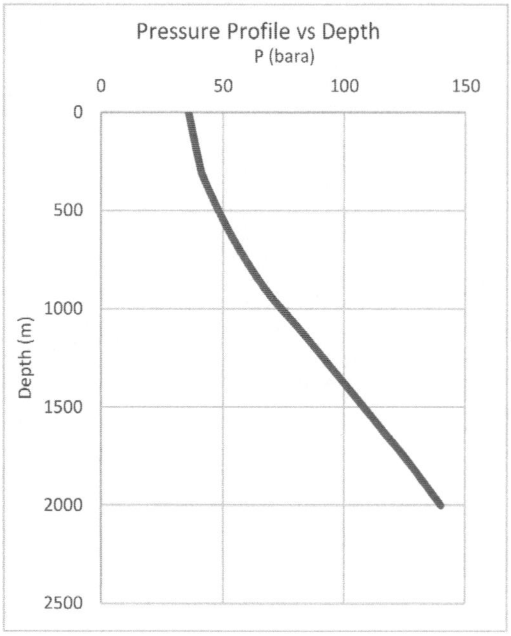

Figure 3: Pressure Profile vs Depth

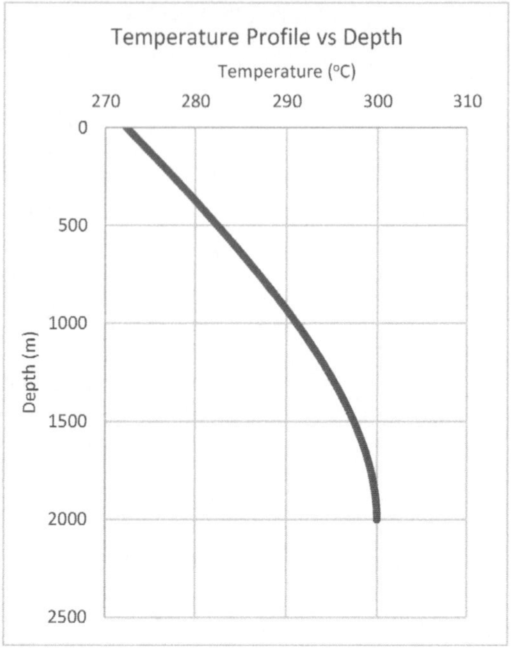

Figure 4: Temperature Profile vs Depth

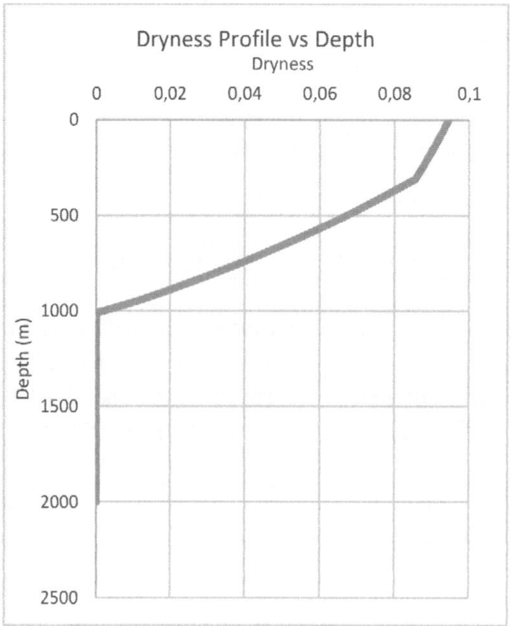

Figure 5: Dryness Profile vs Depth

5.2 Silica Solubility Modelling

Silica deposition is started at the flash point because there is fluid which release from the liquid and formed vapor so that liquid concentration increases. Silica solubility will decrease with decreasing temperature; give impact to deposition of silica through geothermal pipelines. Silica solubility versus temperature is presented in Figure 6, while silica mass rate deposition versus depth is presented Figure

7. Deposition of mass and thickness is lowest at a depth of 1030 meters which then gradually increase to the surface.

Thickness of silica deposition is presented in Figure 8. Thickness of silica deposition is gradually increasing with decreasing of depth. But in depth of 500 meters, thickness of silica deposition is suddenly decreasing due to increasing diameter of casing. Then, thickness of silica deposition is gradually increasing to the surface.

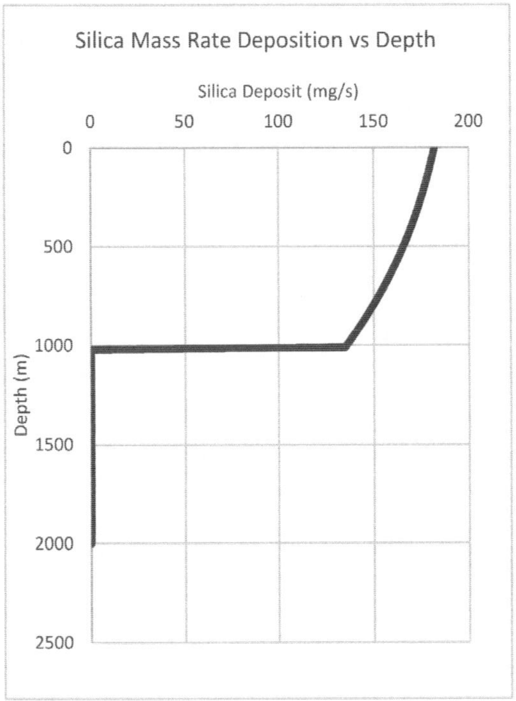

Figure 6: Silica Mass Rate Deposition vs Depth

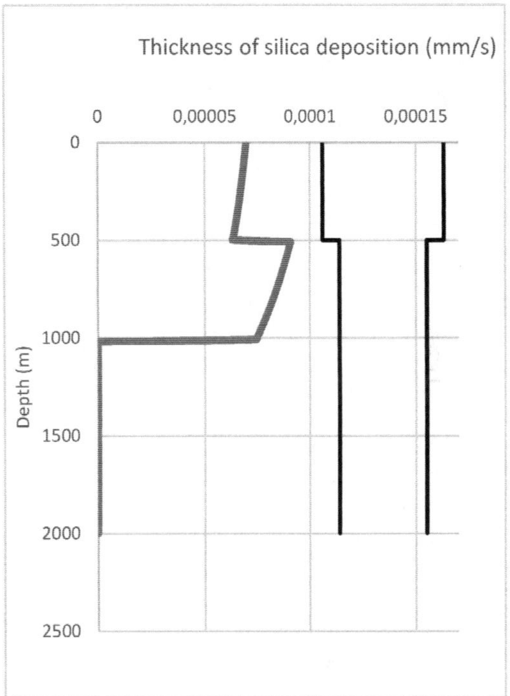

Figure 7: Thickness of Silica Deposition vs Depth

5.3 Sensitivity Analysis of Well X

Sensitivity analysis was performed on well X by varying bottom-hole pressure, bottom-hole temperature and silica concentration in solution to determine depth of flash point and also thickness of silica from the flash point into the surface. Depth of flash point is very important to estimate silica scaling deposition. In this case, silica scaling starts to precipitate in geothermal well when solubility of silica at flash point is equal to silica concentration in solution.

Sensitivity analysis of varying bottom-hole pressure with constant mass flow rate (100 kg/s) is presented at Table 5 and the result of calculation of thickness of silica with varying of bottom-hole is presented in Figure 9. From Table 5 could be concluded that increasing of bottom-hole pressure will cause depth of flash point become shallower. Moreover, decreasing of depth of flash point will cause silica scaling deposition depth will be decreasing (become shallower) too.

Table 5: Pressure variation to depth of flash point

Pressure (bara)	Depth of flash point (m)
130	1220
140	1010
150	780

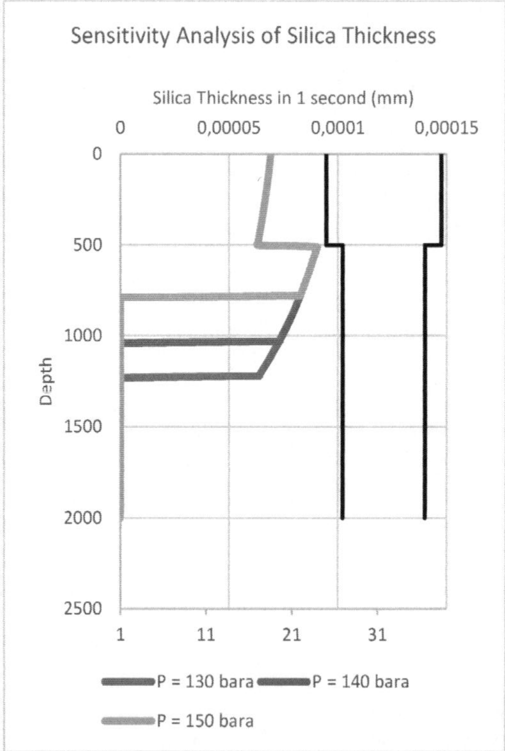

Figure 9: Sensitivity Analysis of Silica Thickness (Bottom Hole Pressure)

Sensitivity analysis of varying bottom-hole temperature with constant bottom-hole pressure (140 bara) is presented at Table 6. From Table 6 could be concluded that decreasing of bottom-hole temperature will cause depth of flash point become shallower. Moreover, decreasing of depth of flash point will cause silica scaling deposition depth will be decreasing (become shallower) too. Sensitivity analysis with varying of bottom hole temperature are undertaken to predict silica scaling thickness. Thickness of silica scaling is presented Figure 10.

Table 6: Bottom-hole temperature variation to depth of flash point

Bottom hole temperature (°C)	Depth of flash point (m)
275	630
300	1010
325	1620

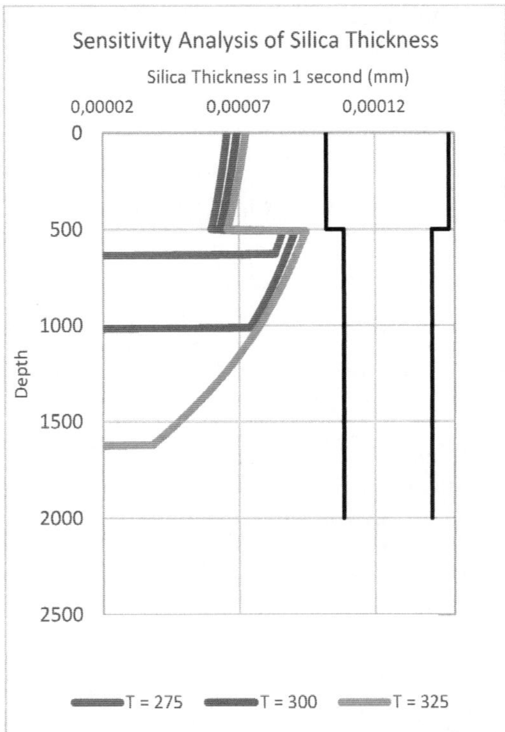

Figure 10: Sensitivity Analysis of Silica Thickness (Bottom Hole Temperature)

Moreover, we also undertake sensitivity analysis where silica concentration in solution is not equal to solubility of silica at flash point. We analyze in several parameters such as silica scaling deposited before reach flash point depth or after reach flash point depth. Depth of silica deposition with varying of silica concentration in solution is presented at Table 7. Thickness of silica at varying of silica concentration is presented at Figure 11.

Table 7: Silica concentration in solution variation to depth of deposition

Silica Concentration in solution (mg/kg)	Depth of deposition point (m)
1600	490
1650	810
1700	1170

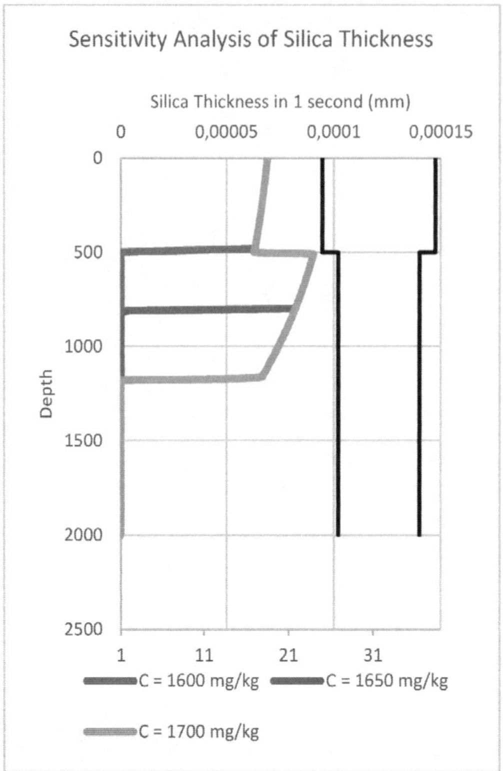

Figure 11: Sensitivity Analysis of Silica Thickness (Silica Concentration)

6. Conclusion

A mathematical model has been successfully developed by integrating concept of pressure drop and heat transfer which were developed by Hasan-Kabir with solubility-temperature correlation for silica. This mathematical model has been also theoretically implemented in well X to estimate depth of flash point, flow regime, amount of silica mass deposition, thickness of silica deposition after a certain time interval. In addition, there are several conclusions were reached during the process of implementing the mathematical model and the sensitivity analysis by the following:

1. Decreasing of geothermal well temperature will cause decreasing of silica solubility in solution so that precipitation of silica scaling through the geothermal wells increase.
2. Increasing bottom hole pressure will cause decreasing of depth of flash point so that depth of silica deposition become shallower.
3. Mass flow rate has no effect on depth of flash point.

7. Recommendations

There are several recommendations for further studies:

1. It is necessary to test the validity of the mathematical model that has been developed on real geothermal field data where there are silica scaling problems.
2. The mathematical model needs to be developed with variation of bottom hole pressure and mass flow rate which is matched with well production history.
3. Prediction of silica scaling mass and thickness will be additional information for workover program to clean silica scaling in geothermal well.
4. Downhole sampling has to be undertaken to collect accurate fluid composition and chemistry and also determine silica concentration in geothermal fluid.
5. Real thickness of silica scaling is necessary to determine by using caliper log.

References

[1] Polii, J., H., Abdurrachim. (2013), "Model Development of Silica Scaling Prediction on Brine Flow Pipe," *Proceedings, 13th Indonesia International Geothermal Convention & Exhibition 2013*, Jakarta, Indonesia, June 12 – 14, 2013.

[2] Hasan, A. R., Kabir, C. S. (2009), "Modeling Two-Phase Fluid and Heat Flows in Geothermal Wells," *Paper SPE Presented at the 2009 SPE Western Regional Meeting*, San Joss, California, USA, 24-26 March 2009.

[3] Vargaftik, N. B., Volkov, B. N., and Voljak, L. D. (1983), "International Tables of the Surface Tension of Water," *J. Phys. Chem. Ref. Data*, Vol 12, No.3, Moscow Aviation Institute, Moscow, USSR.

[4] Sutopo. (2005), "Modelling Silica Deposition in a Porous Medium Column," *Proceeding World Geothermal Congress 2015*, Antalya, Turkey, 24 – 29 April 2005.

[5] Brown, K. (2011), "Thermodynamics and Kinetics of Silica Scaling," *Proceedings International Workshop on Mineral Scaling 2011,* Manila, Philippines, 25-27 May 2011.

[6] Ciptadi, S., Patangke, S., (2001), "Evaluasi Potensi Silica Scaling pada Pipa Produksi Lapangan Panas Bumi Lahendong – Sulawesi Utara," *Proceeding of The 5th INAGA Annual Scientific Conference & Exhibitions*, Yogyakarta, 7 – 10 March, 2001.

[7] Adityawan, S. K., Saptadji, N. M., Ashat, A., (2013), "Development of Mathematical Model to Predict Sulfide Scaling in Geothermal Wells," *35th New Zealand Geothermal Workshop: 2013 Proceedings*, Rotorua, New Zealand, 17 – 20 November 2013.

[8] Hasan, A. R., Kabir, C. S. (2002), "Fluid Flow and Heat Transfer in Wellbores," Richardson, Texas: Textbook Series, SPE.

[9] Chadha, P. K., Malin, M. R., (1993), "Modelling of Two-Phase Flow Inside Geothermal Wells," *Applied Mathematical Modelling*, Vol 17, May 1993.

IOP Conf. Series: Earth and Environmental Science **42** (2016) 012014 doi:10.1088/1755-1315/42/1/012014

Volcanostratigraphy for supporting geothermal exploration

S Bronto[1], J Y Sianipar[2], A K Pratopo[2]

[1]Geological Agency, Ministry of Energy and Mineral Resources
57th Diponegoro Rd., Bandung 40122, Indonesia
[2]Geothermal Technology Master Program, Faculty of Mining and Petroleum Engineering, Institut Teknologi Bandung, Bandung 40132, Indonesia

Email: sutiknobronto@gmail.com

Abstract. Volcanostratigraphy is stratigraphy related to volcanism and its products. This includes stratigraphy for a general scoping of a regional area and detailed analysis in a local area. On the basis of Indonesian Stratigraphic Code 1996, from low to high rank, volcanostratigraphic units are Hummock (*Gumuk*), Crown (*Khuluk*), Brigade (*Bregada*), Super Brigade (*Manggala*), and Arc (*Busur*). For more detailed stratigraphic study these ranked units can be classified into genetic rock units based on source location, processes and absolute age. Genetic processes include transportation and cooling or deposition mechanisms. These lead to physical- and chemical-properties of the volcanic rocks and provide the history of volcanism and potentially geothermal processes in an area. For many areas in Indonesia the understanding of the volcanostratigraphy is an important dataset for geothermal exploration.

1. Introduction

Based on the investigation of Geological Agency, Ministry of Energy and Mineral Resources Indonesia has 29 Gwe potential geothermal energy that is estimated to be associated with active or dormant volcanic system in volcanic arcs having 6,000 km length [1]. One of important factors in geothermal exploration associated with a volcanic terrane is defining the area's volcanostratigraphy. This is the stratigraphy relating to volcanism and its products, to understand the volcanic history and characteristics of the volcanic deposits or rocks. Principally, volcanostratigraphy deals with the spatial and temporal arrangement, relationship and origin of volcanic strata, or volcanic rock/deposit bodies in an area [2]. The terminology for volcanostratigraphic units in Indonesia was originally proposed in mid-1996 [3] which was then accepted by Wirakusumah et al. (1996) [4], and finally adopted in Stratigraphic Code of Indonesia [5]. A volcano is both the place or opening from which molten rock or gas, and generally both, issues from the earth's (or a planet's in the case of planetary geology) interior onto its surface, and the hill or mountain built up around the opening by accumulation of the rock material [6] (Figure 1).

Molten rock and gas are often referred to as magma, and the location of the opening from which magma emanates onto the surface of the earth is called a crater having diameter (Ø) < 2 km, or a caldera (Ø ≥ 2 km) if it is circular in nature or a fissure if it is linear. Based on planetary research volcanoes are presence in other planets. As such, a volcano is redefined as an opening, or rupture, in a planet's surface or crust which allows hot magma, ash and gases to escape from below the surface.

Figure 1. Mayon Volcano in Philippines as an example of a volcano that is an opening, or rupture, in a planet's surface or crust which allows hot magma, ash and gases to escape from below the surface. (Source: Philippines Institute on Volcanology and Seismology)

2. Scope of Volcanostratigraphy

Volcanostratigraphy can be applied to the regional and local setting of an area. In a regional area it covers more than one volcano whereas in local area it is used for a single volcano. Regional volcanostratigraphy in eastern most Java, for example, volcanic systems are divided into two groups, namely Tertiary paleovolcanoes and Quaternary volcanoes (Figure 2). The second group includes Iyang-Argopuro volcanic complex, and Ijen volcanic complex. The Ijen volcanic complex is comprised of Ijen composite volcano and some volcanic cones appearing after Ijen caldera forming event. In both cases, the volcanism may have developed over time producing old to young stages. Figure 3 shows the development of Soufriere St. Vincent from old to young stages. In Indonesia, there are many single composite volcanoes that have developed from one to two or more conical volcanoes. For instance, Merapi and Slamet volcanoes in Central Java, Lawu volcano in the boundary between Central and East Java, Lamongan Volcano in East Java, and Gamalama volcano in Ternate island, North Maluku.

On representative exposures of volcanic units very detailed volcanostratigraphy is necessary for measuring the thickness of units (Figure 4) and collecting samples for physical and chemical analysis. This may be useful for correlating volcanostratigraphic sections in an area, either having a single volcanic cone or comprising more than one volcanic cones or volcanic complexes.

Paleosol or erosional surface may be observed in the measured section. This indicates that the volcano was inactive during a certain period (dormancy), or volcanic deposits had not been deposited in that area. Figure 5 presents three paleosoil layers in Krakatau, Setu Patok and Mount Patiayam volcanic areas.

Before applying volcanostratigraphy, geologists need to understand basic volcanology, e.g. volcanic geology concepts, genetic volcanism, volcanic landscapes, variation of volcanic rocks, volcanic facies, paleovolcanoes, and super-imposed volcanism. On the basis of a principal in geology, i.e. the present is the key to the past, a volcanic geology concept emphasizes that there is a continuous processes from magmatism through volcanism to sedimentation.

Genetically, magma erupting to the surface creating volcanism. Effusive eruptions produce either lava domes/plugs or lava flows as extrusive igneous rocks. Explosive eruptions yield pyroclasts/hydroclasts with varying intensity depending on volcanic explosivity index [7]. Based on the deposition mechanism the pyroclastic/hydroclastic deposits are divided into falls, flows, surges and pyroclastic density currents

[8, 9, 10]. Another important volcanic material results from slope failure, known as Mount St. Helens type, Bezymianny type and Bandai-san type, are called volcanic debris avalanches or rockslide-avalanches [11, 12, 13, 14].

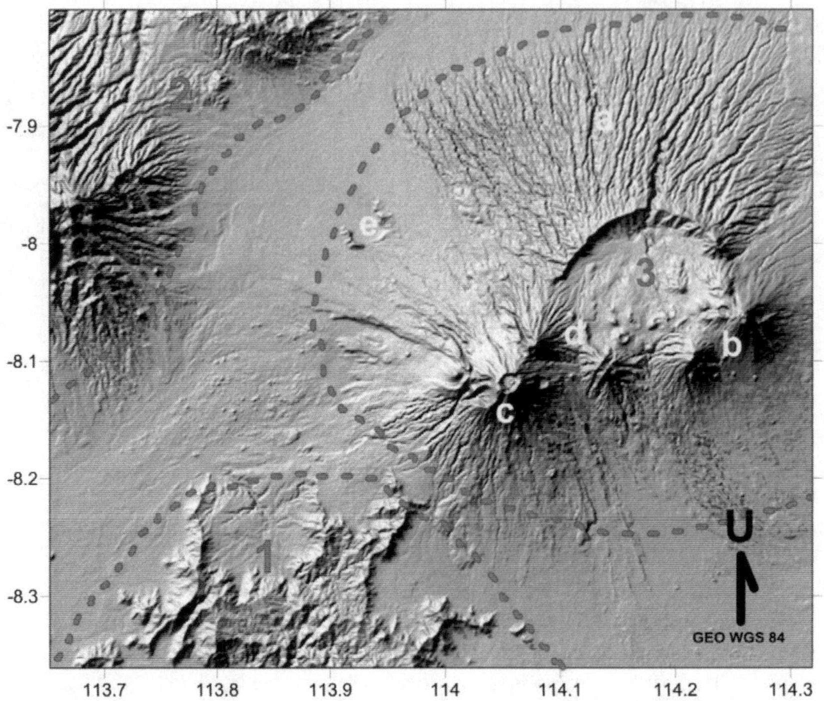

Figure 2. Regional volcanostratigraphy in eastern most Java. Volcanic systems are divided into two groups, namely Tertiary paleovolcanoes (1) and Quaternary volcanoes (2-3). The second group includes Iyang-Argopuro volcanic complex (2), and Ijen volcanic complex (3). The Ijen volcanic complex is comprised of Ijen old composite (a), Marapi volcano (b), Gadung-Raung volcano (c), some smaller cones in the caldera depression (d), and a parasitic cone (e). (Source: Satellite image by US Geological Survey)

Figure 3. Volcanostratigraphy of a single volcano comprised of old (1) and young (2) volcanic cones of Soufriere St. Vincent. (Source: http://www.mnialive.com/articles/video-montserrat-20-years-after-the-first-eruption-by-martinique-television)

(a)

(b)

(c)

Figure 4. Very detailed (measured sections) volcanostratigraphy on the wall of Rakata volcano (a) in Sunda Strait, Galunggung volcano (b) in Tasikmalaya-West Java, and Setu Patok maar deposits (c) in Cirebon-West Java. Photo by: Sutikno Bronto

The magma related to surface volcanism that remains beneath the surface forms intrusive bodies, such as dikes, sills and cryptodomes in the near-surface to intermediate depth and these would overlie a plutonic body or batholith, depending on size, at deeper depth [15]. These shallower intrusions are commonly located over the deeper solidified magma body (e.g. pluton) and are referred to as subvolcanic intrusive rocks. These rocks may be connected to the surface volcanic bodies such as lava plugs/domes by volcanic necks. Ejected magmatic material are referred to as pyroclastic deposits or tephra. Magma at the surface or the near-surface interacting with surface or ground water results in violent explosions due to the formation of steam and produces an explosive volcanic material called hydroclastics.

(a)

(b)

(c)

Figure 5. Paleosol beneath: (a) The pumice-rich 1883 Krakatau deposit, Rakata Island; (b) Setu Patok maar deposits, and (c) Mt. Patiayam, Muria Penninsula, Central Java. Lower part of the Patiayam hydroclastic deposit is greenish geothermal altered mudstone, whereas the upper part is brownish gravely mudstone. (Photo by: Sutikno Bronto)

Soon after deposition of the loose volcanic material, it immediately erodes and is redeposited as syn-eruptive volcaniclastics [16], such as lahars. Texturally and compositionally, this is a non-epiclastic reworked deposit resembling pyroclastic-hydroclastic deposits, but structurally may be comparable with sedimentary siliciclastic rocks. Volcanic vents (craters or calderas) may situated on land or in the ocean floor, and the erupted material can be deposited separately, or continuously from land to the sea, especially for island volcanic activities.

Volcanic processes form many different types of volcanic landscape. Polygenetic volcanoes may form composite, compound, caldera, somma and shield types, while monogenetic volcanoes form lava domes or pyroclastic cones [17]. Among volcanologists, genetic terms of volcanic deposits/rocks are very familiar, although descriptive names should be mentioned. The genetic names of volcanic material are closely related to volcanic facies classification, namely vent facies, proximal (or near vent) facies, medial facies and distal facies [18].

Volcano stratigraphy is also applied to paleovolcanoes. These are volcanoes that were active in the past, but are now extinct or even been eroded away, so their physical features are not as clear as recent active volcanoes [19]. In Indonesia, generally, paleovolcanoes were active during Tertiary or even pre-Tertiary ages. In some cases paleovolcanoes having Paleogene age were superimposed with Neogene ones and Quaternary volcanoes.

3. Volcanostratigraphic Units

Volcanostratigraphic classification is a systematic grouping of volcanic rock/deposit bodies or strata, which enables one to make a simpler description, arrangement and determine mutual relationship among volcanic rock/deposit strata. There are two types of volcanostratigraphic units, i.e. formal and non-formal names. Formal names are volcanostratigraphic units that follow the Indonesian Stratigraphic Code requirements. Non-formal names do not fulfil this code requirement. Naming of informal volcanostratigraphic units should not create confusion in relation to the formal names. Ranks of volcanostratigraphic formal units, from high to low, are respectively Arc, Super Brigrade, Brigade, Crown, and Hummock.

A Crown (in Indonesian: Khuluk) is a basic unit in the volcanostratigraphic classification. This consists of rocks/deposits produced from one eruption point (Figure 1) or more in the case of a composite volcanic body such as, Merapi Crown (Figure 6). Composite volcanoes may erupt various rocks/deposits as products of explosive and effusive activities along with their subvolcanic intrusions. Thus, a Crown unit is applicable for single composite volcanic body which reflects a construction period resulting from a variety of volcanic activities. The Crown unit is mappable at a scale of 1:50,000 or larger.

Figure 6. Merapi Crown in Central Java, consisting of volcanic rocks/deposits accumulation erupted from a composite volcanic body. (Photo by: Oystein Lund Andersen)

A Hummock (in Indonesian: Gumuk) is a part of a Crown formed as eruptive material at the volcanic Crown in the summit crater or on flanks. However, a Crown does not always has a Hummock. In the summit crater the Hummock is known as a child of volcano or volcano children (in Indonesia: gunung api anak), whereas on flanks of the Hummock the eruptive material forms a parasitic cone(s). The volcanic body forming the Hummock is smaller than Crown and usually consists of homogeneous volcanic rocks/deposits. These include lava cones or lava domes, pyroclastic cones (cinder/scoria cones, tuff cones, tuff rings), and maars. Monogenetic volcanoes also imply a single stage of volcanism. Due to its smaller size and homogeneous composition, an excentric volcano can be considered as a Hummock. Excentric volcanic eruptions are eruptions from vents near or beyond the base of the main volcanic cone [6]. The Hummock is mappable at a scale of 1:50,000 or larger.

A Brigade (in Indonesian: Bregada) is a volcanostratigraphic unit covering the distribution of volcanic rocks/deposits resulting from two or more volcanic Crown units, or that is related with a caldera formation, e.g. Sumbing-Sindoro Brigade and Tambora Brigade (Figure 7). Compound volcanoes are included in a Brigade, and these indicate double or multiple construction periods of composite volcanoes. The presence of a caldera system reflects a destruction period of a previous composite volcano. Thus, a Brigade always has a volcanic Crown(s) or a caldera system, and is mappable at the scale 1:100,000 or larger.

IOP Conf. Series: Earth and Environmental Science 42 (2016) 012014 doi:10.1088/1755-1315/42/1/012014

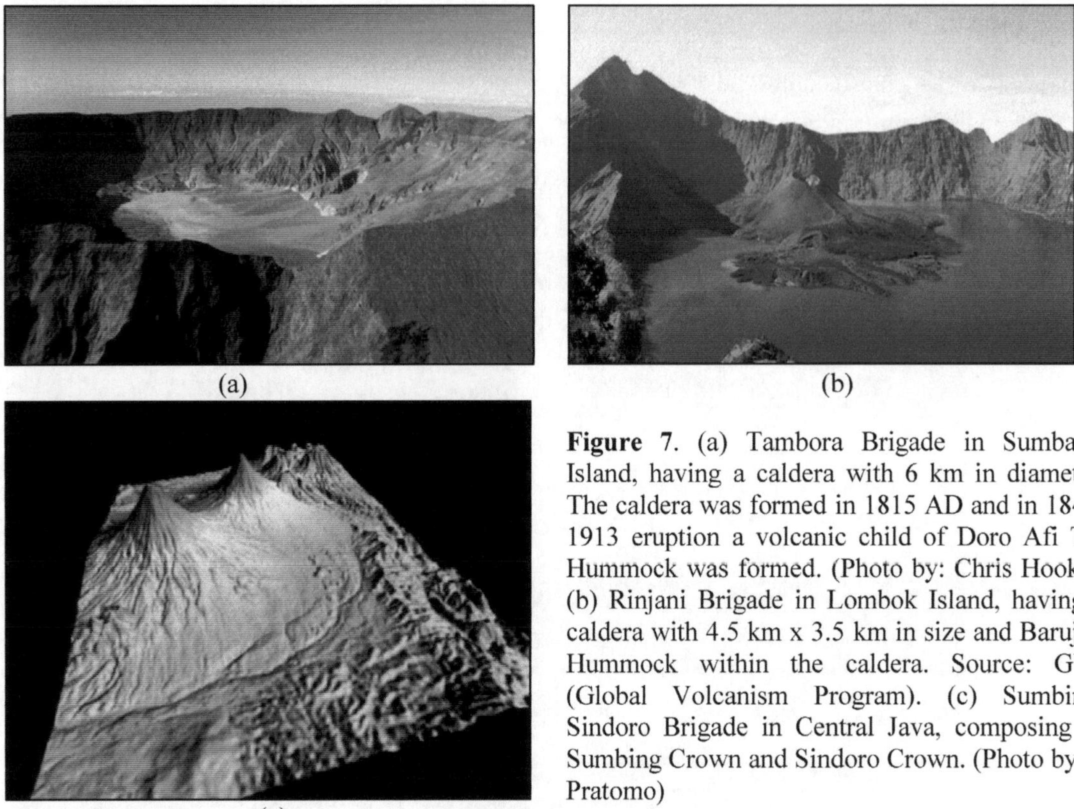

(a) (b) (c)

Figure 7. (a) Tambora Brigade in Sumbawa Island, having a caldera with 6 km in diameter. The caldera was formed in 1815 AD and in 1847-1913 eruption a volcanic child of Doro Afi Toi Hummock was formed. (Photo by: Chris Hooke). (b) Rinjani Brigade in Lombok Island, having a caldera with 4.5 km x 3.5 km in size and Barujari Hummock within the caldera. Source: GVP (Global Volcanism Program). (c) Sumbing-Sindoro Brigade in Central Java, composing of Sumbing Crown and Sindoro Crown. (Photo by: I. Pratomo)

A Super Brigade (in Indonesian: Manggala) is a volcanostratigraphic unit covering the distribution of volcanic rocks/deposits resulting from more than one caldera at one or more volcanic cones. In other words, each (composite) volcanic cone has one or more caldera(s), e.g. Kintamani Super Brigade in Bali Island and Tengger Super Brigade in East Java (Figure 8). Krakatau volcano in Sunda Strait, which has had two prehistoric calderas [20] and the 1883 caldera is also included into Super Brigade unit. A volcanic complex having many volcanic cones, craters and even calderas is also included in the Super Brigade. For example, Dieng Super Brigade in Central Java and Bandung Super Brigade in West Java. A Super Brigade has always two or more Brigade and is mappable at the scale of 1:100,000 or larger.

An Arc (in Indonesian: Busur) is a volcanostratigraphic unit composing many Crowns, Brigades and Super Brigades which have similar tectonic setting. The Arc is mappable at a scale of 1:1,000,000 or larger. The naming of Arc to Quaternary Indonesian volcanoes has been defined, i.e. Sunda Arc, Banda Arc, North Sulawesi–Sangihe Arc and Maluku Arc.

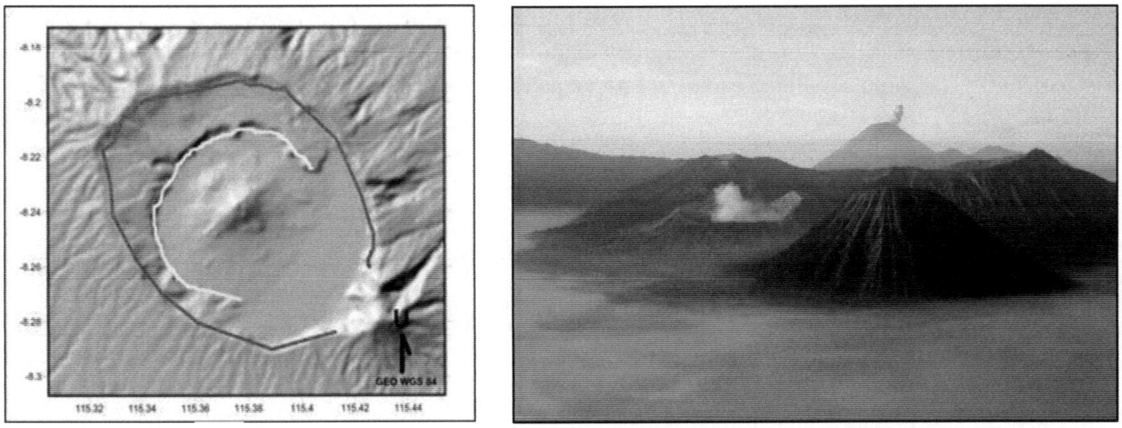

a b

Figure 8. (a) Air photo of Kintamani Super Brigade in Bali Island, showing two caldera features with a new cone of Batur Crown in the center; (b) Tengger Super Brigade in East Java, comprised of Ngadisari Caldera and Sandsea (Lautan Pasir) caldera with Hummocks of Bromo, Batok and Kursi in the center and Semeru Crown in the background. (Photo by: Andika Permana)

4. Nomenclature of A Volcanostratigraphic Unit

Volcanostratigraphic units from Hummock to Super Brigade, or even Arc, provide a general view of volcanism on a volcanic geological map. In more detailed geologic mapping, e.g. 1:50,000 scale or smaller, rock units based on genetic terms are necessary. These, together with their succession, are very useful in analyzing the character and evolution of each volcano. Table 1 presents an example correlation of volcanostratigraphic units in a volcanic geological map, e.g. Geological map of Soputan volcano, North Sulawesi [21].

The nomenclature of a volcanostratigraphic unit is based on volcanic source, genetic term of volcanic rocks/deposits and chronological events. These use three characters in succession as a combination between a letter and a numeral. Letters refer to the name of the volcanic source (in capital letter) and genetic name of a volcanic rock/deposit (in small letter). Numerals indicate chronological events. For a volcanic source that has no name, the closest and easily recognize geographic name can be applied.

In geothermal exploration, lahars, either explosion lahars or rain-generated lahars, and other reworked deposits may be less important compared with other volcanic rocks/deposits. This is due to their deposition environment that located in the distal area of the source volcano, far away from the volcanic center.

By developing study on physical volcanology, the terminology of pyroclastic density currents [10, 22] should be added to the primary rocks/deposits in Table 1. Other problems are volcanic avalanche (gv) and phreatic explosion (fr) deposits. These are included in secondary rocks/deposits. It may be correct if the avalanche was triggered by non-magmatic processes, such as occurred at Mt. Unzen (Mayuyama) in 1792 [14]. However, for volcanic debris avalanches triggered by either magmatic (known Bezymianny Type) or geothermal (known Bandai Type) processes should be considered as primary rock/deposit units, e.g. the 1980 Mount St. Helens rock-slide avalanche [12] and the 1772 Papandayan avalanche [23, 24]. In addition, volcanic avalanches or glowing avalanches due to dome collapse, such as occurred on Merapi volcano, are also considered are primary rock units.

In volcanic areas, a phreatic explosion can be initiated by a magmatic intrusion into older rocks containing a significant amount of meteoric water (e.g. an aquifer) or a long-lived geothermal system producing very high pressure steam beneath the surface. Both processes yield hydroclastic deposits. Although not directly, this geothermal phenomenon is also related to magmatic processes under the reservoir of geothermal fluid. So, phreatic explosion deposits related to geothermal-magmatic system should be considered primary deposit units. Phreatic deposits classified as secondary deposits are those produced by secondary explosions of hot lava flows or pyroclastic flows entering into a body of surface water, such as a river, lake and sea. Secondary explosions producing secondary phreatic deposits may also occur during a heavy rain fall on top of hot pyroclastic flow deposits, lava flows and lava domes. Usually, the two first events occur on slopes or flanks and along river valleys, while the third one located in volcanic crater.

Table 1: Example correlation table for volcanic map units

AGE		ACTIVITY PERIOD	VOLCANOSTRATIGRAPHIC UNITS				ERUPTION CENTER	ROCK UNIT										REMARKS
								PRIMARY ROCKS/DEPOSITS							SECONDARY ROCKS/DEPOSITS			
RELATIVE	ABSOLUTE		Mg	Br	Khu	Gm		k	l	j	a	s	ll	f	gv	lh	fv	

Explanation:

Mg	Super Brigade	k	lava dome	ll	explosion lahar
Br	Brigade	l	lava flow	f	phreatomagmatic explosion
Khu	Crown	j	pyroclastic airfall	gv	volcanic avalanche
Gm	Hummock	a	pyroclastic flow	lh	rain-generated lahar
		s	pyroclastic surge	fv	phreatic explosion

5. Role of Volcanostratigraphy in Geothermal Exploration

Volcanostratigraphy presenting various volcanic rock/deposit units and chronologically arranged is useful to understand the genetic processes of volcanism. This is an important strategic step in geothermal exploration [25]. Volcanostratigraphic units from Hummock through Crown to Brigade and Super Brigade could represent intensity and timing of volcanism. The Hummock unit indicates low intensity and short life volcanism. On the other hand, Brigade and Super Brigade units imply very high intensity and very long-lived volcanism. Low intensity volcanism and short-life time may reflect a small sized magma reservoir which could be the heat source for a geothermal system. This, together with a small recharge area and minimal groundwater supply would cause a limited life to the geothermal system. However, a large group of Hummocks distributed over a very wide area, as monogenetic cones, could produce a long-lived geothermal system. Very high intensity and very long-lived volcanism yielding Brigade or even Super Brigade units, mainly a caldera or multiple caldera systems, suggest a large magma reservoir(s). This, together with very large recharge area and voluminous groundwater supply may able to produce very long-lived geothermal system.

Volcanic source of the youngest volcanostratigraphic units may indicate the position of the heat source or magma at depth. Principally, heat energy flows conductively depending on characteristics of the volcanic rocks or deposits in the absence of open structure(s) which would be conduits for convective heat transfer. Massive coherent lavas are a very good conductor. Those cover subvolcanic intrusive rocks, volcanic necks, lava domes and lava flows beneath the surface. It is thought that intrusive rocks containing rich in heavy metallic elements, e.g. iron and magnesium in basic- and ultra basic rocks, would be better conductor than that having poor ones. Whereas, volcaniclastic deposits with high porosity and permeability, and rich in liquid (meteoric-and or magmatic water) will play in convective role through their porosity and permeability. These two end member cases should cover the distribution of volcanic geothermal system to be encountered.Genetically, potential volcaniclastic deposits for geothermal reservoir are pyroclastics, hydroclastics, autoclastics, and cataclastics because relatively they are loose material and commonly are good in porosity and permeability. Pyroclastic and hydroclastic deposits are produced by magmatic- and phreatic eruptions, respectively. Autoclastic rocks or blocky lavas are formed during super cooling processes of either lava domes or lava flows. Cataclastic deposits are yielded by deformation due to tectonic- or volcanic activities, namely fault breccias, volcanic debris avalanches or rockslide avalanches. Voluminous hydroclastic deposits may suggest an area containing ample meteoric water.

The study of volcanostratigraphic units, their volume, physical- and chemical-characteristics and stratigraphic successions will assist the interpretation of potential candidates for reservoir and cap rocks/deposits. These are important to estimate the location and size of a potential geothermal system.

6. Conclusion

Volcanostratigraphic classification is a systematic grouping of volcanic rock/deposit bodies or strata. Ranks of volcano-stratigraphic formal units, from high to low, are respectively Arc, Super Brigrade, Brigade, Crown, and Hummock. A Crown is a basic unit in the volcanostratigraphic classification. Hummock unit may reflect a short-lived geothermal system, but Brigade and Super Brigade units could imply a very long-lived geothermal system. These are important parameters in understanding the genetic processes related to volcanism and the potential size and viability for development of a geothermal system.

Acknowledgement

We wish to thank to Dr. Suryantini, a senior lecture at Geothermal Institute ITB, who had motivated in writing up this volcanostratigraphic paper. We would also like to acknowledge the Scientific Committee of the 5th IIGW 2016 for presenting this paper in the workshop. We are grateful to Mr. Joe Lovenitti for his critical and constructive comments which improve the manuscript.

References

[1] Sukhyar R and Danar A 2010 *Energi Panas Bumi di Indonesia: Kebijakan Pengembangan dan Keputusan Investasi*, Badan Geologi K-ESDM, Bandung, 179.

[2] Martodjojo S 1975 *Stratigraphic Code of Indonesia*, The Commission for the Stratigraphic Code of Indonesia, IAGI, Jakarta, 19.

[3] Bronto S 2003 *Asal muasal penamaan satuan stratigrafi gunung api*, Berita IAGI, no. 11.40/Nopember 2003, 4-5.

[4] Martodjojo S and Djuhaeni 1996 *Sandi Stratigrafi Indonesia*, Komisi Sandi Stratigrafi Indonesia, IAGI, Jakarta, 25.

[5] Wirakusumah A.D, Bacharudin R, Sujanto and Bronto S 1996 *Satuan Stratigrafi Gunung Api Indonesia*, Direkt. Vulkanologi, Bandung, 8.

[6] Macdonald G.A, 1972 *Volcanoes*, Prentice-Hall, Englewood Cliffs, New Jersey, 510.

[7] Newhall C G and Self S 1982 *The Volcanic Explosivity Index (VEI): An Estimate of Explosive Magnitude for Historical Volcanism*. J. Geophys. Res., 87, 1231-1238.

[8] Fischer RV and Schmincke H U 1984 *Pyroclastic Rocks*, Springer-Verlag, Berlin, 472.

[9] Cas R A F and Wright J V 1987 *Volcanic successions: modern and ancient*, Allen and Unwin, London, 528.

[10] Branney M J and Kokelaar P 2002 *Pyroclastic Density Currents and Sedimentation of Ignimbrites*, Geological Society Memoir, no. 27, 143.

[11] Ui T, Takarada S and Yoshimoto M 2000 *Debris Avalanches*, in H. Sigurdsson, Editor-in- Chief, *Encyclopedia of Volcanoes*, Academic Press, San Diego, 617-626.

[12] Voight B, Glicken H, Janda R J and Douglass P M 1981 *Catastrophic rockslide-avalanche of May 18*. In: P W Lipman and D R Mullineaux (Eds.) *The 1980 eruptions of Mount St. Helens*. Washington, U.S. Geological Survey Prof. Paper 1250, v. 98: 347-377.

[13] Siebert L 1984 *Large volcanic debris avalanches: Characteritics of source areas, deposits and associated eruptions*, Jour. Volc. Geoth. Res., v. 66: 367-395.

[14] Siebert L, Glicken H dan Ui T 1987 *Volcanic hazards from Bezymianny- and Bandai Types Eruptions*, Bull. Volcanol., v. 49: 435 – 459.

[15] Marsh B D 2000 *Magma Chambers*, in H. Sigurdsson, editor-in-chief, Encyclopedia of Volcanoes, Academic Press, San Diego, 191 - 206.

[16] McPhie J, Doyle M and Allen R 1993 *Volcanic Textures. A guide to the interpretation of textures in volcanic rocks*, Centre for Ore Deposit and Exploration Studies, University Tasmania, Tasmania, 196.

[17] Siebert L, Simkin T and Kimberly P 2010 *Volcanoes of the world*, 3rd ed., University of California Press, Los Angeles, 551.

[18] Bronto S 2006 *Fasies gunung api dan aplikasinya*. Jurnal Geologi Indonesia, 2 (1), 59-71.

[19] Bronto S 2013 *Geologi Gunung Api Purba*, Cetakan kedua, Badan Geologi, K-ESDM, Bandung, 184.

[20] Bronto S 1983 *The nature of the Krakatau ash flow deposits*, Sympos. 100th Krakatau 1883-1983, Jakarta, 23-27 Agust. 1983, 23.

[21] Kartadinata M N, Irawan W, Solihin A, and Mulyana A R 1998 *Geological map of Soputan volcano, North Sulawesi*, Volcanological Survey of Indonesia, Bandung.

[22] Druitt T H 1998 *Pyroclastic density currents*, in : J.S. Gilbert and S. J. Sparks (eds), 1998. The Physics of Explosive Volcanic Eruptions, Geological Socciety, London, Special Publication, 145, 145-182.

[23] Neumann van Padang M 1951 *Catalogue of the Active Volcanoes of the World Including Solfatra Fields*. Part I Indonesia, International Volcanology Association, Via Tasso 199, Napoli, Italy, 271.

[24] Kusumadinata K Ed 1979 *Catalogue of references on Indonesian volcanoes with eruptions in historical time*, Volc. Surv. Indon., Bandung, 820.

[25] Sumotarto U 2015 *Eksplorasi Panas Bumi*, Penerbit Ombak, Yogyakarta, 125.

Identifying Non-Volcanic Geothermal Potential in Amohola, Southeast Sulawesi Province, by Applying the Fault and Fracture Density (FFD) Method

Kiddy Nahli, Fajar Mulyana, Gian E. Tsani, M. Aulia Alwan, M. Hilman Darojat and Rezky N. Hendrawan

Faculty of Geological Engineering, Padjadjaran University, Bandung, West Java, Indonesia Jl. Raya Jatinangor, Km. 21. Phone: (+62-22 7796545)

kiddynahli@gmail.com

Indonesia has numerous volcanic areas that lead to a significant geothermal potential. The geothermal potential could be produced by volcanic and non-volcanic processes. To date, geothermal possibilities by volcanic process are the most commonly known and explored in several areas in Indonesia, whereas, geothermal potential by non-volcanic process is rarer. Therefore, Indonesia is likely has a promising non-volcanic geothermal also. Non-volcanic geothermal systems could be identified by using Fault and Fracture Density (FFD) method, lineaments analysis from SRTM and topography data, as well as using earlier research models by previous researchers. By using those methods, the existing of fractures and possibility level of area that being recharge and discharge zones for geothermal might be predicted. Thus, the result could be the next target of exploration. This paper describes the application of those methods for the Amohola region and surrounding areas in Southeast Sulawesi Province. In general, the study area has complex geological characteristics. Consequently, the methods could be applied to identifying the presence of non-volcanic geothermal system in this study area.

1. Introduction

Geologically, Sulawesi is situated at the intersection of three major plates, which are the Eurasia, Indo-Australian and Pacific. Consequently, these tectonic activities result in the formation of volcanoes in Sulawesi, and in particular, both volcanic and non-volcanic hosted geothermal areas.

Sulawesi is well-known for its complex tectonic setting resulting in faults and fractures (Sompotan, 2012), among other geologic feature, e.g., volcanoes. A discovery of geothermal manifestations in the Amohola region and surrounding areas [4] in a region with no volcanic activity has spurred consideration that a non-volcanic geothermal system may exist in this area. In geothermal systems, faults, fractures or contacts between intrusive and surrounding rocks may become conduits for thermal water [3].

The simple method to indicate highly fractured is Fault and Fracture Density (FFD) method. In addition, this method is possible to assess structure density areas that formed by interconnection of faults and fractures [3]. By applying this method to a non-volcanic geothermal system, the relationship between fractures and surface geothermal manifestations could be identified.

2. Geological Setting

The focus area of this study is the Amohola region and surrounding areas in Southeast Sulawesi Province with coordinates of 122°36'2.29" - 122°47'27.69" E and -4°16'22.98" - -4°16'22.98" S. Geologically, the study area is comprised of the East Sulawesi Ophiolite Belt that contains an ophiolite complex and pelagic sediment on the east and southeast arms of Sulawesi. This belt consists of mafic and ultramafic rocks as well as pelagic and mélange sediments in several locations. Ultramafic rocks are dominant in the southeast arm and mafic rocks are dominant in northern Sulawesi, particularly along the north shore of Sulawesi Southeast arm [2].

The most important geologic structures for the occurrence of geothermal manifestation in the Amohola region are faults with patterns that strike NW-SE and NE-SW. The tectonic setting of the region generate normal faulting which locally may have increased permeability at the structural intersection of oblique and normal faults. Yushantarti and Rezky [4] reported that this increased rock permeability at the fault intersections creates a pathway for the upward movement of hydrothermal fluid from depth to the surface.

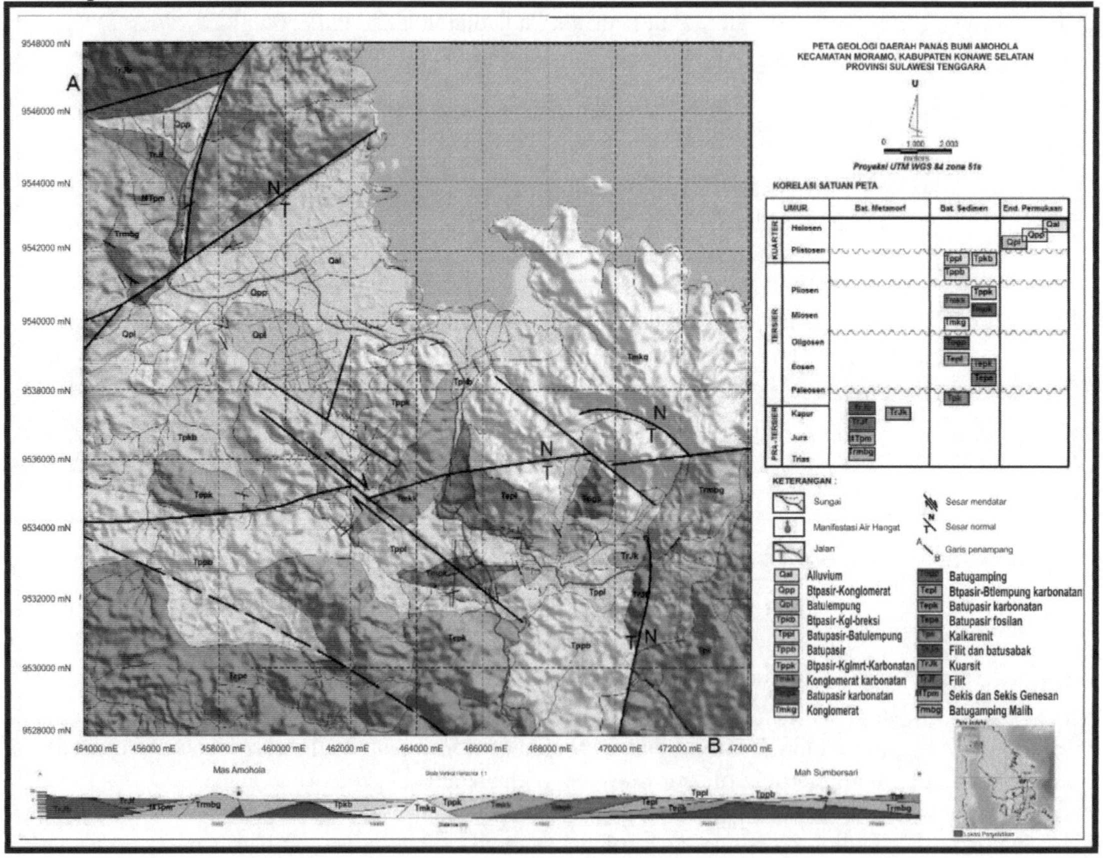

Figure 1. Geothermal Geological Map of Amohola Region [4]

3. Methodology

The FFD method has been applied in this study. Firstly, ASTERGDEM (Advanced Spaceborne Thermal Emission and Reflection Global Digital Elevation Map) image data was used for remote sensing image analysis. The data was processed by using GlobalMapper software to obtain lineament patterns.

Shading features from ASTERGDEM image are used in lineament analysis. Identification of lineaments is needed to support the analysis and estimation of the extent of lineaments at the macroscopic scale. Azimuth of lighting direction is divided into 4 quadrants, which are 0°, 90°, 180° and 270° with similar lighting elevation. The next step is making a grid for determining density and total lineament length data for each grid. Furthermore, exact value according to total lineament length is added in each center of grid. Then the grid values are contoured. The contour map will now reflect fractures density. High density contours indicate high density fracturing and could be interpreted as correlating to high permeability.

In addition to the FFD analysis, another literature study of the geochemistry data was conducted to identify the type of thermal water present in study area.

4. Discussion

In this study, ridge and valley lineaments data are used as a primary dataset, while geological and geochemistry data based on Yushantarti and Rezky, [4] are used as secondary dataset.

The lineament data are processed into rose diagrams and the resulting lineaments direction for the ridges is relatively NNW – SSE and various directions (Nm– S, NW – SE and NE – SW) are indicated for the valleys (Fig. 2). Additionally, the lineament data is also used to create a FFD map (Fig. 2). There are nine different zones that reflecting high permeability value. In this study, 2 km/km² FFD contour is applied as a cut off value to define an anomalous area.

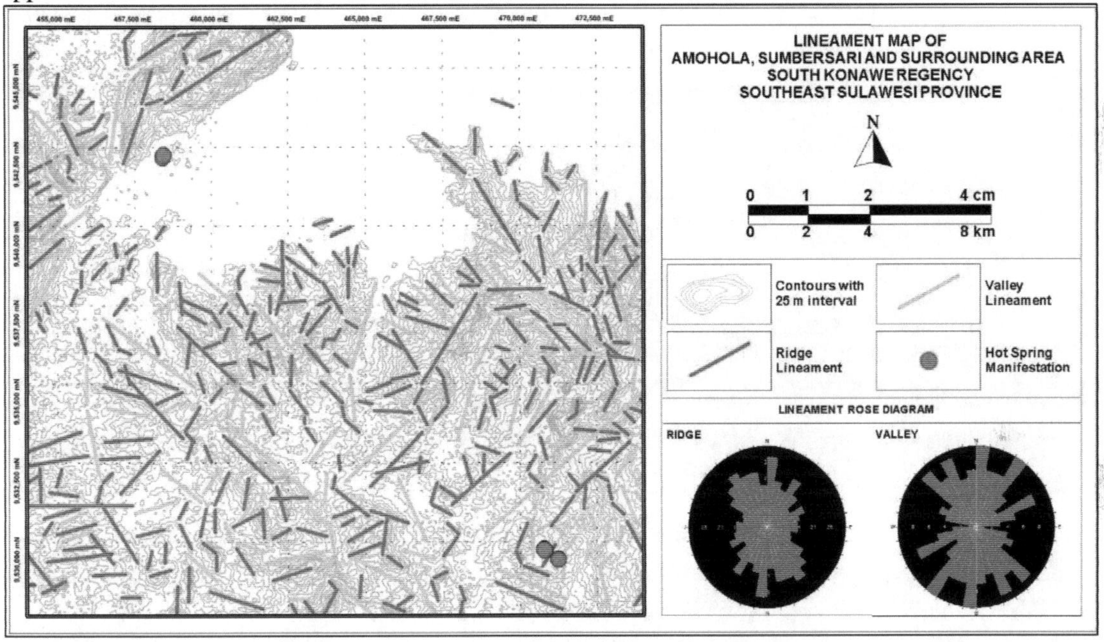

Figure 2. Lineament Map of Study Area

Of the total nine zones identified, four zones are located east of study area, two zones lie in the center and the rest occur to the NW and SW of the study area. By contrast, there is no zone at all in N and NE of the study area owing to those areas are a marine region, hence, lineament interpretation could not be done.

Based on topography, zone 1, 5, 6, 7 and 9 are situated on the slope areas, while zone 2, 4 and 8 are located on the hills as well as zone 3 occurs surrounding one hill (Fig. 3).

From the nine distinct zones (Fig. 3), only two are associated with warm or hot springs, which are in Amohola region (zone 1) and Sumbersari (zone 7). Those remaining seven zones not associated with warm or hot springs are dominated by graben blocks in a normal fault environment. Moreover, the control of erosion and rock weathering might have responsibility to this, in spite of excessively fractures existence in seven zones.

In addition, Amohola and Sumbersari geothermal manifestations emerge from permeable rocks. In particular, Amohola springs appear on altered limestone unit (Trmbg), whereas Sumbersari springs come up on calcarenite unit (Tpk) (Fig. 1). It might be assumed that manifestation related to travertine associated with calcarenite unit.

4.1 Geothermal Possibility Area and Its Association with Geological Setting

Two zones that are interpreted as having possibility as a geothermal prospect area are associated with normal faults in the NW and SE portions of the study area.

Amohola springs are not located in one of the high permeability zones identified by the FFD method. However, these springs are situated on a normal fault zone with a NE – SW trend and its hanging wall facing SE. As a result, thermal manifestations could appear on the surface. As mentioned previously, Amohola springs issue through altered limestone [4]. The altered limestone was faulting as leaking normal fault that triggered the manifestation existence.

On the other hand, Sumbersari springs are exactly located in a FFD method designated high permeability zone (zone 7). This zone is also associated with a normal fault striking N – S and the hanging wall in W – NW slip. Based on fault movement, we refer that fault as an oblique fault. These springs appear on travertine. According to the discussion above, it could be concluded that Sumbersari springs emerge on the surface is caused by high fractures intensity and strengthened by its spring location that situated exactly on the fault zone.

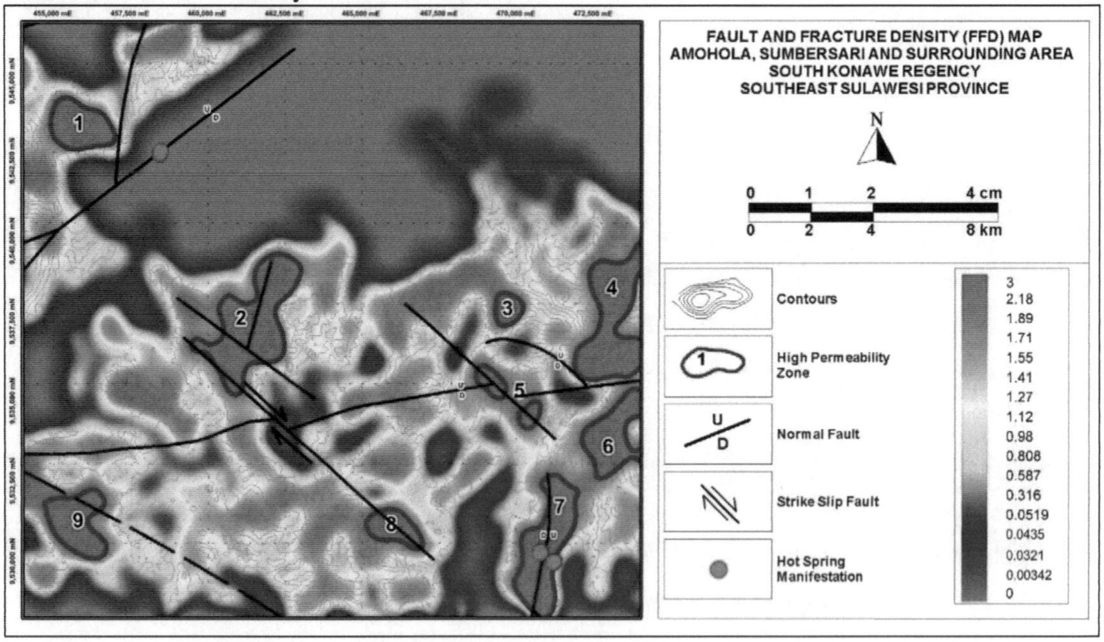

Figure 3. Fault and Fracture Density Map of Study Area

4.2 Geochemistry

Amohola springs consist of four separate springs, one that is hot and three that are warm [4]. The hot spring has pH 6.59 and 50°C water temperature at 26.8° C air temperature. In addition, the hot water in this spring is clear (colorless), odorless and tasteless.

In comparison, the warm springs have pH in the range of 6.35 – 6.62 with temperature ranging from 37.5°C to 46.3°C. Warm water in Amohola springs have various characteristics, such as color (white to colorless) and scent (sulfurous to odorless).

In terms of Sumbersari springs, there are five distinct manifestations that are categorized as warm springs with pH ranging from 6.65 to 6.94 and temperature from 33.3° C – 48.2° C. All springs in Sumbersari have similar properties which are clear (colorless), odorless and tasteless [4].

Warm water at Sumbersari springs is categorized as bicarbonate water. This type of chemical signature may indicate that the water has a surface water characteristic or may indicate mixing with with surface water dominance. The presence of bicarbonate leads to an estimation that the springs are associated with the thermal water flow that contain gas, in particular CO_2 which condenses in a shallow aquifer.

According to the previous research [4], Amohola and Sumbersari springs are at different geothermal system. The geothermal system at Amohola springs is considered to be upflow or at the margins of an upflow. On the other hand, geothermal system at Sumbersari springs is interpreted as outflow.

5. Conclusion

Study area consists of nine high permeability zones based on the FFD method and seven of them have no associated warm or hot springs. It is postulated that the lack of a geothermal signature for the position of seven zones, which are located on normal fault graben (Fig. 1 cross-section). As such, these seven zones are thought to be far from a heat source.

The two remaining zones (Amohola and Sumbersari) have warm and hot springs associated with fractures. In comparison, these two hot springs have a different surface occurrence. At the Amohola springs, the warm or hot water appears to be correlated with a normal fault, whereas in the Sumbersari hot springs are associated with both a normal fault and a high permeability zone defined by the FFD method.

It could be assumed that both hot springs have a different genesis. This study results are strengthen by the the result of previous work [4] who report that geochemically, geothermal system between Amohola and Sumbersari is also different.

To obtain more convincing interpretation, a further study with another method such as drilling or geophysics survey is needed. From those methods, subsurface data in details could be procured for better analysis.

6. References

[1] Soengkono S 1999 Analysis of Digital Topographic Data for Exploration and Assessment of Geothermal System *21st New Zealand Geothermal Workshop*

[2] Sompotan A 2012 *Struktur Geologi Sulawesi* (Bandung: Perpustakaan Sains ITB) p 20

[3] Suryantini and Wibowo H H 2010 Application of Fault and Fracture Density (FFD) Method for Geothermal Exploration in Non-Volcanic Geothermal System; a Case Study in Sulawesi-Indonesia *Jurnal Geoaplika* **5** 27-37

[4] Yushantarti A and Rezky Y 2014 Penyelidikan Terpadu Geologi dan Geokimia Daerah Panas Bumi Amohola, Kabupaten Konawe Selatan, Provinsi Sulawesi Tenggara *Prosiding 2014* (Bandung: Pusat Sumber Daya Geologi)

IOP Conf. Series: Earth and Environmental Science **42** (2016) 012016 doi:10.1088/1755-1315/42/1/012016

Characteristic of geothermal fluid at East Manggarai, Flores, East Nusa Tenggara

Mochamad Iqbal[1], Niniek Rina Herdianita[2], Dikdik Risdianto[3]

[1]Geological Engineering, Bandung Institute of Technology

[2]Applied Geology Research Group, Bandung Institute of Technology

[3]Center for Geological Resources, Bandung

Email: iqbalduri@gmail.com

Abstract. The research area is located in East Manggarai and its surrounding area, Flores. In the study area there are two geothermal systems, i.e. Mapos geothermal system which is associated with Anak Ranakah volcano and Rana Masak geothermal systems which is associated with Watuweri volcano. The difference within these systems is shown by the relative content of conservative elements of Cl, Li and B. Geothermal surface manifestations in Mapos include 4 hot springs having temperatures of 34,3-51,4°C and bicarbonate and sulphate-bicarbonate waters; the discharge area in Rana Masak consist of 3 hot springs with temperatures of 38-46,6°C and chloride and chloride-bicarbonate water. Stable isotopes $\delta^{18}O$ and δD analyses showed that the geothermal fluid derived from meteoric water. The Mapos geothermal system is a high temperature system having reservoir temperature of 250-270°C with natural heat loss of 230 kW. The Rana Masak geothermal system is a low temperature system having reservoir temperature of 120-140°C with natural heat loss of 120 kW.

1. Introduction

East Manggarai is located at Flores Island, East Nusa Tenggara. The research is conducted around Ruteng and Borong to find the geothermal potential (Figure 1). This research is integrated geothermal survey conducted by CGR (Center for Geological Resources) to discover energy resources for electricity in the eastern Indonesia. The previous study is reconnaissance was also conducted by CGR in 2013.

The aim of this research is to identify the characteristic of the geothermal reservoir in the research area, particularly temperature and the depth of reservoir based on water geochemistry study from the surface manifestations. Hopefully this research can be used for further data as a geothermal power development in East Manggarai.

Geological studies conducted by CGR in 2014 mention that the research area consists of several morphology i.e. volcanic cones, volcanic slope, lava dome, and undulating hills. The stratigraphy of research area consists of volcanic rocks such as lava, pyroclastic and volcanic breccia. There is basement in the research area which consists of sedimentary rocks having Tertiary age (Tmk) which underlie all volcanic products aged Quaternary. The geological map of the research area is presented in Figure 2 made by Center for Geological Resources [1]. The lava is dominantly andesite with the characteristic: grey, porphyritic, pyroxene and plagioclase phenocryst, mafic groundmass, in some-

places have sheeting joint structure. Volcanic breccia is the product of volcanic eruption in the research area with the component consisting of igneous rock. The pyroclastic rock formed from a

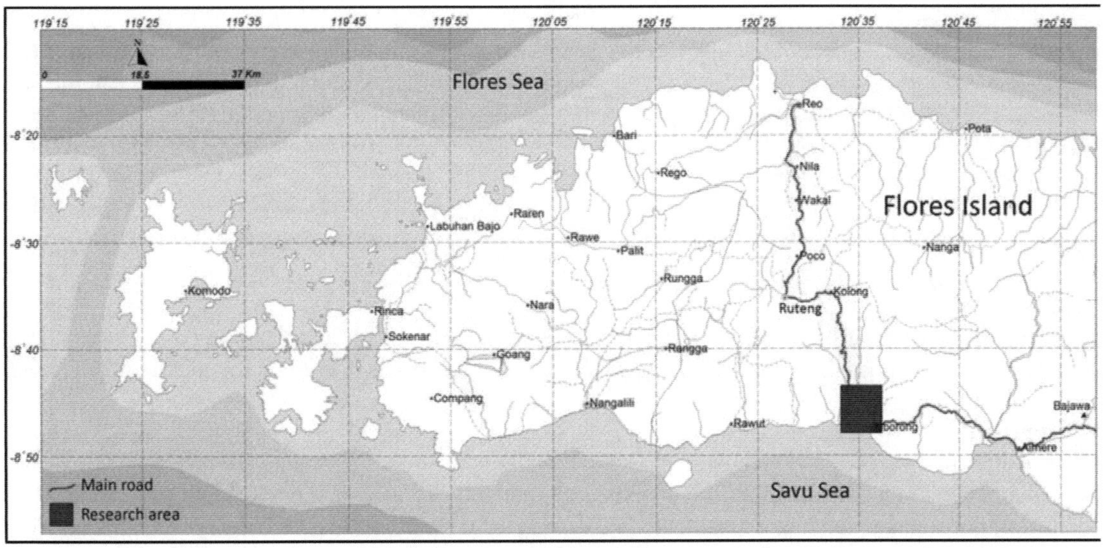

Figure 1. Research area.

pyroclastic flow with quartz and plagioclase in composition. Sedimentary rock (Tmk) consists of interbedded sandstone, claystone and tuff, grey to white, in some places is calcareous and altered. Strike of the bedding is N40° – 50°E with dip 20°-35° to the SE. There are several faults relatively trending NNW-SSE and NE-SW in the research area (Figure 2).

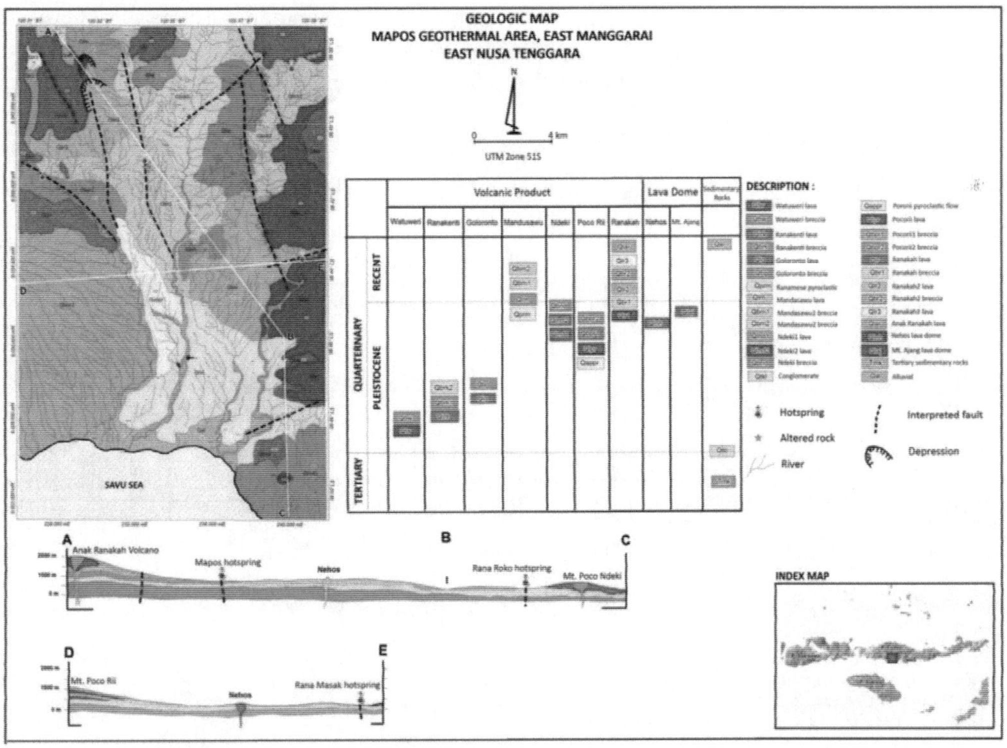

Figure 2. Geological map of research area (made by Center for Geological Resources, 2014).

2. Methodology

In this study data have been collected in the field, including data on geology and geochemistry (hot water and cold water samples), laboratory analysis, and interpretation. Analysis of water samples has been conducted at CGR laboratory focused on the chemistry and stable isotope from the water samples. Water chemistry analysis has been conducted on 7 hot waters and 2 cold water samples. The purpose of chemistry analysis is to determine the element content in the water i.e. Na^+, K^+, Li^+, Fe^{3+}, Ca^{2+}, Mg^{2+}, As^{3+}, Al^{3+}, Cl^-, SO_4^{2-}, HCO_3^-, F^-, CO_3^{2-}, B, SiO_2, and NH_4^+. Analysis of stable isotope δD (deuterium) and $\delta^{18}O$ (oxygen-18) were conducted on all hot water samples and 6 samples of cold water. The purpose of stable isotope analysis is to know the origin of geothermal fluid by using graph δD vs $\delta^{18}O$ (see Section 3.4).

Ion balance calculations are also performed to assess the eligibility samples. Ion balance calculation is following this equation: $(\sum anions - \sum cations)/((\sum anions + \sum cations)$, and the result should not be greater than 5%.

3. Result and analysis

There are 7 surface manifestations in research area appearing as hot springs (Figure 3). The manifestations characteristics are presented in Table 1. Chemistry and isotope analysis is performed to all hot water samples. Cold water sample has also taken from the wells and spring in the research area. This is useful for isotope analysis to know the origin of geothermal fluid. There are 6 cold water samples taken in the research area and its vicinity, 2 were analyzed chemistry and all samples are isotope analyzed. Water chemistry and isotopic data are shown in Table 2. Result of water chemistry analysis then can be used to define the type of geothermal water and geoindicators using ternary diagram. Geoindicator diagram can be used to determine how many reservoirs in the research area.

The first step that must be done to analyze data on the geothermal area is to assess the feasibility of the sample by calculating the ion balance. Ion balance below 5% is assessed as eligible data for further analyses [2]. All ion balance of the sample is below 5%, so the samples in the study area could be used for further analysis (see Table 2).

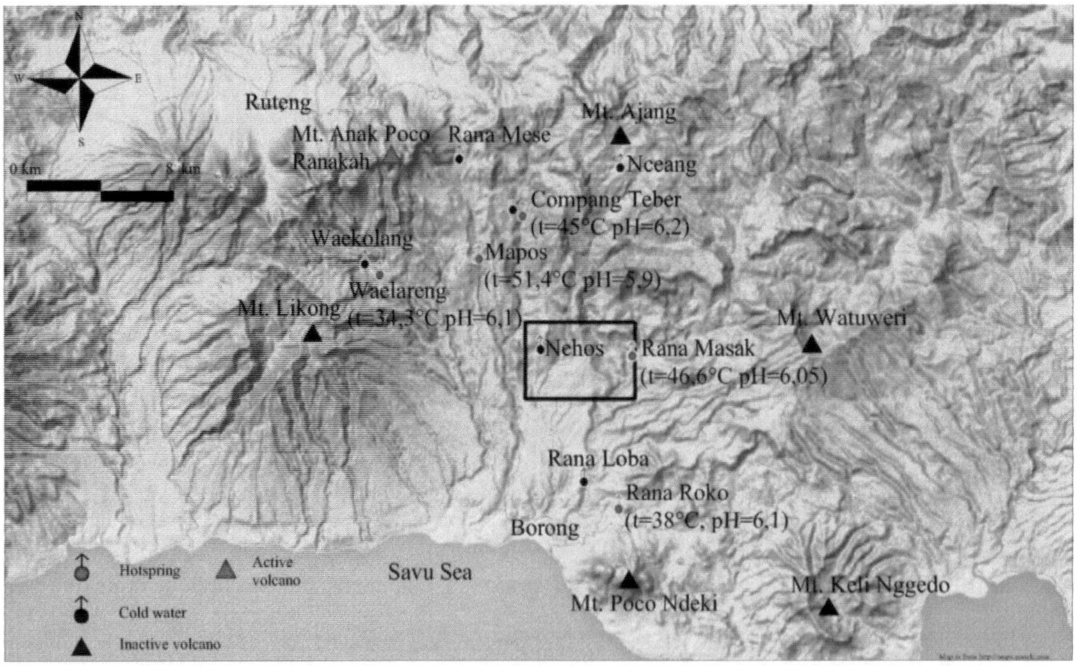

Figure 3. Location of hot spring manifestations and cold water (image from http://maps.google.com, August 2014).

Table 1. Location and characteristic of surface manifestation in research area.

No	Manifesta-tion	Coordinate (Zone 51S)		Eleva-tion	T air	T water	Debit	pH	Conduc-tivity	Sampli-ng Date	Description
		m	m	(masl)	(°C)	(°C)	(L/s)		(umhos /cm)		
1	Mapos-1 hot spring (APMP-1)	232.674	9.039.323	724	28,6	51,4	1,4	5,9	1.300	8 March 2014	Hot water pool with 3,6 m diameter, clear, tasteless – a bit acid, H_2S smell, have continuous gas bubbles, has residual silica around the spring, has mud deposit that contain Sulphur in the flow.
2	Mapos-2 hot spring (APMP-2)	232.680	9.039.324	727	29,2	50,1	0,4	6,1	1.284	8 March 2014	Clear, tasteless - a bit acid, slightly smell of Sulphur, has mud deposit that contain Sulphur in the flow.
3	Compang Teber hot spring (APCT)	234.991	9.041.552	650	26,8	45,0	0,5	6,2	2.160	11 March 2014	Clear, odorless, a bit acid, has iron oxide deposit and thin sinter carbonate.
4	Waelareng hot spring (APWL)	227.323	9.038.454	707	29,3	34,3	0,5	6,1	964	12 March 2014	Clear, a bit acid, odorless.
5	Rana Masak-1 hot spring (APRNM-1)	240.836	9.034.332	243	32,0	46,6	0,6	6,0	16.216	9 March 2014	Hot water pool with 4,8 m diameter, clear, odorless, salty, have continuous bubbles, has wide (100x100m) sinter carbonate, has salt deposit and iron oxide deposit.
6	Rana Masak-2 hot spring (APRNM-2)	240.845	9.034.331	242	29,0	43,0	0,4	5,9	16.210	9 March 2014	Hot water pool with 4,8 m diameter, clear, odorless, salty, have continuous bubbles, has wide sinter carbonate, has salt deposit and iron oxide deposit.
7	Rana Roko hot spring (APRNK)	240.144	9.026.390	65	27,0	38,0	0,5	6,1	6.200	10 March 2014	Hot water pool with 0,76 m diameter, clear, odorless, salty, have continuous bubbles, has sinter carbonate, has salt deposit and iron oxide deposit.

Table 2. Results of chemistry and stable isotope analysis of hot water manifestation and cold water in research area.

Code		APMP-1	APMP-2	APRNM-1	APRNM-2	APRNK	APCT	APWL	ADNH	ADRL	ADRNM	ADCT	ADWK	ADNC
pH	unit	5,9	6,1	6,0	5,9	6,1	6,2	6,1	7,2	7,1	-	-	-	-
Conductivity	(umhos/cm)	1.300	1.284	16.216	16.210	6.200	2.160	964	102	889	-	-	-	-
SiO_2	(mg/L)	27,55	26,75	29,99	31,21	136,66	61,52	135,22	60,61	114,97	-	-	-	-
B	(mg/L)	1,27	1,49	103,97	92,07	10,79	0,57	0,66	0,09	0,57	-	-	-	-
Al^{3+}	(mg/L)	0,03	0,05	0,04	0,04	0,03	0,03	0,04	0,03	0,03	-	-	-	-
Fe^{3+}	(mg/L)	0,13	0,12	0,00	7,59	2,16	1,08	0,00	0,00	0,00	-	-	-	-
Ca^{2+}	(mg/L)	152,50	161,00	547,10	486,10	360,90	322,10	71,80	7,67	34,60	-	-	-	-
Mg^{2+}	(mg/L)	31,85	32,94	118,83	115,46	96,69	11,68	34,46	6,37	17,71	-	-	-	-
Na^{+}	(mg/L)	127,60	128,70	2311,20	1995,30	466,20	117,80	66,90	5,52	135,00	-	-	-	-
K^{+}	(mg/L)	11,15	10,58	308,41	264,64	11,52	5,97	9,32	2,86	9,88	-	-	-	-
Li^{+}	(mg/L)	0,10	0,10	8,38	7,76	0,13	0,13	0,02	0,04	0,03	-	-	-	-
As^{3+}	(mg/L)	0,05	0,05	0,76	1,37	0,00	0,01	0,01	0,00	0,00	-	-	-	-
NH_4^{+}	(mg/L)	1,16	1,14	5,58	6,86	1,65	2,17	0,98	0,89	1,30	-	-	-	-
F^{-}	(mg/L)	0,19	0,12	1,00	0,10	0,44	0,14	0,00	0,01	0,21	-	-	-	-
Cl^{-}	(mg/L)	18,26	20,47	4562,40	4169,30	1050,44	8,76	2,40	1,17	55,43	-	-	-	-
SO_4^{2-}	(mg/L)	470,50	492,75	282,70	272,70	306,53	942,44	32,80	1,70	39,69	-	-	-	-
HCO_3^{-}	(mg/L)	299,70	297,98	683,17	924,20	544,72	189,56	503,11	68,63	390,00	-	-	-	-
CO_3^{2-}	(mg/L)	0,00	0,00	0,00	0,00	0,00	0,00	0,00	0,00	0,00	-	-	-	-
$\delta^{18}O$	‰	-6,44	-6,28	-4,64	-4,71	-4,86	-6,07	-6,17	-5,56	-5,31	-6,54	-6,16	-6,12	-6,83
δD	‰	-34,31	-34,47	-26,76	-26,16	-24,16	-30,28	-31,31	-26,51	-26,65	-35,34	-33,07	-31,48	-38,12
Sum of Cation (meq)		16,15	16,70	147,02	129,09	46,73	22,49	9,62	1,28	9,38	-	-	-	-
Sum of Anion (meq)		15,23	15,73	145,84	138,45	44,97	22,98	9,00	1,19	8,79	-	-	-	-
Ion Balance (%)		3%	3%	0%	3%	2%	1%	3%	3%	3%	-	-	-	-

Description: APMP-1: Mapos-1 hot spring; APMP-2: Mapos-2 hot spring; APRNM-1: Rana Masak-1 hot spring; APRNM-2: Rana Masak-2 hot spring; APRNK: Rana Roko hot spring; APCT: Compang Teber hot spring; APWL: Waelareng hot spring; ADNH: Nehos cold water; ADRL: Rana Loba cold water; ADRNM: Rana Mese cold water; ADCT: Compang Teber cold water; ADWK: Wae Kolang cold water; ADNC: Nceang cold water; - : not analyzed.

3.1. Types of Geothermal Water

The types of water are classified by major anion, which refer to the relative content of Cl^- (chloride), $SO4^{2-}$ (sulphate), and HCO_3^- (bicarbonate), as illustrated in the ternary diagram (Figure 4). The study area has four types of water:

3.1.1. Chloride

Belonging to this type of water are APRNM-1 and APRNM-2. This manifestation has a pH of 5.9-6 and tasted a bit salty. Around these hot springs there is a wide travertine deposit occurs. This water is not a chloride water type coming from the reservoir. High chloride content is derived from a mixture of water formation. That statement is also supported by the content of Ca, Na, and K that is quite high.

3.1.2. Chloride-bicarbonate

Belonging to this type of water is APRNK. It is characterized by clear water with a pH of 6.1 and there are deposits of thin sinter carbonate near the discharge. This type of water formed by the mixing of the water in the reservoir that contains high Cl with groundwater which is has high bicarbonate content (dilute waters).

3.1.3. Bicarbonate

Belonging to this type of water is APWL, characterized by clear water and odorless, with pH close to neutral (6.1). This type of water is formed by adsorption of CO_2 gas and condensation of vapor to the groundwater (steam condensates/steam heated waters).

3.1.4. Sulphate-bicarbonate

Belonging to this type of water are APMP-1, APMP-2, and APCT. Characterized by clear water with pH 5.9-6.2 with a slight smell of sulfur, they present mud deposits around the discharge. This hot spring is formed as a result of vapor and gas on the upflow zone that contains H_2S and CO_2 condense with meteoric water thus forming condensates water (steam condensates).

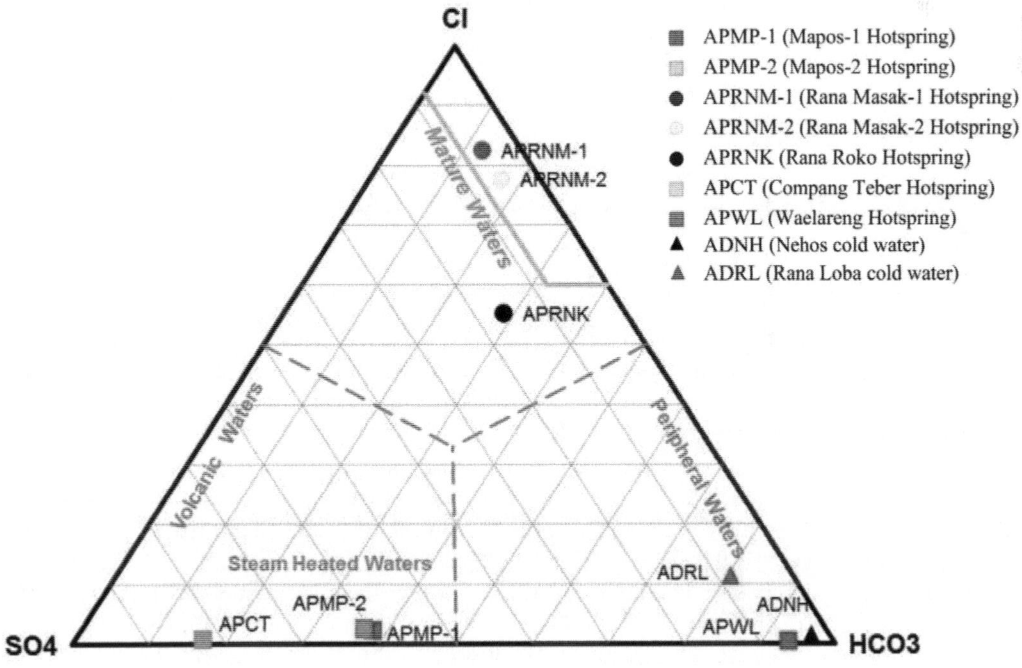

Figure 4. Ternary diagram Cl-SO4-HCO3 to classified types of geothermal water.

3.2. Geoindicator

Cl-Li-B diagram is used to identify similarities the geothermal fluid reservoir [2]. Based on the Cl-Li-B diagram (Figure 5) and the geographic location of the manifestations, the research area has two reservoirs, which are reservoir Mapos and Rana Masak.

3.2.1. Mapos Reservoir

Reservoir Mapos is estimated to have a heat source derived from active volcano Mt. Anak Ranakah. It can be seen from the characteristic of the manifestation that contain sulfur, and high content of sulphate in the springs that indicates the influence of H_2S gas coming from active volcano. This reservoir has 3 sulphate-bicarbonate manifestations (APMP-1, APMP-2 and APCT), and one bicarbonate water manifestation that is APWL. The content of B/Cl in this reservoir ranges from 0.07 to 0.28 due to the fluid interaction with the surrounding rocks and steam heating.

3.2.2. Rana Masak Reservoir

Reservoir Rana Masak is estimated has heat source derived from Mt. Watuweri, it can be seen from the geographical location of manifestations and hydrology pattern. This reservoir has chloride water discharge (APRNM-1, APRNM-2) and chloride-bicarbonate water (APRNK). The reservoir shows B/Cl value between 0.01-0.02, which is caused by fluid-rock-steam-interaction.

Plotting the manifestation on the ternary diagram Na-K-Mg (Figure 6) shows all the water can be classified as immature waters, which means the water is already mixed with meteoric water or groundwater.

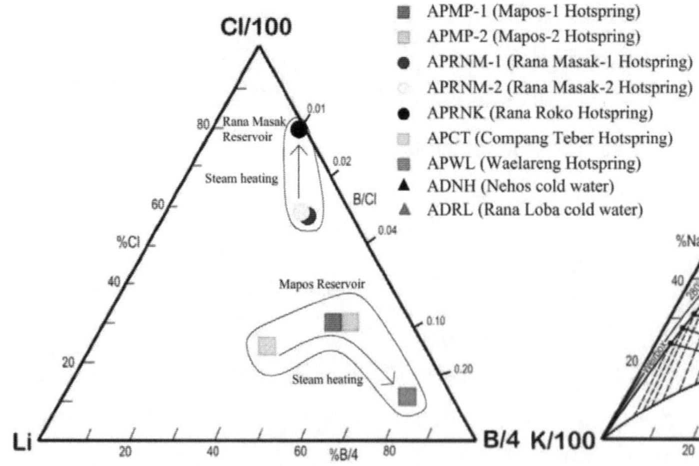

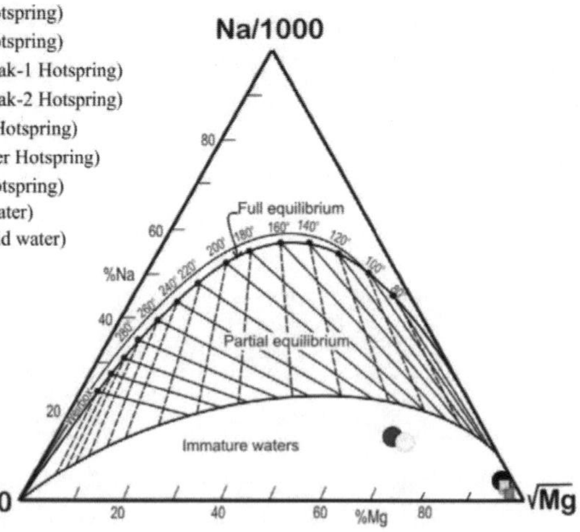

Figure 5. Ternary diagram Cl-Li-B shows 2 reservoirs in research area.

Figure 6. Ternary diagram Na-K-Mg shows water from manifestation is immature waters.

3.3. Reservoir Temperature and Natural Heat Loss

The reservoir temperature has been calculated by using the geothermometer Na/K and K/Mg [3]. This method is applied because it is often used in a geothermal field in Indonesia. It shows as well results that approach to the actual reservoir temperatures. Based on the geothermometer calculation, Mapos reservoir temperature is 250°-270°C, and Rana Masak reservoir temperature is 120°-140°C.

Natural heat loss is the heat that comes out from the geothermal area through the surface manifestations. There are 2 types of natural heat loss in this case. The first one is natural heat loss that comes out through convection that happens to hot spring flow. The second is through evaporation if the manifestation in the form of hot pool that has area. Natural heat loss is calculated in each discharge

IOP Conf. Series: Earth and Environmental Science **42** (2016) 012016 doi:10.1088/1755-1315/42/1/012016

manifestations then summed in each group. Total natural heat loss is sum between convection and evaporation. After calculated, natural heat loss from Mapos Reservoir is around 230 kW and Rana Masak Reservoir is around 120 kW.

3.4. Origin of Geothermal Fluid

Based on the plotting of δD and δ18O contents from each manifestation (Figure 7), geothermal fluid on the research area has content of δD and δ18O which is not much different from the local meteoric water, this shows that the geothermal fluid in research area is derived from meteoric water. Stable isotope variation δD and δ18O local meteoric water follow a linear equation which is calculated based on 6 samples of cold water taken in research area, that is:

$$\delta D = 8\delta^{18}O + 17, R^2 = 1$$

The isotope data plotting is generally showed enrichment δD and δ18O, this is caused by surface evaporation or steam heating. But on the ADNH, APRNM-1 and APRNM-2 the addition is only on δ18O (Figure 7) this is caused by water rock interaction. The reaction between rock and fluid at the depth causing exchange of oxygen with a heavier isotope will be concentrated in the solution phase [2].

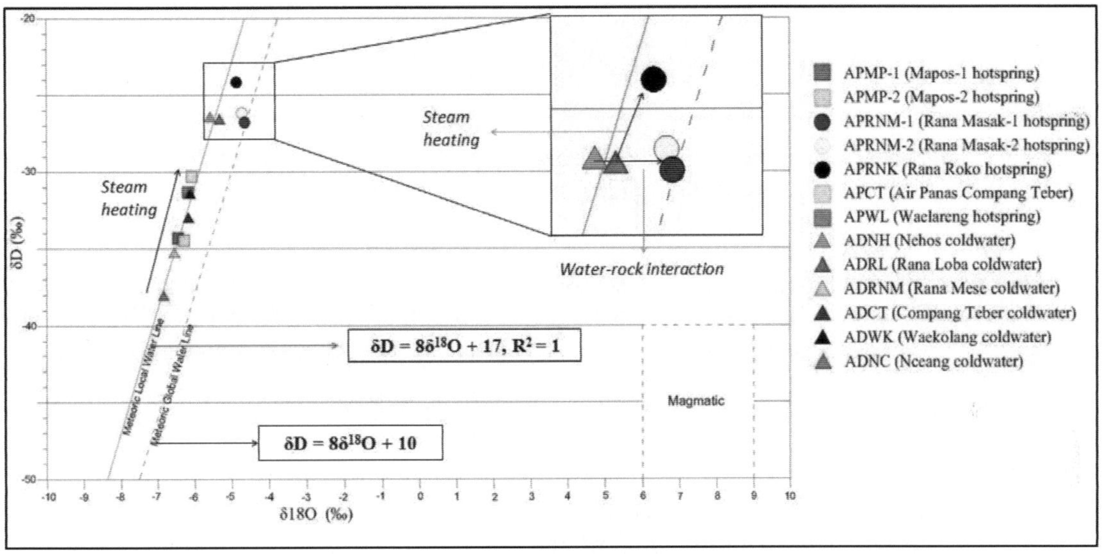

Figure 7. Stable isotope plot for the sample investigated in this study.

3.5. Estimating Depth of Reservoir

Determination depth of reservoir has been performed by using statistical data of temperature reservoir measurements at several drilling locations of geothermal fields in Indonesia presented by Hochstein and Sudarman [4] (Figure 8). From the data, we obtained a graph by using quadratic regression with the following equation:

$$y = -0.0257x^2 + 0.4446x + 31.248$$
$$R^2 = 0.7573$$

with: y = well depth,
x = bottom temperature of reservoir,

The equation above is then applied to the research area. The reservoir Mapos is characterized by temperature of 250°C having depth of about 1.500 m while the reservoir Rana Masak temperature 120°C having depth about 300 m (Figure 8).

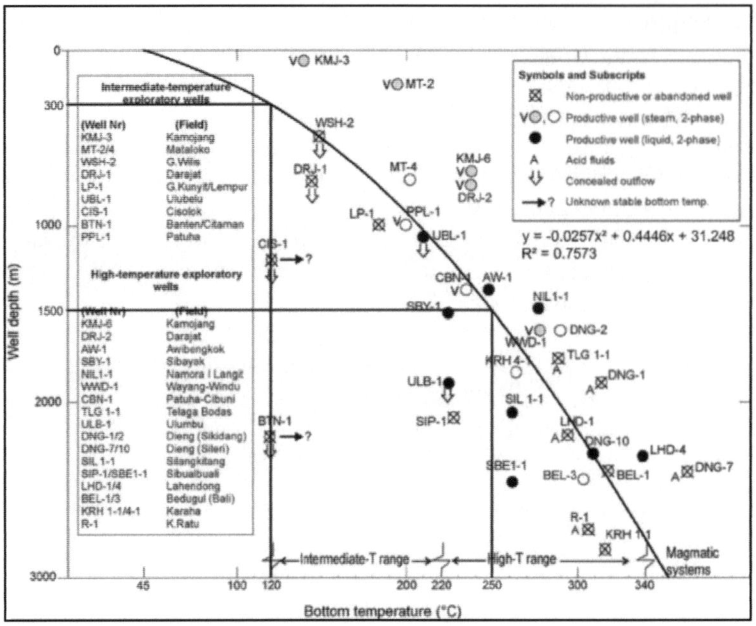

Figure 8. Estimation of the reservoir depth using statistical data of geothermal field in Indonesia. Modified from Hochstein and Sudarman [4].

3.6. Tentative Model of Geothermal System

From all data that has been taken and processed, a made tentative model of the geothermal system of the study area has been developed. Figure 9 shows the tentative model of the geothermal system in research area made by CGR [1].

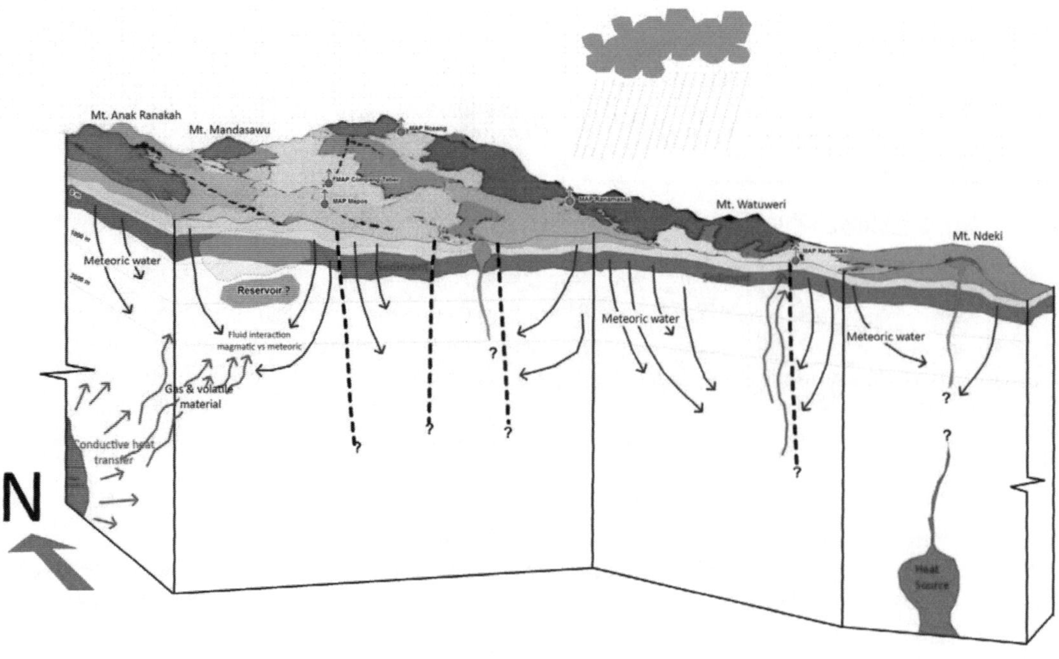

Figure 9. Tentative Model of Geothermal System in the Research Area (made by Center for Geological Resources, 2014).

4. Conclusion

The research area is located at East Manggarai and its vicinity. The geology of the research area consists of volcanic products such as lava and pyroclastic with Quaternary age, and there are Tertiary sedimentary rocks basement.

The water geochemistry shows that the research area has two reservoirs: Reservoir Mapos and Reservoir Rana Masak. Reservoir Mapos has temperature 250-270°C with 1.500 m depth, associated with active volcano Mt. Anak Ranakah and natural heat loss 230 kW. Reservoir Rana Masak has temperature 120-140°C with 300 m depth, associated with Mt. Watuweri and natural heat loss 120 kW. The isotope data show that reservoir water is mixed with meteoric water.

Acknowledgement

The author would like to thank CGR for funding the research including sample analysis data and as well as permission to publish this paper.

References

[1] Center for Geological Resources (CGR) 2014 Report for geothermal integrated survey geology and geochemistry in Mapos area (Bandung: Center for Geological Resources). Unpublished work
[2] Nicholson K 1993 Geothermal Fluids (Berlin, Heidelberg: Springer Berlin Heidelberg)
[3] Giggenbach W F 1988 Geothermal solute equilibria. Derivation of Na-K-Mg-Ca geoindicators Geochim. Cosmochim. Acta 52 2749–65
[4] Hochstein M P and Sudarman S 2008 History of geothermal exploration in Indonesia from 1970 to 2000 Geothermics 37 220–66

IOP Conf. Series: Earth and Environmental Science **42** (2016) 012017 doi:10.1088/1755-1315/42/1/012017

Energy Optimization Modeling of Geothermal Power Plant (Case Study: Darajat Geothermal Field Unit III)

R H M Sinaga[1] and **P S Darmanto**[2]

[1]Engineering Management Study Program, Del Institute of Technology, Laguboti, Toba Samosir 22381, Indonesia
[2]Mechanical Engineering Study Program, Bandung Institute of Technology, Bandung 40132, Indonesia

E-mail: rizal.sinaga@del.ac.id

Abstract. Darajat unit III geothermal power plant is developed by PT. Chevron Geothermal Indonesia (CGI). The plant capacity is 121 MW and load 110%. The greatest utilization power is consumed by Hot Well Pump (HWP) and Cooling Tower Fan (CTF). Reducing the utility power can be attempted by utilizing the wet bulb temperature fluctuation. In this study, a modelling process is developed by using Engineering Equation Solver (EES) software version 9.430.The possibility of energy saving is indicated by Specific Steam Consumption (SSC) net in relation to wet bulb temperature fluctuation from 9°C up to 20.5°C. Result shows that the existing daily operation reaches its optimum condition. The installation of Variable Frequency Drive (VFD) could be applied to optimize both utility power of HWP and CTF. The highest gain is obtained by VFD HWP installation as much as 0.80% when wet bulb temperature 18.5 °C.

1. Introduction

Darajat geothermal power plant is located in West Java, Garut residence, about 35 km southeast from Bandung on 1750 -2000 masl. Darajat unit III was operated in July 2007 for capacity 110 MWe. Since 2009, it has been uprated up to capacity 121 MW, load 110%. There are other geothermal power plants near Darajat Geothermal, those are Kamojang and Wayang Windu plants. Kamojang is located 9 km northeast from Darajat. Meanwhile Wayang Windu about 10 km west from Darajat[8].

Initial design of geothermal power plant was started by choosing suitable thermodynamics cycle, preparing power plant modeling, and then determination of technical specification of every equipment using a simulation. It aims to optimize the net output of a power plant by setting up the thermodynamics parameters[7]. During its operation, power plant systems should be analysed both for full and partial load. Partial load operation is conducted to determine power plant system condition during maintenance of some equipment without shutting down the power plant[9]. A Simulator is required to make decisions for such condition.

This study aims to optimize the overall plant efficiency of Darajat Unit III plant through modeling and simulating the wet bulb temperature fluctuation with particular attention to the refrigeration system. The modeling is validated by Process Flow Diagram (PFD) Darajat Unit III load 110 % (121 MW) for 1% NCG percentage, atmosphere pressure 0.8 bar, wet and dry bulb temperature 15.5°C and 21.77 °C, respectively, and relative humidity of 55 %.

2. Basic Theory

1.1. Heat and Mass Balance

Power plant facility is analysed based on heat and mass balance. There are some assumptions for analysing heat and mass balance[4]:

a. Fluid flows independent on time (steady state)
b. No heat transfer from/to every facility
c. No work produced/consumed, except turbine and pump
d. Kinetic and potential is ignored since enthalpy is far higher than those kinds of energy.

Heat and mass balance is analysed based on the following equations, [6] :

$$\frac{dm_{cv}}{dt} = \sum_{inlet} \dot{m}_i - \sum_{exit} \dot{m}_e \tag{1}$$

$$\left(\frac{dE}{dt}\right)_{cv} = \dot{Q}_{cv} - W_{cv} + \sum_i \dot{m}_i \left(h_i + \frac{v_i^2}{2} + gz_i\right) - \sum_e \dot{m}_e \left(h_e + \frac{v_e^2}{2} + gz_e\right) \tag{2}$$

Where:

$\dot{Q}$ = Heat transfer to or from system (kW)
$\dot{W}$ = Work produced or consumed by facility(kW)
$\dot{m}$ =mass rate(kg/s)
h =enthalpy specific (kJ/kg)
v =fluid velocity$({}^{m}/_{s})$
g =gravitation constant$\left({}^{m}/_{s^2}\right)$
z = high level(m)

1.2. Energy Conversion System

Darajat plant energy conversion system is displayed by figure 1. Point 1 shows fluid outflow from well head which has steam quality 99.98%. Then, fluid flows in to steam scrubber and produces dry steam to be expanded in to turbine. Meanwhile brine flows out from steam scrubber (point 3) to cooling tower before injected into the reservoir. Figure 1 illustrates simple scheme Darajat unit III process flow[1]. Figure 2 shows T-s diagram of Darajat unit III process flow. A complete Process Flow Diagram (PFD) of Darajat unit III is shown in figure 3.

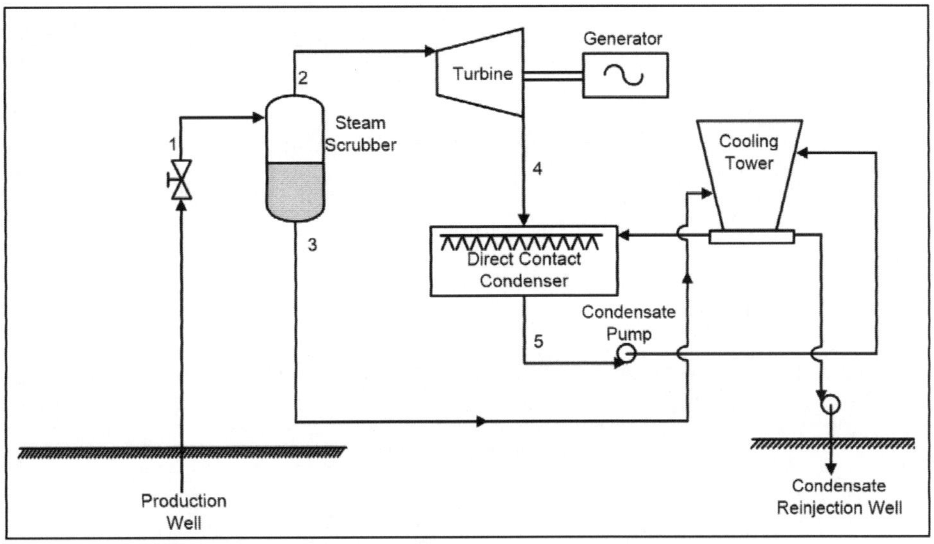

Figure 1. Simple Process Flow Scheme of Darajat Unit III

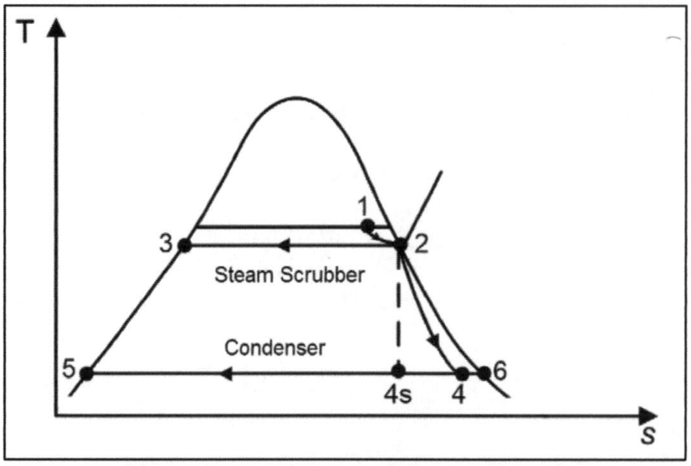

Figure 2. *T-s* Diagram of Darajat Unit III

Based on figure 2, steam mass flow rate in to turbine is expressed as $\dot{m}_{s,2} = x_1 \cdot \dot{m}_{s,1}$. The isentropic process of turbine results $s_2 = s_{4,s}$. . Turbine outlet steam quality (point 4s) is calculated based on equation (3):

$$x_{4s} = \frac{s_{4s} - s_5}{s_6 - s_5} \tag{3}$$

Then, fluid enthalpy at point 4s is:

$$h_{4s} = x_{4s} \cdot h_6 + (1 - x_{4s}) \cdot h_5 \tag{4}$$

Thus, turbine output is calculated by equation (5)

$$\dot{W}_t = x_1 \cdot \dot{m}_{s,1} \cdot (h_2 - h_4) \tag{5}$$

Turbine efficiency is :

$$\eta_t = \frac{h_2 - h4}{h_2 - h_{4s}} \tag{6}$$

Based on PFD, inlet turbine pressure (point 2) is 16.83 Bar where saturation temperature, both steam and NCG enthalpy are $T_2 = 203.9$ °C , $h_{s,2}$ =2795 $\frac{kJ}{kg}$, and$h_{NCG,2}$ =159.7 $\frac{kJ}{kg}$, respectively. Turbine isentropic process results $s_2 = s_6$ =6.402$\frac{kJ}{kg\ K}$. Therefore, enthalpy at point 6 as function of both entropy and pressure, $h_{s,6}$ =2011 $\frac{kj}{kg}$. Turbine isentropic efficiency is 82.5%, therefore steam enthalpy at point 6 is

$$\eta_{turbine} = \frac{h_{s,2} - h_6}{h_{s,2} - h_{6s}} \tag{7}$$

Electricity power is calculated by

$$\dot{W} = \frac{\dot{m}_{s,3} \cdot (h_{mix,2} - h_{mix,6})}{1000} \cdot \eta_{generator} \tag{8}$$

Generator effciency is 98%, thus gross electricity power is $\dot{W}_{gross}$ =121 MW.

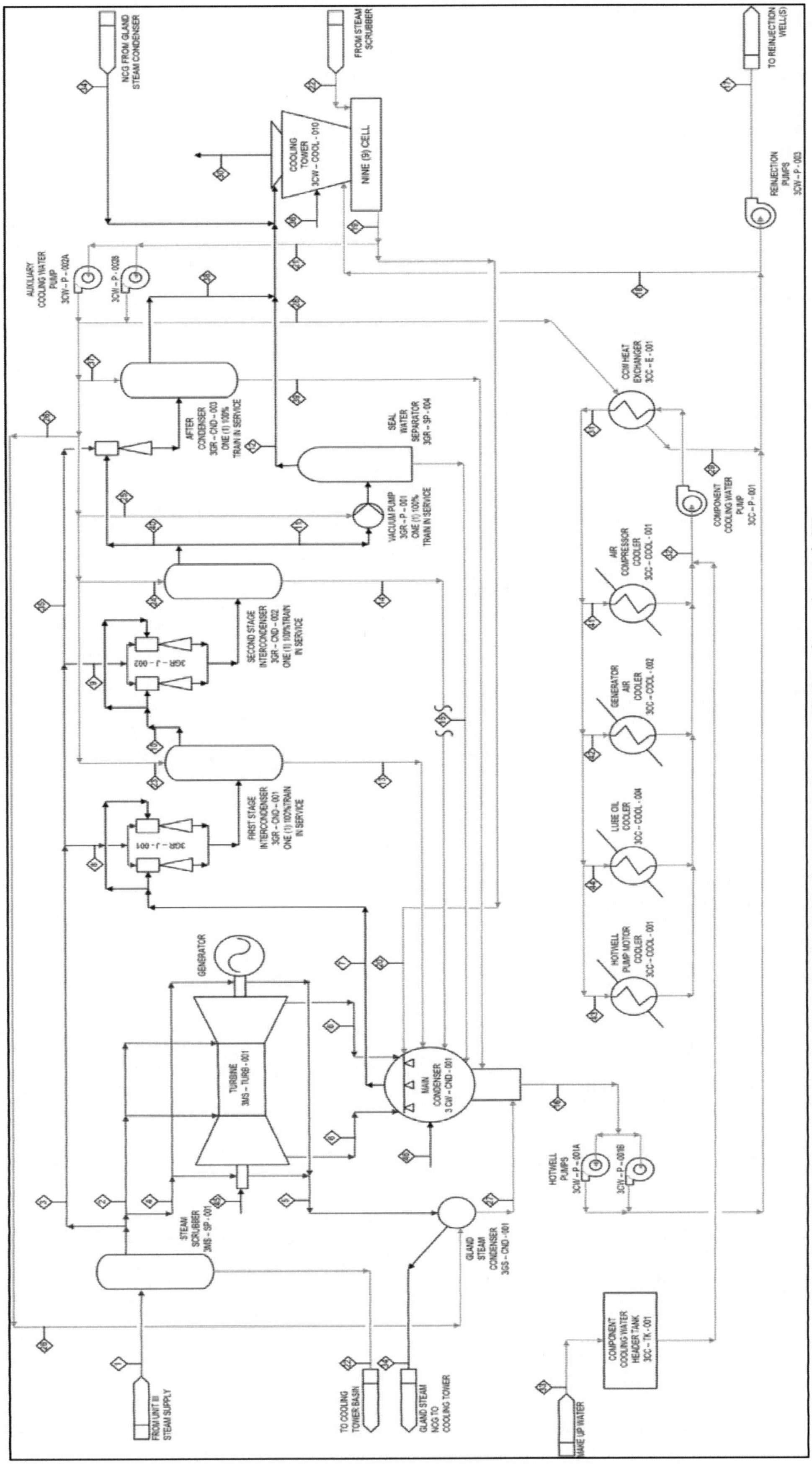

Figure 3. Process Flow Diagram (PFD) of Darajat Unit III

3. Power Plant Modeling

Some main facilities are modelled by heat and mass balances are shown below [4]:

3.1. Main Condenser System

Main condenser scheme is illustrated by figure 4. Mass balance of steam and water, NCG, and air are shown through equation 9, 10, and 11, respectively.

$$2\dot{m}_{s,6} + \dot{m}_{w,13} + \dot{m}_{w,14} + \dot{m}_{w,15} + \dot{m}_{w,20} + \dot{m}_{w,27} + \dot{m}_{w,39} = \dot{m}_{s,7} + \dot{m}_{w,16} \quad (9)$$

$$2\dot{m}_{NCG,6} = \dot{m}_{NCG,7} \quad (10)$$

$$\dot{m}_{a,46} + \dot{m}_{a,20} = \dot{m}_{a,7} \quad (11)$$

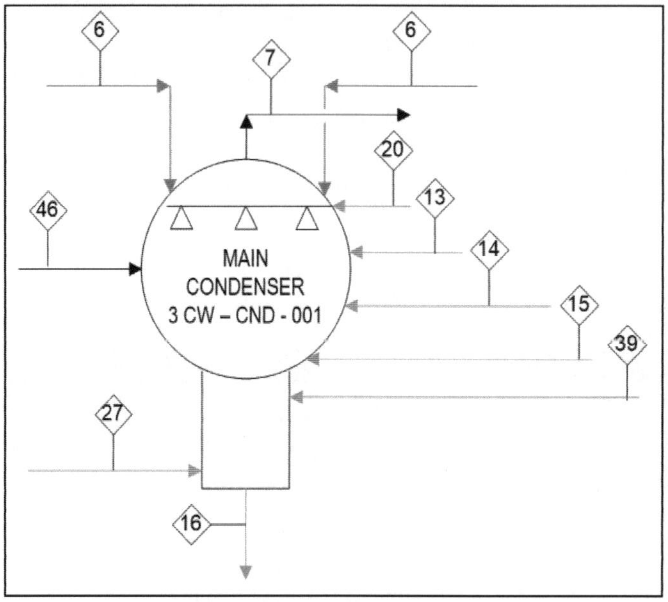

Figure 4 Main Condenser Scheme

Meanwhile, energy balance is shown below

$$2(\dot{m}_{s,6} + \dot{m}_{NCG,6})h_{camp,6} + \dot{m}_{w,13} \cdot h_{w,13} + \dot{m}_{w,14} \cdot h_{w,14} + \dot{m}_{w,15} \cdot h_{w,15} + \dot{m}_{w,20} \cdot h_{w,20} + \dot{m}_{a,20} \cdot h_{a,20} + \dot{m}_{w,27} \cdot h_{w,27} + \dot{m}_{w,39} \cdot h_{w,39} + \dot{m}_{a,46} \cdot h_{a,46} = (\dot{m}_{s,7} + \dot{m}_{NCG,7})h_{camp,7} + \dot{m}_{a,7} \cdot h_{a,7} + \dot{m}_{w,16} \cdot h_{w,16} \quad (12)$$

The calculation of cooling water mass rate $(m_{w,20})$ is determined by comparing water mass flow rate entering cooling tower with cooling air, defined as L/G [3]. It is 1.049 for wet bulb temperature 15.5°C. Therefore, both condensate mass flow rate $(m_{w,16})$ and temperature (T_{16}) can be obtained.

3.2. Cooling Tower

The flowing of water, air, and NCG inside cooling tower is illustrated in figure 5.

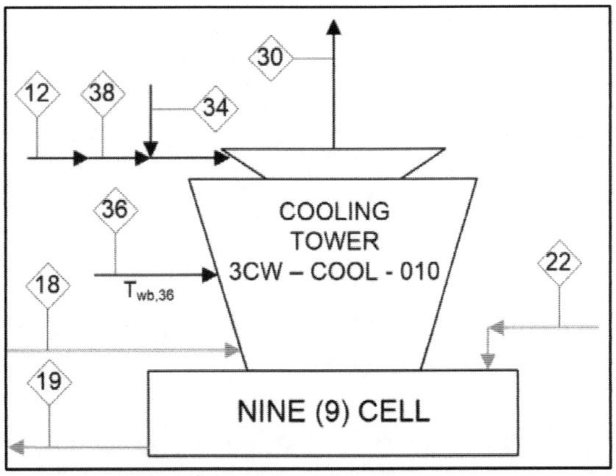

Figure 5. Cooling Tower Scheme

Water/steam, NCG, and air mass balance is shown in equation (13), (14), and (15), respectively:

$$\dot{m}_{s,12} + \dot{m}_{s,36} + \dot{m}_{s,38} + \dot{m}_{w,18} + \dot{m}_{w,22} = \dot{m}_{w,19} + \dot{m}_{s,30} \quad (13)$$

$$\dot{m}_{NCG,12} + \dot{m}_{NCG,34} + \dot{m}_{NCG,38} = \dot{m}_{NCG,30} \quad (14)$$

$$\dot{m}_{a,12} + \dot{m}_{a,34} + \dot{m}_{a,36} + \dot{m}_{a,38} = \dot{m}_{a,30} \quad (15)$$

Energy balance:

$$h_{a,36} + \omega_{36} \cdot Cp_{v,36} T_{wb,36} + \frac{1}{\dot{m}_a}\left(\dot{m}_{w,18} \cdot Cp_{v,18} \cdot T_{18} + \dot{m}_{w,22} \cdot Cp_{v,22} \cdot T_{22} + \dot{m}_{s,36} \cdot h_{s,36}\right) =$$
$$h_{a,30} + \omega_{30} \cdot Cp_{v,30} \cdot T_{wb,30} + \left(\omega_{36} - \omega_{30} + \frac{1}{\dot{m}_a}\left(\dot{m}_{w,18} + \dot{m}_{w,22}\right)\right) \cdot Cp_{v,19} \cdot T_{19} \quad (16)$$

As explained before that L/G is 1.049. Therefore water mass rate entering cooling tower $(m_{w,19})$, inlet $(m_{a,36})$ and outlet air mass rate $(m_{a,30})$ can be obtained.

3.3. Utilities Load

Some kinds of utilities load are shown below:

3.3.1. Hot Well Pump. Hot well pump is used to remove condensate water to cooling tower. The scheme of hot well pump is shown in figure 6.

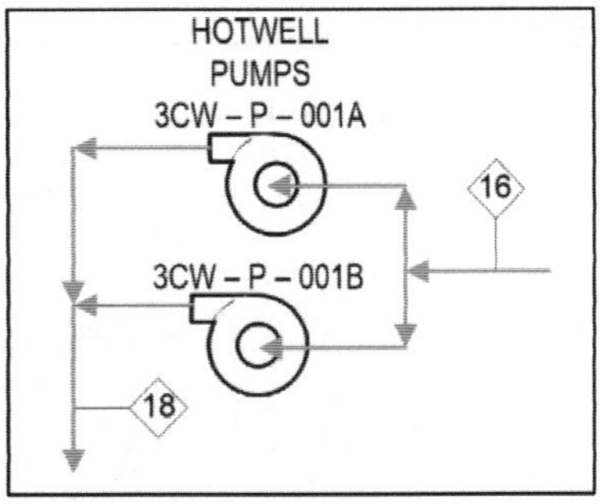

Figure 6 Hot Well Pump Scheme

Hot well pump utility power is calculated by [5]:

$$\dot{W}_{3CW-P-001A} = \frac{\frac{1}{2}\dot{V}_{16} \cdot g \cdot Head}{v_{w,16} \cdot \eta_{HWP} \cdot \eta_{motor}} \quad (17)$$

For given data, hot well pump's power is $\dot{W}_{3CW-P-001A}$ =1077 kW. Based on manual data, the efficiency of pump and pump motor is 84.72 % and 87.85%, respectively. The calculated head pump is 24.44 m. Meanwhile the head of pump for $m_{w,16}$ calculated is 24.44 m.

3.3.2. Fan Cooling Tower. There are nine existing cells inside the cooling tower. The efficiency of each fan and its motor is 76.1% and 74.25%, respectively. Fan utility power is calculated by:

$$\dot{W}_{fan} = \frac{\dot{m}_{air,16} \cdot v_{air,36} \cdot \Delta p}{\eta_{fan} \cdot \eta_{motor}} \quad (18)$$

Variable Δp defines pressure drop inside cooling tower. Total fan utility power is $\dot{W}_{fan}$ =1666 kW. By using the similar approach, total utility power is 4310 kW.

4. Result

Power plant performance is determined by Specific Steam Consumption (SSC) steam supply to produce1 kWh electricity. SSC is calculated by [9]:

$$SSC_{nett} = \frac{\dot{m}}{\dot{W}_{nett} \cdot 1000} \cdot 3600 \quad (19)$$

4.1. Variable Frequency Drive (VFD) Application

Wet bulb temperature fluctuation provides a chance for power plant system optimization by using Variable Frequency Drive (VFD)[2]. It is installed to hot well pump and/or cooling tower fan.

4.1.1. VFD Application for Hot Well Pump (VFD HWP). The lower wet bulb temperature, the lower exit cooling water temperature will be. Therefore cooling water mass rate in to condenser should be reduced to keep constant pressure. VFD HWP should be installed to reduce hot well pump utility power. Fluid flowing capacity in to a pump can be controlled either by an open valve or by a pump rotation[2]. The difference of those methods is displayed in figure 7.

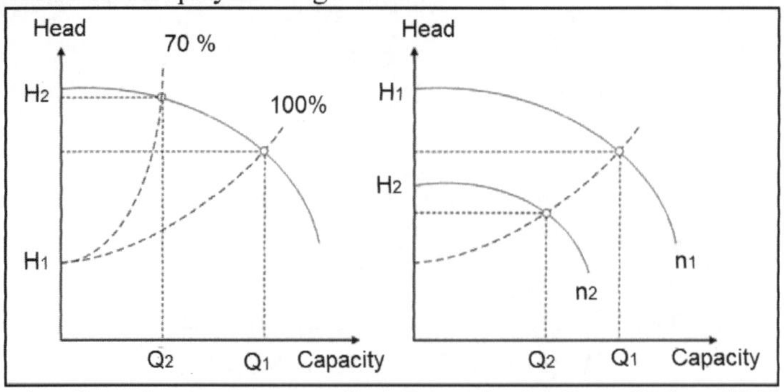

Figure 7 Flowing Capacity Arrangement in A Pump by (a) Open Valve, (b) Pump Rotation
Note : Percentage denotes valve opening

Figure 7(a) shows the increased capacity Q_1 to Q_2 is controlled by open valve only. It yields constant utility power. Meanwhile figure 7(b) shows that increased capacity is controlled by the speed of pump rotation. This method could reduce utility power for lower capacity.

The ratio between pump rotations is determined by

$$\frac{Q_1}{Q_2} = \frac{n_1}{n_2} \cdot \left(\frac{D_1}{D_2}\right)^3 \tag{20}$$

Q_1 : First stage of pump capacity
Q_2 : Second stage of pump capacity
n_1 : Pump speed rotation on first stage
n_2 : Pump speed rotation on second stage
D_1 : Pump diameter on first stage
D_2 : Pump diameter on second stage

Pump constant diameter ($D_1 = D_2$) results eq. (21) :

$$\frac{Q_1}{Q_2} = \frac{n_1}{n_2} \tag{21}$$

Therefore frequency changes as result of stator field rotation of pump motor

$$n_s = \frac{120f}{P} \tag{22}$$

Where f denotes frequency of stator current. Meanwhile, P denotes the number of pole inside electric motor. Another parameter is slip. It is occurred as result of the difference of rotation motor with stator field which formulated as:

$$Slip = \frac{n_s - n_p}{n_s} \tag{23}$$

Where n_p denotes pump rotation (rpm)

4.1.2. VFD Application for Cooling Tower Fan (VFD CTF). It is applied to maintain constant condenser pressure by the adjustment of both cooling water temperature and mass rate in relation to wet bulb temperature. The adjustment is controlled by cooling tower fan. Most of equations that are used both for VFD HWP and VFD CTF are relative same. The only difference is the reduction ratio of fan[2] which is formulated as equation (24)

$$R = \frac{n_{motor}}{n_{fan}} \tag{24}$$

Where

R : Gearbox reduction ratio
n_{motor} : Motor fan rotation speed (rpm)
n_{fan} : Fan rotation speed (rpm)

4.2. Power Plant Analysis

Estimation of cooling water temperature is obtained by cooling tower heat transfer theory. Both cooling water and condensate temperature, for wet bulb temperature, range from 9°C to 20.5°C, displayed in figure 8. It shows that the temperature difference of the two unit processes are relatively constant over the increasing wet bulb temperature since the mass rate of the cooling water is also relatively constant.

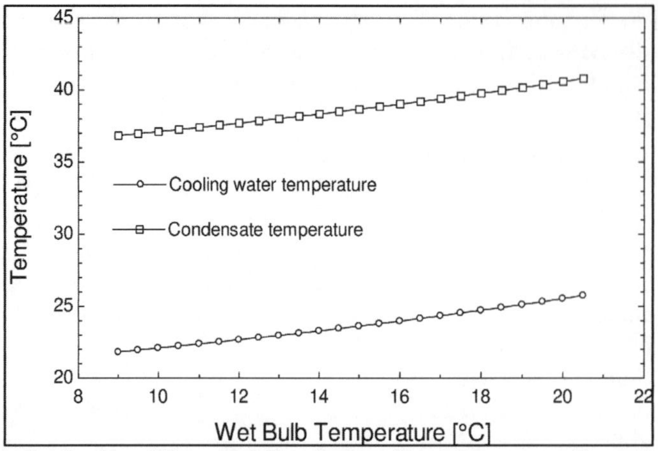

Figure 8. Cooling Water and Condensate Temperature on Constant L/G

The higher cooling water temperature, the higher condenser pressure is. The increase of condenser pressure is shown in figure 9. It displays that condenser pressure increases 23.6% in relation to the increase of wet bulb temperature from 9°C up to 20.5°C

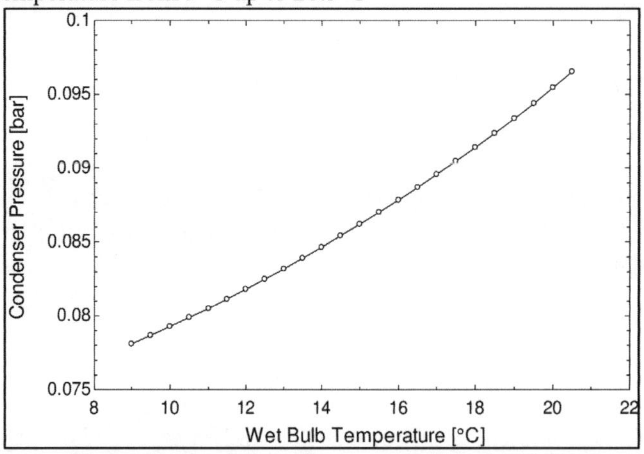

Figure 9. Condenser Pressure on Constant L/G

The increase of condenser pressure will affect electricity generation which is displayed in figure 10. The increase of condenser pressure yields the decrease of 2.9% gross output power.

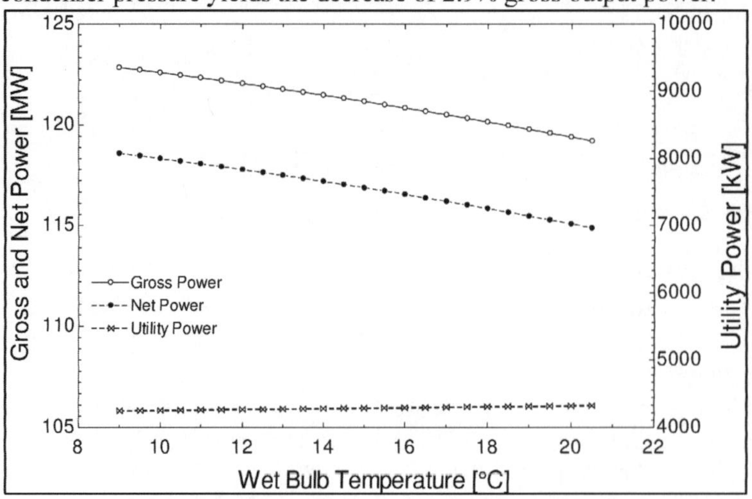

Figure 10 Output Power Plant on Constant L/G

Using similar approach, SSC net increases approximately 3.19% until 3.24%, as shown in figure 11.

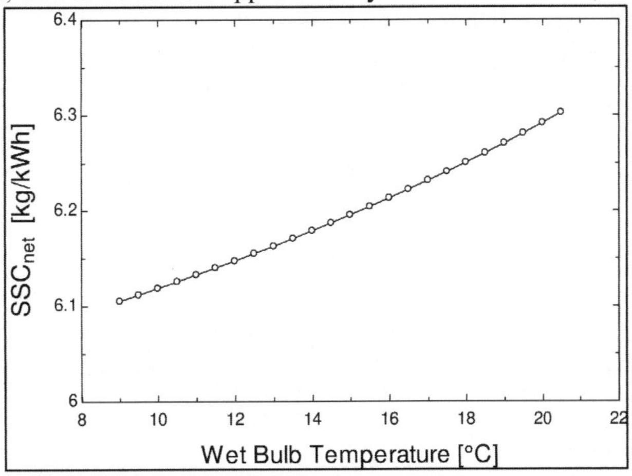

Figure 11 Wet Bulb Temperature Impact to SSC Nett on Constant L/G

Figure 12 shows the impact of VFD HWP to maintain condenser pressure 0.087 bar. Both cooling water and condensate temperature increase directly proportional with wet bulb temperature. The increase temperature is quite different with constant L/G. The increase of bulb temperature yields the less difference of those temperatures where caused by the amount of cooling water mass rate to maintain constant condenser pressure 0.87 bar.

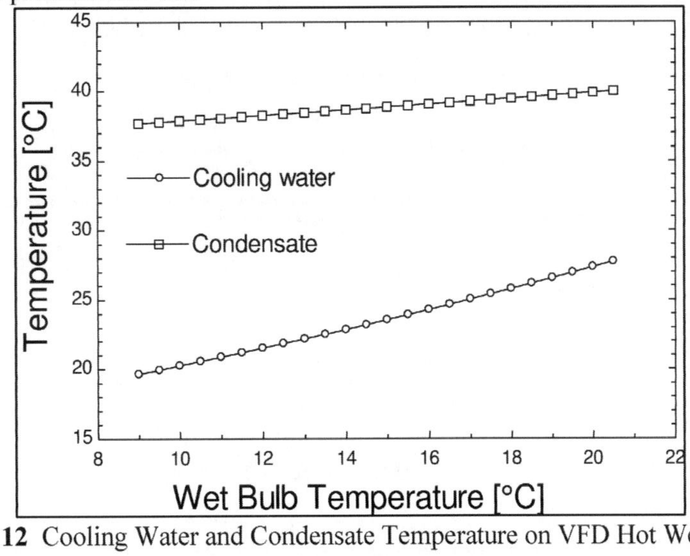

Figure 12 Cooling Water and Condensate Temperature on VFD Hot Well Pump

Both cooling water and condensate mass rate is revealed in figure 13. Their increase are 52.92% and 45.12%, respectively.

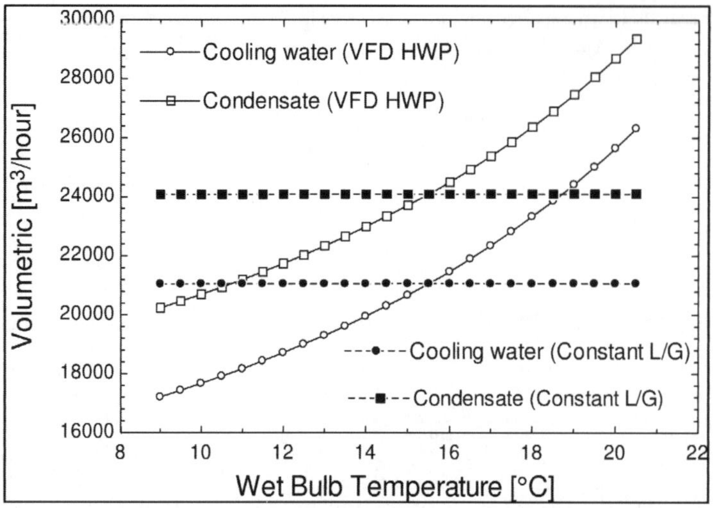

Figure 13 Cooling Water and Condensate Mass Rate on VFD HWP

Based on HWP design, normal capacity of each pump is 11.420$\frac{m^3}{hour}$ which provide energy saving of HWP utility power for a certain wet bulb temperature. Figure 14 shows that pump utility power range from 1930 kW up to 2146 kW for wet bulb temperature below 15.5°C. VFD HWP reduces 11.20% utility power. VFD can be applied up to wet bulb temperature 18.5 °C in case of the ability of HWP to operate until frequency 60 Hz. It is displayed in figure 14 by shaded region.

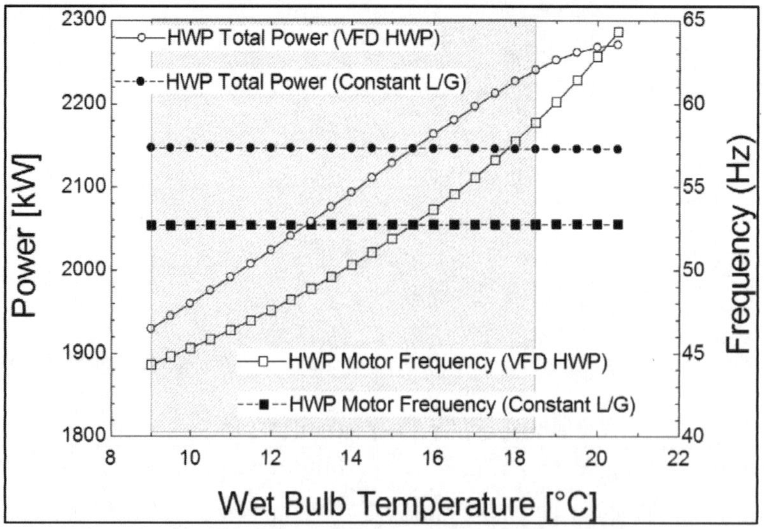

Figure 14 Utility Power and Frequency of Hot Well Pump

VFD can also be installed to cooling tower fan (CTF) electrical motor. Cooling air mass rate is adjusted to keep constant condenser pressure. It is displayed in figure 15.

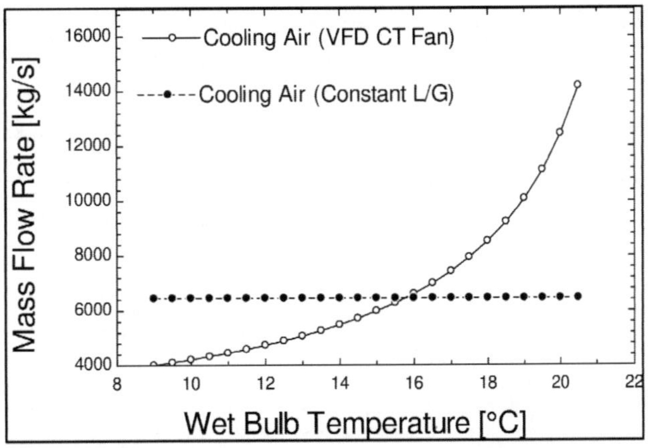

Figure 15 Cooling Air Mass Rate as Wet Bulb Increase

Cooling air mass rate increases 273.93% to maintain exit cooling water temperature 23.9°C. Based on cooling tower design, each cell can absorb cooling air as much as 771.6 kg/s.

Normal cooling air is absorbed by 8 of 9 existing cells. Figure 16 reveals fan quantities to adjust the amount of cooling air as described before. It shows that VFD CTF is applied for wet bulb below16.5°C

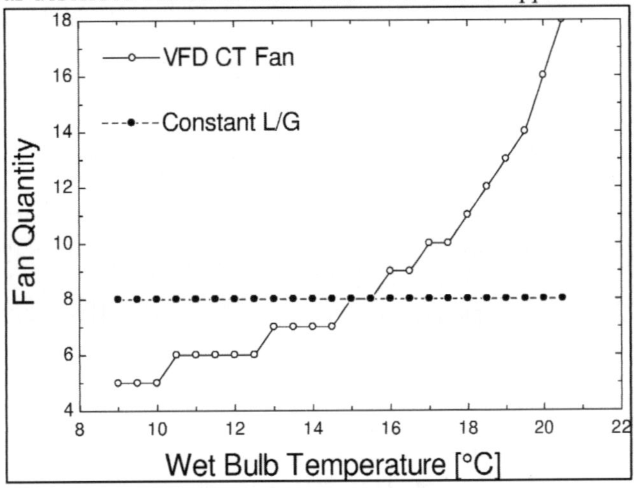

Figure 16 Cooling Tower Fan Quantity

Total fan utility power is represented in figure 17.

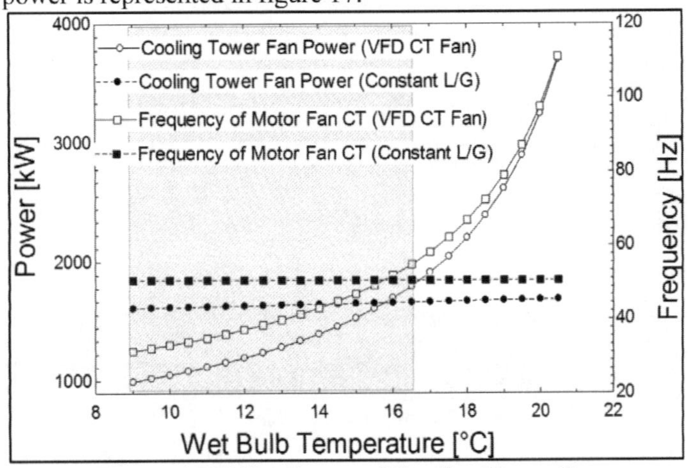

Figure 17 Utility Power of Cooling Tower Fan

The possibility of energy saving of CTF is applied below wet bulb temperature 15.5°C as well as VFD HWP. It reduces energy up to 37.97% when wet bulb temperature 9°C. Meanwhile, VFD CTF can be applied up to wet bulb temperature 17°C in case of motor fan ability to operate until 60 Hz. Both gross and net power of each scenario is represented in figure 18. It shows that both gross and net power constant L/G is higher than other VFD applications

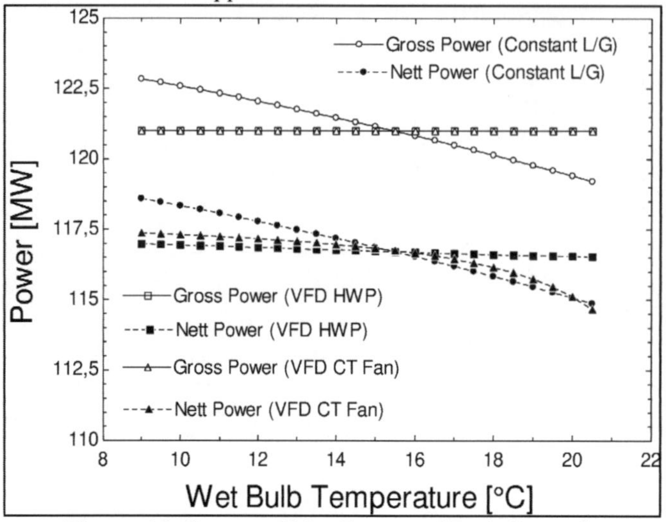

Figure 18 Gross and Nett Power of Each Scenario

SSC net of each scenario is considered to decide power plant performance. SSC net defines the amount of steam to generate 1 MWh net electricity. The less steam supply required to generate 1 MWh electricity, the more efficient the power plant is. SSC net consideration is displayed in figure 19. When the wet bulb temperature is lower than 15.5°C, constant L/G SSC net is lower than others. Otherwise, when the wet bulb more than15.5°C, SSC net VFD HWP is lower than constant L/G.

In case of motor pump ability to operate until frequency 60 Hz, VFD HWP can be applied starts from wet bulb temperature 15.5°C until 18.5°C. It is shown in Figure 19. The highest gain of VFD HWP is obtained when wet bulb temperature 18.5°C, about 0.80% compared to constant L/G.

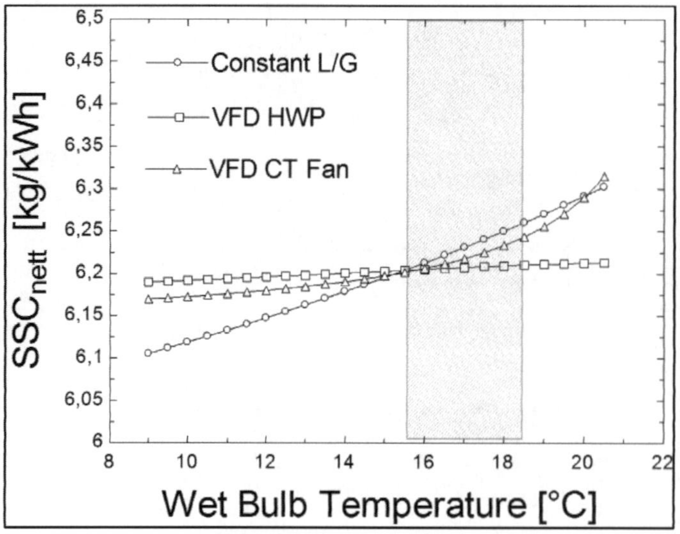

Figure 19 SSC Net

5. Conclusions and Suggestions

5.1. Conclusions
Having developed and simulated some variables, some conclusions are described below:
- Based on consideration of wet bulb fluctuation, the existing daily operation reaches its optmimum performance.
- In case of motor pump ability to operate until frequency 60 Hz, VFD HWP can be applied starts from wet bulb 15.5 °C until 18.5 °C. The highest gain of VFD HWP is obtained when wet bulb temperature18.5 °C, about 0,80% compared to constant L/G.

5.2. Suggestions
- Besides technical consideration, economical approach should be considered for power plant optimization.
- This study mainly focuses on simulation based on full load. Further study should elaborate partial load in order to keep better performance during maintenance.

References
[1] DiPippoR 2007 *Geothermal Power Plants : Principal, Applications, Case Studies, and Environmental Impact*(Darmouth, Massachusets: Elsevier Ltd.)
[2] Effendi H 2009*Main Cooling Water System Optimization Kamojang Unit 2 Geothermal Power Plant, PT. Indonesia Power*(Bandung, ITB) pp 10-45
[3] Freeston D 1996*Geothermal Technology:Teaching The Teachers - Course Stage III.* pp 15-20
[4] Incropera, Frank P 2008*Fundamental of Heat and Mass Transfer*(John Willey and Sons)
[5] Maloney J O 2008*Perry's Chemical Engineers' Handboook*(The McGraw-Hill Companies, Inc) pp 12-30
[6] Moran M J and Shaphiro H N 2000*Fundamental of Engineering Thermodynamics*(John Willey and Sons) Chapter 4 pp 164-209
[7] Nugroho A 2011*Optimization of Electrical Power Production from High-Temperature Geothermal Fields with Respect to Silica Scaling Problem*(Reykjavik, Iceland: The United Nations University) pp 3-47
[8] Rejeki S, Rohrs D, Nordquist G, and Fitriyanto A 2010*Conceptual Model Update of the Darajat Geothermal Filed, Indonesia*(Bali, Indonesia : World Geothermal Congress) pp. 1-2
[9] Sulistyardi H B2010*Basic Design of Lumut Balai 2x55 MW Geothermal Power Plant, Indonesia.* (Geothermal Training Program, The United Nations University) pp 615-624

Experimental Study of Isothermal Plate Uniformity for Blood Warmer Development using Geothermal Energy

J Hendrarsakti[1,2] and Y Ichsan[1]

[1]Faculty of Mechanical and Aerospace Engineering, ITB Jalan Ganeca No 10, Bandung, Indonesia
[2]Geothermal Master Program, ITB Jalan Ganeca No 10, Bandung, Indonesia

*E-mail: jooned@ftmd.itb.ac.id

Abstract. This research was conducted to assess the direct use of geothermal energy for blood warmer. The heating plate was made form aluminium plates with dimensions of 100 x 200 mm and then fed from the hot water heater. Tests were conducted in the laboratory where geothermal source water is replaced with the heat generated from the heater. The hot water from the heater in the temperature range 55°C - 60°C flowed into vertical chamber. Setting the temperature of the hot water heater is done by changing the flow of hot water coming out of the heater. Results showed that the value of a standard deviation of plate temperature was about 0.42 °C, so it can be said isothermal accordance with design requirement and objective. The test data used for the analysis of the manufacture of the heating plate in the blood warmer to regulate the discharge of hot water at intervals of 21.47 mL/s to 24.8 mL/s to obtain a temperature of 37.20 °C – 40.15 °C. Geothermal energy has the potential for blood warmer because blood warmer is part of the energy cascade in a temperature range of 40°C to 60°C

1. Introduction

Indonesia has a tremendous opportunity in the field of geothermal energy. Potential reserves of geothermal energy in Indonesia amounted to 40% of the total reserves of geothermal energy worldwide, equivalent to 28,000 MWe [1]. Geothermal potential is spread from Aceh to Papua. To take advantage of this opportunity, it is necessary to conduct in-depth study so that the development of geothermal utilization for power generation is also balanced with an assessment of the direct utilization of geothermal fluid, such as for agriculture, fisheries, tourism and even for chemical processes.

Based on data from the Indonesia Ministry of Energy and Mineral Resources in 2010, the utilization of geothermal energy for electricity in Indonesia has reached 1189 MWe which are spread throughout Indonesia [1]. Direct use of application geothermal energy is still limited to the hot water bath used at tourist spots, such as Ciwidey, Cipanas, and Garut. In addition, the utilization of geothermal resources directly used for drying tea industry and aquaculture, but use is still limited

This research was conducted as based Lindal diagram shown in Figure 1 that there is a potential source of geothermal heat at a temperature of 40 °C to 60 °C to heat the blood warmer. A blood warmer, itself is a device that is used to heat the blood during the blood transfusion process. In blood warmer, heater used consists of two vertical plates as one example is presented in Figure 2.

At the time of transfusion, blood preheated for blood stored in the refrigerator at a temperature of 1 °C - 6 °C in order to avoid lysis. Van Der Walt and Russell (2015) argued that ideally the blood is heated to body temperature, 37 °C, but at temperatures between 32 °C and 37 °C is still acceptable [3].

Heating the blood also may not exceed a certain temperature because it can result in either hemolysis damage or destruction of red blood cells due to disruption of membrane integrity of red blood cells and the release of hemoglobin [3].

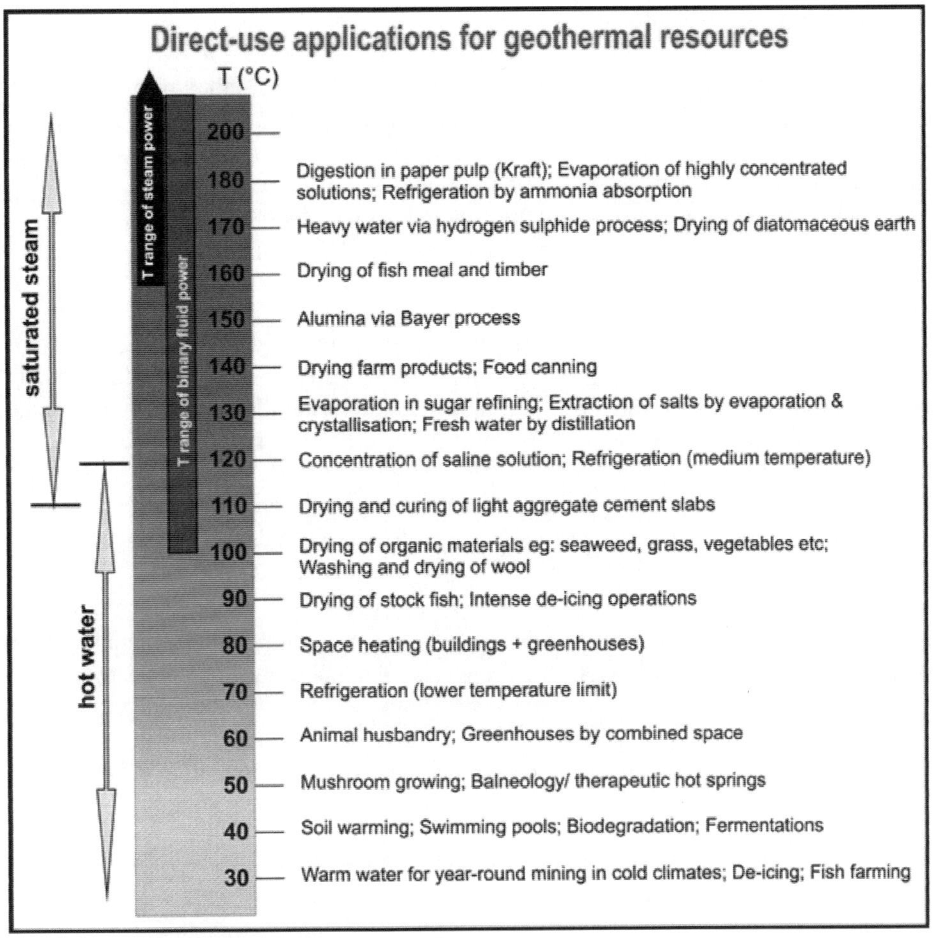

Figure 1. Lindal Diagram. [2]

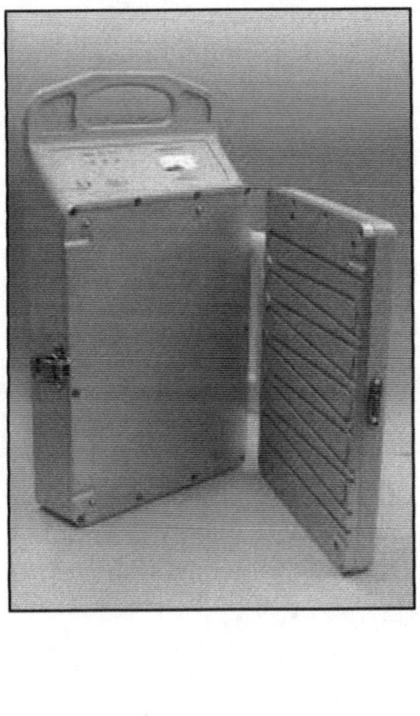

Figure 2. Example of Blood warmer: *Thermofluid TF251.*

The purpose of this research is to study and manufacture isothermal plate as the heating plate in the blood warmer. This research was carried out on a laboratory scale using a water heater in the form of heat generated from the heater. The current study was also to determine the geothermal potential as a source of heating in heating appliance blood

2. Experimental Apparatus and Procedure

2.1 Experimental Apparatus

Tests were conducted on the chamber by flowing hot water from the heater as shown in Figure 3. In order the results of this test can be used in applications, the design requirement for blood warmer that must be met in this test are:

1. Chamber positioned vertical to resemble blood warmer position.
2. Aluminum plates with no coated with insulator was used to heat the blood.
3. The temperature of the plate uniform with a standard deviation of less than 1 °C

2.1. Experimental Procedure

Tests were conducted to determine temperature distribution in the plate with a variety of a certain slope. Heater in Figure 4 is placed outside the room so that the exhaust gases can be neutralized by air. Furthermore, the hose is connected from the tap water supply to the heater. Then the system was put under the test plate to facilitate installation of the heater hoses from the heater to the plate. The procedures in the testing are as follows:

1. The water pump is turned further water-supply valve to the heater is opened. Valve governing effluent water heater is closed. Heater is heated with a gas-fueled torch to the water in the heater reaches a certain temperature (about 55-60 °C).
2. The next step is to organize the entry and exit discharge constant heater. To generate a constant flow of the heater valve that regulates water heater opened out with specific openings, simultaneously valve that regulates incoming water heater is also arranged so that the water level at heater remains constant.

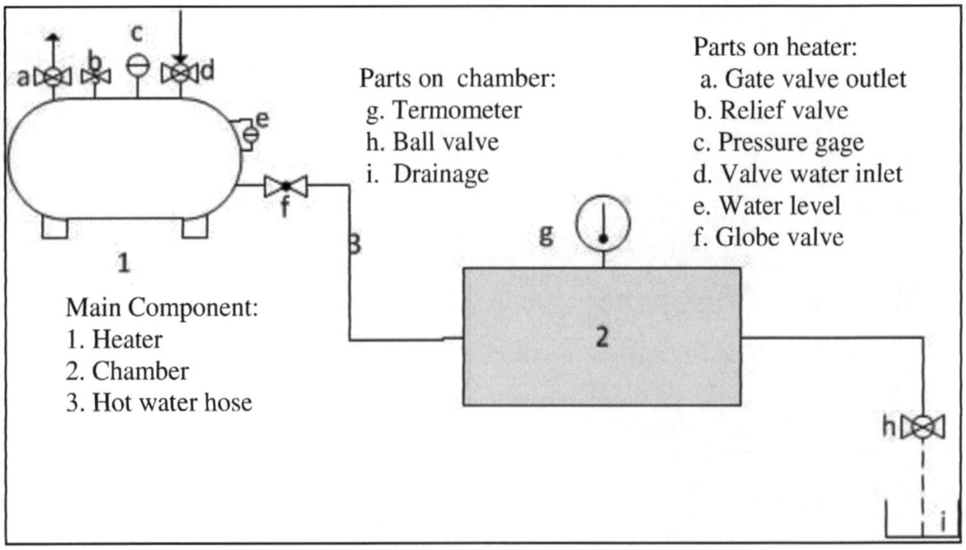

Figure 3. System apparatus used in this study.

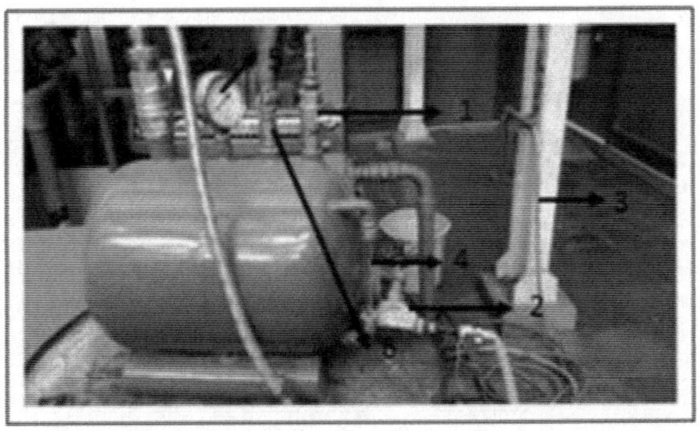

Figure 4. Heater used for the study.

3. If the water flow in the range of 12.00 to 13.33 mL/s (when measured in the range of 18-20 seconds) with a water temperature of 55-60 °C then the next step drain the hot water from the heater to the isothermal plate system.
4. The hot water flows using a hydraulic hose isolated 0.5 inch diameter. Hoses connected with niple on a test system plate. Valve that regulates the water out on a plate covered with a test system thus hot water will fill the test system until it reaches a certain volume then marked on the glass.
   To maintain the volume of hot water in the system test plate to remain constant as described in Figure 5, then the flow of water entering the system plate must be equal to the flow of water out of the system test plate. It is thus necessary setting the valve on the test plate system manually. Hot water coming out of the test plate system is accommodated in a container before being discarded. Temperature readings on the plate in Figure 6 that are not insulated using a digital thermometer at eight different points. Before testing each thermometer calibrated beforehand with the mercury thermometer by inserting temometer on water in order to obtain a correction factor for each digital thermometer. Digital thermometer readings done by making a video starts when the water flow ouof the system constant test plate with a duration of five to eight-minute video.

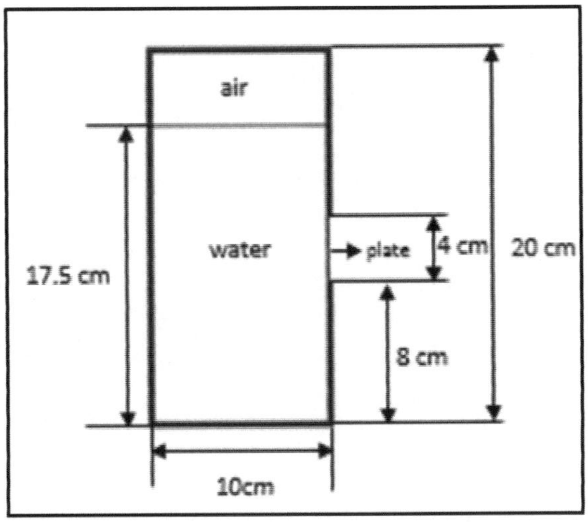

Figure 5. *Position chamber with testing plate*

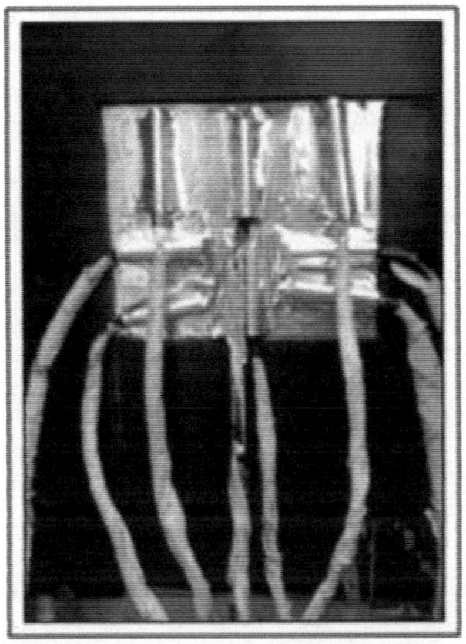

Figure 6. Testing plate with temperature sensors

IOP Conf. Series: Earth and Environmental Science **42** (2016) 012018 doi:10.1088/1755-1315/42/1/012018

3. Results and Discussion

3.1 Temperature Distribution on the Plate

Temperature measurement performed at eight different points with position sensors as shown in Figure 7 to determine the temperature distribution in the plate. This is necessary because blood plate warmer temperatures should be uniform (design requirement).

The sensor is placed on a square plate with dimensions of 4 cm x 4 cm as shown in Figure 7. Selection of a square-shaped plate as a square is a symmetrical shape so as to facilitate analysis of the influence of symmetry of the position sensor plate temperature uniformity. The distance between the point of the sensor is 1.5 cm so that the sensor number 1, 3, 6, and 7 are not exactly on the edge of the plate, but the position of the sensor is 0.5 cm from the edge of the plate. Table 1 shows the temperature distribution plate with a standard deviation of 0.42 °C, thus the temperature distribution in the plate can be said to be isothermal.

Table 1. *Temperature distribution (°C) on plate*

T1	T2	T3	T4	T5	T6	T7	T8	Tmax	Tmin	Tavg	St Deviation
50.10	50.40	50.10	50.30	49.10	49.70	49.70	50.10	50.40	49.10	49.93	0.42

3.2 Effect of Water Flow rate Against Plates Temperature

Figure 8 shows the relationship between the discharges of hot water that flowed from the heater to the average temperature of plate by Equation (1). This test result is useful for designing blood warmer. Hot water comes from geothermal arranged debits resulting in a certain temperature on the plate. Determination of the hot water discharge that can be streamed using Equation (1) flow test results from the heater. Equation 1 is useful for determining the flow of hot water to be flowed into the chamber to get a certain temperature on the plate.

$$T = 63.906e^{-0,025Q} \quad (1)$$

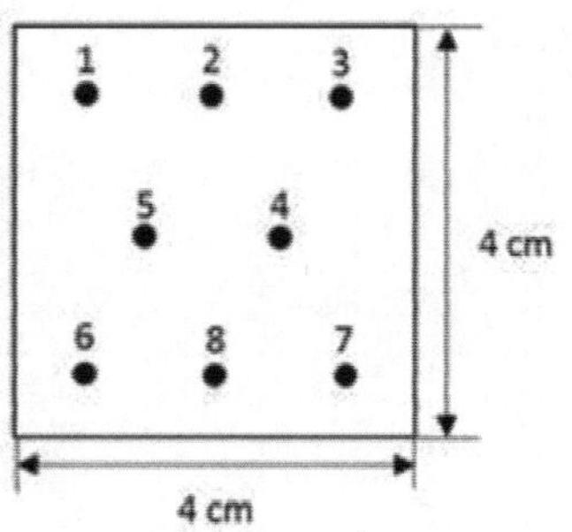

Figure 7. Sensor position on plate

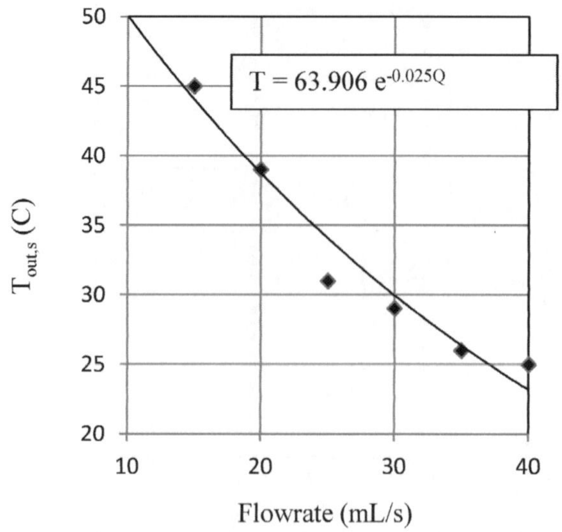

Figure 8. The influence of hot water flow rate to plate temperature.

3.3 Designing Blood Warmer

In designing the heating plate in the blood warmer, the data needed are type of plate material, plate dimensions, the effective area of the heating plate, the upper and lower temperature limits the heating plate, and the discharge of blood flow in the hose. The heating plate to heat the blood during transfusion be modeled referring to the heating plate designed of Biegler Protherm II in Figure 9 [5].

Due to data limitations, some assumptions used to design the plate. The assumptions used are:

1. Material aluminum plate (k = 180 W/mK)
2. Heating plate dimensions 200 x 100 x 2 mm
3. The area of effective heating plate 80 x 160 mm
4. Based on data obtained from Biegler Protherm II plate temperature can be set at intervals of 37 ± 0.5 °C - 41 ± 0.5 °C
5. Steady state fluid conditions
6. Constant material properties

In Figure 10, dimensions shown in the schematic drawing of the plate, the effective area of the heating plate, the area of insulation, and water level inside the chamber. In Figure 10 the heating plate has dimensions of 10 cm x 20 cm with an effective heating area 8 cm x 16 cm and height of the water in the chamber is maintained at a height of 18 cm. A tube of blood will be positioned wrapped around the front of the plate surface area of effective heating.

3.4 Calculation of the hot water discharge temperature limits of the upper and lower plates

In designing the heating plate of blood warmer, the temperature of the upper and lower plates should be known. The purpose of calculating the temperature of the upper limit and lower limit on the plate is to determine the amount of discharge of hot water to be flowed into the chamber. The following step will be given calculation of discharge as the steps shown in Figure 11 so that the plate reaches a temperature of 37 ± 0.5 °C - 41 ± 0.5 °C.

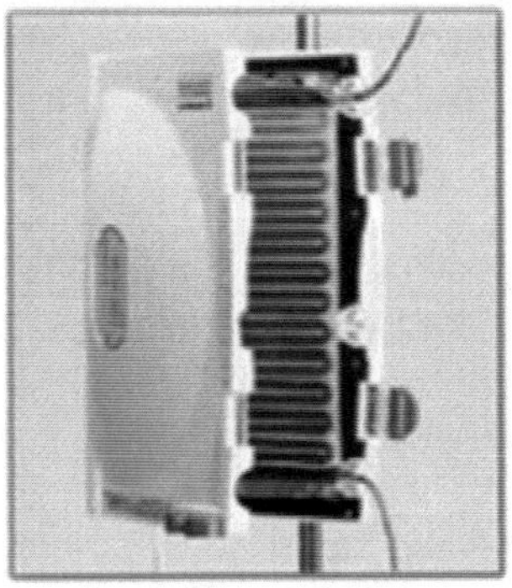

Figure 9. Biegler Protherm II [5].

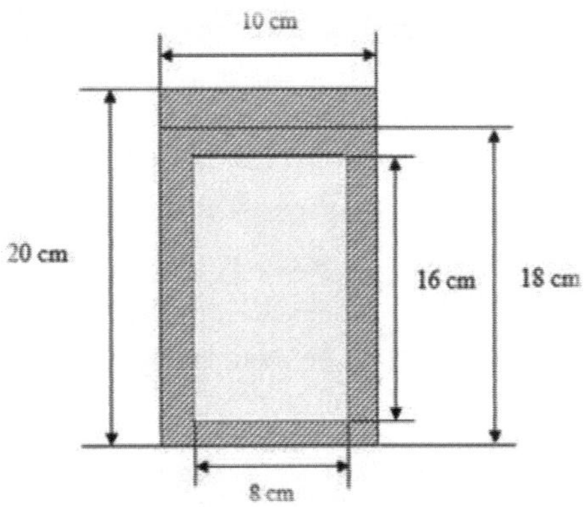

Figure 10. Schematic of plate designed.

Determination of the temperature upper limit and lower limit based on the blood warmer on the market, in this case refers to design of Biegler Protherm II. Calculation of the hot water discharge flow in chamber involves the calculation of the internal flow convection in the chamber. Thus in calculations involving dimensionless number Nusselt internal flow [6]. While the water temperature data inlet and outlet obtained from measurements of the temperature of the hot water heater.

A detailed explanation Figure 11 is described in the following steps:

1. The first step is to determine the hot water discharge (Q) flowing on the hot plate with the equation in Figure 8, $T = 63.906e^{(-0.025Q)}$

2. Determine the temperature of hot water coming out of the heater using equation $T = 69.39e^{(-0.016Q)}$. This equation shows the relationship discharge of hot water from the hot water heater with temperature
3. Calculation of convection on the internal flow in the chamber
4. Calculation of the outside temperature plateThe calculation result discharge hot water must flow into the chamber to reach temperatures lower limit (37 ± 0.5 °C) plates are shown in Table 2.

Outside temperature plate $T_{s,out}$ = 37.2 ° C which is in the interval below the minimum limit of the desired temperature of the heating plate.Thus the discharge of hot water from the heater should be drained into the chamber was 24.80 mL/s.

The next step is to determine the hot water discharge in order to reach the upper limit temperature of 41 ± 0.5 ° C. Calculations are performed the same as in the previous phase to obtain the results as shown in Table 3. The first line in Table 3 calculation $T_{s,out}$ with the use of debit from the equation in Figure 6. After the calculation, $T_{s,out}$ = 42.26 °C then there is a difference by 2.26 °C or 5.35% of the desired temperature is 40 °C.

In the second line of iterations performed by determining the discharge current plate temperature = 40 to 5.35% x 40 = 37.86 °C. By using step 2 was obtained debit of 20.91 mL/s. After calculation until at step 4 $T_{s,out}$ obtained results, $T_{s,out}$ = 40.56 °C. There are differences in value by 1.38% from the desired temperature at 40 °C. This value is outside the upper limit of the temperature interval desired plate so do one more iteration. The second iteration can be seen in the third row of Table 3. Determination of discharge is by Equation 1 when the temperature plate = 37.86 to 1.38% x 37.86 = 37.33 °C. By using step 2 was obtained debit of 21.47 mL/s. After calculation until at step 4, $T_{s,out}$ obtained results, $T_{s,out}$ = 40.15 °C. Thus it can be concluded that the discharge of hot water to be flowed into the chamber to get a plate temperature upper limit was 21.47 ml/s.

Discharge calculations that have been done on the heater can also be done on a geothermal heat source at temperatures 50 °C - 60 °C. By using the hot water temperature data on the geothermal heat source, then the discharge must flow into blood warmer can be known.

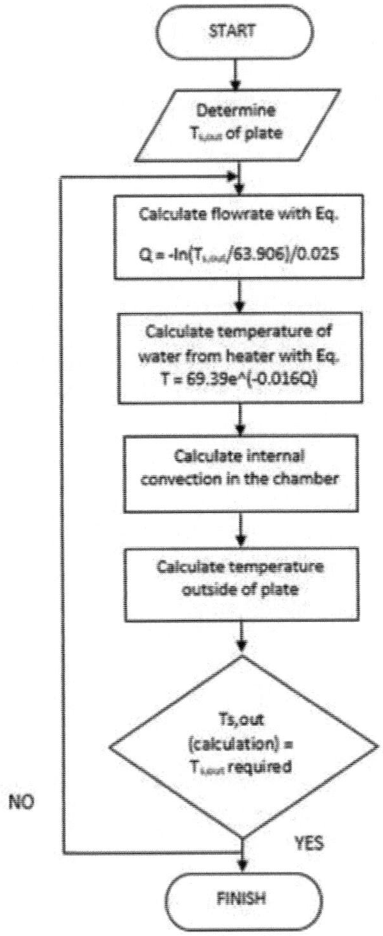

Figure 11. Flowchart diagram calculation of outer plate temperature

Table 2. Calculation of the lower limit temperature of plates

Debit (mL/s)	$\dot{m}$ (kg/s)	$T_{m,i}$ (°C)	$T_{m,o}$ (°C)	$\overline{T_m}$ (°C)	$T_{s,in}$ (°C)	Re_D	$\overline{Nu}$	$\bar{h}$ (W/m²K)	$T_{s,out}$ (°C)
24.80	0.025	46.65	41.90	44.27	37.30	513.64	14.61	92.60	37.20

Table 3. Calculation of the upper limit temperature of plates

Debit mL/s	$\dot{m}$ (kg/s)	$T_{m,i}$ (°C)	$T_{m,o}$ (°C)	$\overline{T_m}$ (°C)	$T_{s,in}$ (°C)	Re_D	$\overline{Nu}$	$\bar{h}$ (W/m²K)	$T_{s,out}$ (°C)
18.71	0.018	51.43	46.70	49.00	42.30	421.58	12.49	79.87	42.26
20.91	0.020	49.66	44.90	47.28	40.60	456.79	13.29	84.76	40.56
21.47	0.021	49.21	44.50	46.85	40.20	465.56	13.49	85.94	40.15

IOP Conf. Series: Earth and Environmental Science **42** (2016) 012018 doi:10.1088/1755-1315/42/1/012018

3.5 Calculation of blood flow

Another calculation needs to be done to design the blood warmer is debit of blood in the hose at plate. It is necessary to know the time of the blood reaches body temperature.

The blood before transfusion prior stored at a temperature of 1 °C to 6 °C, therefore it must be heated in order to reach body temperature. Setting the blood flow is important to know the blood of warm-up time at the plate until it reaches body temperature when the blood out of the blood warmer.

In this calculation using the following assumptions:

1. Steady state
2. Constant properties
3. The outside temperature of hose equal to the temperature of the plate
4. The outer surface of the plate is well isolated

A Blood hose with a length of 60 cm wound on the heating plate would produce 10 pieces windings. The heating plate is set at a temperature of 41 ° C. Calculation of blood flow and related temperatures can be conducted using Equations (2)-(6).

$$\bar{h}A_s\Delta T_{L,M} = \dot{m}c_p\Delta T \quad (2)$$

$$\Delta T_{lm} = \frac{\Delta T_i - \Delta T_o}{\ln(\frac{\Delta T_i}{\Delta T_o})} \quad (3)$$

$$A_s = \pi D_o L \quad (4)$$

$$\Delta T = T_{m,o} - T_{m,i} \quad (5)$$

Blood convection coefficient calculated by Equation (6)

$$\overline{Nu_D} = 3.66 \text{ (Ts constan)} \quad (6)$$

The time it takes the blood to reach temperatures of 37 ° C can be calculated by Equation (7)

$$t = \frac{\text{Volume of blood}}{\text{Debit}} = \frac{\frac{\pi}{4}Di^2L}{Qb} \quad (7)$$

Data calculations of blood flows are as follows:

1. Surface temperature of hose is constant, $T_s = 41$ °C
2. The temperature of blood into the chamber, $T_{m,i} = 6$ °C
3. Temperature blood out chamber, $T_{m,o} = 37$ °C
4. The diameter of the blood hose, $D_i = 6$ mm
5. The outer diameter of the blood hose, $D_o = 9$ mm
6. The length of the blood hose, L = 60 cm
7. Blood properties k = 0,52 W/m K ; cp = 3617 J/kg K ; $\rho = 1060$ kg/m^3 [7]

With Equation (2), blood convection coefficient can be found to be $\bar{h} = 317.2$ W/m^{2}K. Calculation of mass rate of blood through the hose obtained using Equation (2) $\dot{m} = 4.57 \times 10^{-4}$ kg/s. Thus obtained blood flow through the hose at $Q_b = \dot{m}/\rho = 4.31 \times 10^{-7}$ m^3/s = 0.43 mL/s = 25.8 mL/min. The time for the blood takes to reach temperatures of 37 °C by Equation (7) is about 39 seconds. Setting the blood flow in order to reach the debit at 25.8 mL/s was conducted by turning the hose clamp the blood. Blood flow calculation results are shown in Table 4.

Step calculations of hot water discharge and the discharge of blood can be applied to the use of geothermal directly on a small stream and a temperature between 50-60 °C. However, considering that the geothermal heat source generates heat that is unstable will require further studies regarding the stability of the temperature so as to produce a uniform temperature on the plate.

Table 4. Blood flow calculation

$\dot{m}$ (kg/s)	Debit (mL/min)	Time (seconds)
4.57 x 10^{-4}	25.80	39

4. Conclusions

Based on the studies that have been done, it can be concluded as follows:

1. Blood Warmer can be application use of geothermal direct use.
2. Configure the plate at an angle of 0 ° can be used as a heating plate in the blood which can produce a warmer temperature of 37.20 °C - 40.15 °C by regulating the discharge of hot water from the heater of 21.47 mL/s - 24.80 mL/s.
3. To reach 37 °C, the blood flow in the discharge of 28.30 mL/min and heating takes 39 seconds.

5. Nomenclature

$\bar{h}$ = average coefficient of blood convection (W/m^2K)
A_s = Surface area of blood tube (m^2)
$\Delta T_{L,M}$ = Logaritmic mean temperature (K)
$\dot{m}$ = Mass rate of blood (kg/s)
c_p = Heat capacitance of blood (J/kgK)
T_s = Temperature surface (°C)
$T_{m,i}$ = Temperature inlet (°C)
$T_{m,o}$ = Temperature outlet (°C)
$\overline{T_m}$ = Temperature average between inlet and outlet (°C)
ΔT =Difference temperature between temperature outlet and inlet (K)
ΔT_i =Difference temperature between temperature surface and inlet (K)
ΔT_o =Difference temperature between temperature surface and outlet (K)
$T_{s,out}$ = Outside temperature of plate (°C)
T = Plate temperature
$\overline{Nu_D}$ = Nusselt number for blood in the hose
Re_D = Reynolds number for water in the chamber
Q = a discharge of hot water that flowed into the chamber.
Qb = Flowrate of blood (mL/s)
L = Length of blood tube (m)
D_i = Inner diameter of blood tube (m)
D_o = Outer diameter of blood tube (m)
t = Total time the blood flows on the plate (s)

References

[1] http://www.esdm.go.id/berita/55-siaran-pers/3021-pengembangan-energi-panas-bumi-di-indonesia.html accessed on 26/01/2016
[2] http://www.nova.org.au/technology-future/feeling-heat-geothermal-energy accessed on 2/02/2016
[3] T Poder, W Nonkani, and E Leponkouo, (2015), "Blood Warming and Hemolysis: A Systematic Review With Meta-Analysis", Transfusion medicine Reviews, 29, p. 172 – 180
[4] F Anugrah, (2015), "Design, Manufacture and Testing Heater Capacity 2 Bar", Undergraduate Thesis, ITB
[5] http://www.cedar.wales.nhs.uk/sitesplus/documents/1091/CEP10014%20IV%20fluid%20 warmers%20MR.pdf accesed on 27/11/2015
[6] F Incropera & D Dewitt, (2011), "Introduction To Heat Transfer 6th Edition", New York.

[7] http://www.itis.ethz.ch/virtual-population/tissue-properties/database/heat-capacity/ accessed on 26/02/2016
[8] J Cutnell and K Johnson, (1998), "Physics, Fourth Edition",Wiley,p: 308

Electricity generation from hydrothermal vents

Y Aryadi[1], I S Rizal[2], M N Fadhli[3]

[1,2,3]Department of Petroleum Engineering, [3] Department of Mechanical Engineering, Institut Teknologi Bandung, Bandung, Indonesia

E-mail: *ujaraziz@gmail.com

Abstract. Hydrothermal vent is a kind of manifestation of geothermal energy on seabed. It produces high temperature fluid through a hole which has a diameter in various range between several inches to tens of meters. Hydrothermal vent is mostly found over ocean ridges. There are some 67000 km of ocean ridges, 13000 of them have been already studied discovering more than 280 sites with geothermal vents. Some of them have a thermal power of up to 60 MWt. These big potential resources of energy, which are located over subsea, have a constraint related to environmental impact to the biotas live around when it becomes an object of exploitation. Organic Rankine Cycle (ORC) is a method of exploiting heat energy to become electricity using organic fluid. This paper presents a model of exploitation technology of hydrothermal vent using ORC method. With conservative calculation, it can give result of 15 MWe by exploiting a middle range diameter of hydrothermal vent in deep of 2000 meters below sea level. The technology provided here really has small impact to the environment. With an output energy as huge as mentioned before, the price of constructing this technology is low considering the empty of cost for drilling as what it should be in conventional exploitation. This paper also presents the comparison in several equipment which is more suitable to be installed over subsea.

1. Introduction

Energy plays an important role in the history of human development. Firewood, coal, oil, natural gas and so on are considered as the energy resources for warmth, coking, automobile and other human activities until now. However, environmental problems associated with the use of the energy also began to appear, and tended to be more serious. One of the ways to solve the contradiction between environmental problems and energy demand is searching for new and clean energy resource.

Geothermal is a kind of sustainable energy and even when it is developed well, it can become a renewable energy. Geothermal come from subsurface which sometimes the manifestation of the energy can emerge to the earth surface. On land, the type of manifestation can be in kind of fumarole, geyser, hot spring, hot lake, steaming ground, or many others. These types of manifestation can also be found on seabed and one type of them is hydrothermal vent.

Thermal energy from geothermal manifestation can be exploited using various technologies depend on the characteristics of the manifestation itself. Organic Rankine Cycle (ORC) is a method for exploiting heat energy by using organic fluid instead of steam directly from evaporated water. In this paper, this ORC method, then, is used in model of electricity generation form thermal energy supplied by hot water flowed through hydrothermal vent.

IOP Conf. Series: Earth and Environmental Science **42** (2016) 012019 doi:10.1088/1755-1315/42/1/012019

2. Hydrothermal Vents

Hydrothermal vent is the result of seawater percolating down through fissures in the ocean crust in the vicinity of spreading centers or subduction zone (places on Earth where two tectonic plates move away or towards one another). The cold seawater is heated by hot magma and reemerges to form the vents. Seawater in hydrothermal vents may reach temperatures of over 340 ^{0}C. Hot seawater in hydrothermal vents often does not boil because of the extreme pressure at the depths where the vents are formed.

There are some 67 000 km of ocean ridges, 13 000 of them have been already studied discovering more than 280 sites with geothermal vents. Some of them with a thermal power of up to 60 MWt [2] but there are some of them really gigantic as the one in Rainbow with an output of 5 GWt [2].

Scientist first discovered hydrothermal vents in 1977 while exploring an oceanic spreading ridge near the Galapagos Islands. To their amazement, the scientists also found that the hydrothermal vents were surrounded by large numbers of organisms that had never been seen before. These biological communities depend upon chemical processes that result from the interaction of seawater and hot magma associated with underwater volcanoes [1].

3. Organic Rankine Cycle

Organic Rankine Cycle is conceptually similar to Steam Rankine Cycle in that it is based on the vaporization of a high pressure liquid which is in turn expanded to a lower pressure thus releasing mechanical work. The cycle is closed by condensing the low pressure vapour and pumping it back to the high pressure. Therefore, the ORC cycle involves the same components as a conventional steam power plant (a boiler, a work-producing expansion device, a condenser and a pump). However, the working fluid is an organic compound characterized by a lower ebullition temperature than water and allowing power generation from low heat source temperatures.

Low-temperature geothermal ORC plants are also characterized by relatively high auxiliary consumption: the pumps consume from 30% up to more than 50% of the gross output power. In conventional development of ORC, the main consumer of power is the brine pump that has to circulate the brine over large distance and with a significantly high flow rate. The working fluid pump consumption is also higher than in higher temperature cycles, because the ratio between pump consumption and turbine output power ("back work ratio") increases with decreasing evaporation temperature.

4. Heat Exchanger

When installed over submarine, we need to do some modifications to the system of ORC. For the boiler, it is modelled here to use the external part of one extreme of the submarine that is the mouth of the vent itself. A heat exchanger in the form of a conic coil, then, is installed exactly at the top of the vent in a way that the hot water flows naturally through the coil, touching as much surface of the tubes as possible. This is the evaporator of the cycle. In the other extreme of the submarine, another heat exchanger is installed in the cold water of the surrounding sea, also in the shape of a conic coil.

Key characteristics regarding heat transfers are the efficiency (or pinch point) and pressure drops. Each heat exchanger in the cycle is sized according to these two parameters. Different type of heat exchangers can be used. In conventional use, the most common used heat exchangers are shell and tube (mainly in larger-scale systems) and plate heat exchangers (mainly in small-scale systems, due to their compactness).

Related to submarine geothermal exploitation, the design of heat exchanger should be modified to match with the site where it will be installed. The efficiency of the heat exchanger (boiler and condenser) is directly proportional to the touching area of geothermal fluid from the earth and the heat exchanger itself. Relation between the amounts of heat transferred in conduction is shown in the following equation:

$$\frac{Q}{t} = \frac{\mathrm{KA}(T_{hot} - T_{cold})}{d} \quad (1)$$

where (K) is conductivity of the barrier, (A) area, (T) temperature, and (d) the thickness of barrier.

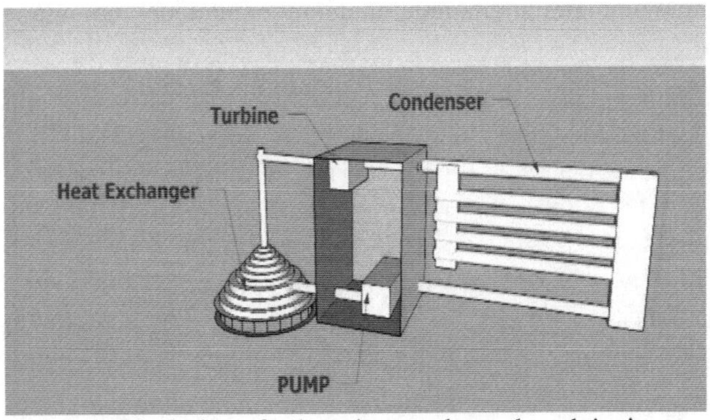

Figure 1. Schematic design of submarine geothermal exploitation machinery

5. Pump

The low pressure organic fluid come out from the condenser, then need to be pumped to become high pressure liquid before it goes through the boiler. A pump as the tools for pumping the fluid has big consumption of energy when used in ORC system compared to Stem Rankine Cycle system. The ratio between pump electrical consumption and expander output power is called Back Work Ratio (BWR):

$$BWR = \frac{\dot{W}_{pp}}{\dot{W}_{exp}} \qquad (2)$$

The pump efficiency is a crucial parameter in low temperature cycles and in trans critical cycles. Figure 2 give us a description about relation between fluid type and BWR value. From that figure, it can be concluded:

1. The higher the critical temperature of the working fluid, the lower the BWR.
2. BWR increases with T_{ev}, and gets significantly high when operating the cycle close to the critical point.

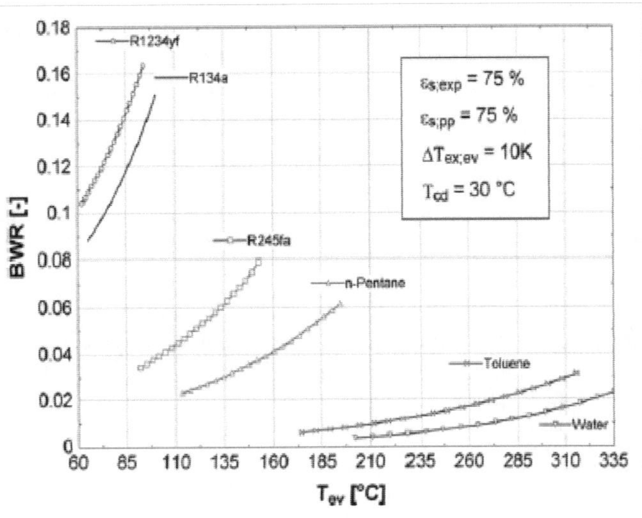

Figure 2.BWR as a function of evaporation temperature for different fluids [3].

6. Turbine

The performance of an ORC system strongly correlates with that of the expander. The choice of the technology depends on the operating conditions and on the size of the system. Three main types of machines can be distinguished: the turbo and positive displacement types [3].

The main feature of a turbine designed for the expansion of an organic Rankine fluid, typically rated less than 15 MW, is the overhung configuration, meaning an expander wheel, composed of maximum 3 stages supported by bearings on a single side. The main reason for this technological choice rather than a double-support (horizontally split or barrel casing) machine is the necessity to limit costs and ease maintenance, favouring at the same time the operation of the plat: the double support design implies two end shaft seal, doubling seals maintenance and eventual troubles. This technology currently represents the most common turbine technology for this application [4].

A second technology, with a minor share of the market, is the radial inflow expander: typically a high-speed integrally geared machine derived from gas expansion application and applied on some medium power geothermal binary plants.

A new type of turbine, developed by Exergy, defined the radial outflow turbine, combines the simplicity of the axial turbine with the efficiency of a radial inflow turbine (figure 3.).

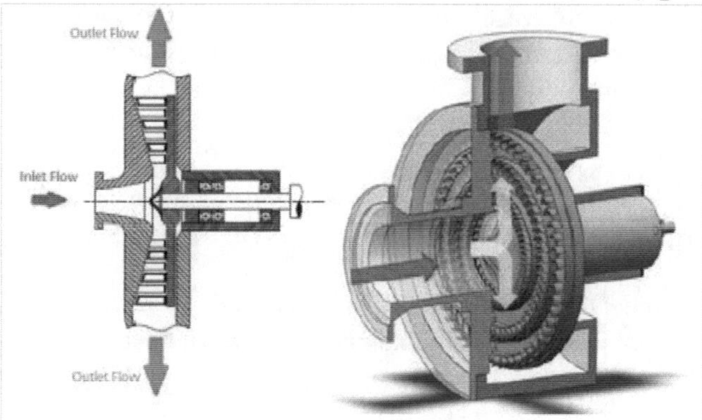

Figure 3. Cross sections of a typical radial outflow turbine [4].

This radial outflow turbine is composed of a series of radial stages, alternating one stator and one rotor blade ring per stage, fitted on one disk side all contained in a casing together with the discharge volute. The mechanical coupling between turbine disk and generator is completed by a system of seals and bearings supporting the overhung shaft [4].

Related to the utilization of this turbine in submarine geothermal manifestation, there are several advantages which make this type of turbine is more suitable, those are as follows:

- it has low vibration profile,
- it is the most efficient stages develop more enthalpy head, at the same pressure ratio,
- pressure difference favours expansion without major 3D effects,
- It has relatively low rotating speed, meaning direct coupling with the generator,
- It has limited overhung mechanical structure.

7. Energy Output Calculation

In calculation of energy output from hydrothermal vent, we use the assumption that the fluid discharged from the vent is at its saturation temperature. This is inferred to the maximum temperature of the fluid. Figure 4 show the maximum temperature that the water can attend in a hydrothermal vent and the maximum efficiency that can be achieved. In thermal cycle, the maximum efficiency which can be achieved is the Carnot Efficiency, that is$(1-\frac{T_1}{T_2})$. With T is the temperature of the cold and hot water respectively, in Kelvin. The Carnot efficiency presented in figure 4 is using the saturation temperature of the fluid from hydrothermal vent as (T_2) and the low temperature of sea water surrounding the vent as (T_1).

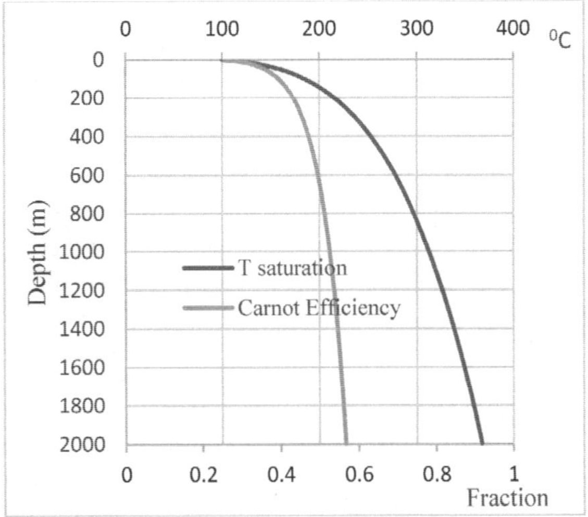

Figure 4. Saturation temperature and maximum efficiency (Carnot) as a function of depth.

Using the value given in figure 4, we can calculate an approximate electrical output of a given vent flowing at the highest temperature that it can have at certain depth. Assuming machine total efficiency is equal to 80%, thermal recovery efficiency 10%, and fluid velocity 1 m/s, the sensitivity analysis of energy output from several vent diameters and depths is described in figure 5.

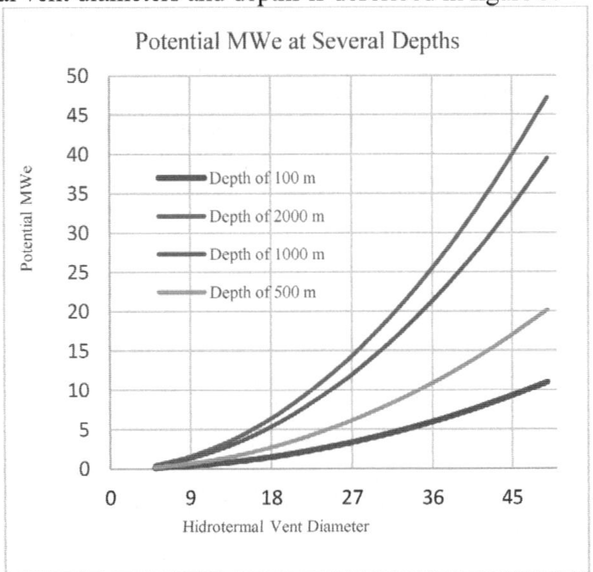

Figure 5. Potential MWe from hydrothermal vent at several depths at any given vent diameter.

The basic calculation of the diagram above is that by calculating the discharged mass of fluid which contains thermal energy. Multiplying the area of the vent with assumed fluid velocity will result the volumetric flowrate of the fluid. Then, when it multiplied again with the density and the enthalpy calculated from steam table, the resulting value is the total thermal energy output of the vent. Finally, the total electricity potential of the vent is achieved by multiplying the thermal energy with machinery efficiency and thermal recovery efficiency.

IOP Conf. Series: Earth and Environmental Science **42** (2016) 012019 doi:10.1088/1755-1315/42/1/012019

Hydrothermal vent is mostly found in depth of 2000 m. With a diameter ranging from several inches to tens of meters [1], choosing a middle value that is 27 inches will result a potential electricity from hydrothermal vent over 15 MWe.

This value of energy is big enough compared to the energy achieved from conventional development. In conventional development, the utilization of geothermal energy need of drilling several wells which is costly and not natural friendly.

8. Conclusion

Many surveys carried out around the world has given a description about the existence of hydrothermal vent over World Ocean. It is found that there are over 280 sites of hydrothermal vent exist from 13000 km studied of 67000 km oceanic ridges. With a big thermal produced from those vents, they become an object to be developed as future energy resources.

Detailed calculation of energy output using conservative value of overall efficiency give 15 MWe for one exploited hydrothermal vent.

With several constraints related to the environment, the technology is needed to be developed well for more appropriate environmentally. ORC system is best method to be used in this case. With well operated system, it will not discharge any material to the environment. So, the biota surrounding the exploited vent will not get any disturbance.

More research and development are needed to be done further. The challenge related to the utilization of energy from deep sea should be faced. But, considering the energy output that can be generated, this becomes more interesting compared with conventional development of geothermal energy onshore which need to drill several wells.

References

[1] Bosche, N. E. S., Arriaga, M. C. S., Samaniego, F. V., Delgado, V. (2005). "*Fundamental Characteristics of Hydrothermal Submarine Systems and Preliminary Evaluation of its Geothermal Potential in Mexico*". Proceeding World Geothermal Congress.

[2] Hiriart, G., Prol-Ladesma, R. M., Alcocer, S., Espíndola, S. (2010). "*Submarine Geothermics; Hydrothermal Vents and Electricity Generation*". Proceedings World Geothermal Congress.

[3] Quoilin, S., Broek, M. V. D., Declaye, S., Dewallef, P., Lemort, V. (2013). "*Renewable and Sustainable Energy Reviews*". Elsevier.

[4] Spadicini, C., Centemeri, L., Danieli, M., Rizzi, D., Xodo, L. (2015)."*Geothermal Energy Exploitation with the Organic Radial Outflow Turbine*". Proceedings World Geothermal Congress.

Design of Tomato Drying System by Utilizing Brine Geothermal

W Afuar[1], B Sibarani[1], G Abdurrahman[1], J Hendrarsakti[1,2]

[1]Geothermal Master Program ITB, Bandung, Indonesia
[2]Faculty of Mechanical and Aerospace Engineering ITB, Bandung, Indoensia

Email : waldyafuar@gmail.com

Abstract. Cultivation of tomato plants in Indonesia has been started since 1961.Tomatoes generally will rot in three days if left on storage. Moreover, low quality tomatoes have cheaper price. After harvested, tomatoes need to be treated by drying process so it can last longer. Energy for drying tomatoes can be obtained by utilizing heat from geothermal brine. Purpose of this research is to design a tomato drying system by extracting heat of geothermal brine from separator with certain flow rate to heat up water by using a heat exchanger. Furthermore, this water will be used to heat up the surrounding air which is circulated by blower system to heat up the tomatoes chamber. Tomatoes drying process needs temperature range of 50-70°C to evaporate water content from 95.7% to 26%. After that treatment, the tomatoes are expected to have better durability. The objective of this study is to determine the quantity of hot brine which is needed for drying tomatoes and to design a drying system so that tomatoes can last longer.

1. Introduction

Most geothermal fields in Indonesia are high temperature - liquid dominated systems with high flow rates. By using single flash method on those conditions, brine from separator will be injected directly back to reservoir. However, brine as residual product with temperature 175-190^{0}C (based on cumulated data from geothermal fields in West Java) can be utilized for direct use application such as for tomato drying process. One of these by using the potential of brine is for tomato drying. This paper will discuss drying tomatoes by using hot brine from the separator.

2. Methods

This study was conducted by collecting information from research of Andritsos and Ehiem [2009], who explained information about geothermal direct use, characteristic of tomato and fruit dryer. Based on those data, some assumptions were collected to calculate mass flow rate of brine needed for drying tomatoes.

3. Drying Tomatoes

In Indonesia, tomato (*Lycopersiumesculentum*) is a kind of fruit that can be harvested a whole year long. Cultivating tomatoes has been prioritized since 1961 to fulfill daily needs[4]. Tomatoes grow in lowland and highland areas, but they cannot survive for a long period of time after harvested. After three days in storage, tomatoes respiration and transpiration will cause increase of CO_2 and H_2O content [4]. It will trigger the growth of bacteria that will cause tomato rotting.

The quality of tomatoes really depends on its freshness. Low water content of tomatoes discourage the growth of bacteria. In order to make the tomatoes last longer after harvesting, preservation method is needed to drying tomatoes while retaining flavor. Dried tomatoes are usually consumed in Nigeria, USA, Greece as processed food and sweetmeat which packed with nutrients that deliver health benefits. In Indonesia, tomatoes are usually not dried after harvesting. It causes tomatoes to become rotten and thrown away before them can be sold.

The drying process requires heat to decrease the moisture content from a tomato. Normally, drying process is conducted by exposing tomato to the sun. On the other hand, artificial drying is done by using engineered equipments. This study will be conducted by using artificial drying system with heat taken from geothermal brine. The heat will use to increase the working fluid's temperature. In this process, fresh air will be heated by working fluid (water)and then circulated into the drying room. At this system, temperature, cleanliness, and the drying process can be controlled to get the optimum, continuous, and faster dryness.

Once harvested, the tomatoes are sorted, washed, and cut into several pieces. This treatment can accelerate the decline of tomato's moisture content by evaporation. During the drying process, texture, color, and flavor of the tomatoes will change. However, their nutritional value still can be preserved by selecting the optimum drying conditions[2].

The desirable moisture content of tomatoes is 26% [5].This condition is very stable so that does not easily change the taste and resistant to microorganisms, especially fungi.

4. Tomatoes Drying System By Using Brine

A drying process using hot air requires heat for increasing the air temperature in order to reduce the relative humidity of the air. These sources included solar energy, fuel, or geothermal. In this study, the heat comes from residual fluid of geothermal energy. The heat will heat up fresh water, then the heat from circulated fresh water will heat up the air inside the drying room. Hot air in the drying room will absorb the moisture of the tomatoes. The room temperature oftomatoes dryeris based on the standard condition which is determined by a number of cumulated data from geothermal fields in West Java.

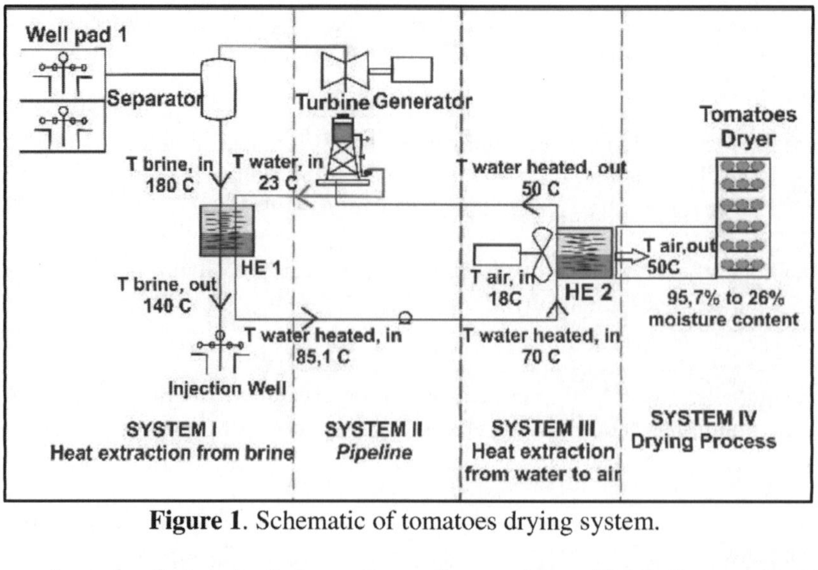

Figure 1. Schematic of tomatoes drying system.

Figure 1 is a schematic of a tomato drying system using geothermal brine. It consists of four systems, heat extraction from brine (system I), fresh water circulation inside pipeline of approximately 2 km long (system II), heat extraction from water to air (system III), and drying process (system IV). Fresh water is heated by geothermal brine through first heat exchanger (HE1). Then, hot water is transferred to the location of tomatoes dryer. The heat extracted from the heated water in pipeline is used to warm the air inside the drying room through a blower in the second heat exchanger (HE2). Next, hot air will absorb moisture from the tomatoes in drying room. After that, the hot water will continue to circulate from HE2 towards HE1 through a pipeline. Those process are shown in Figure 2.

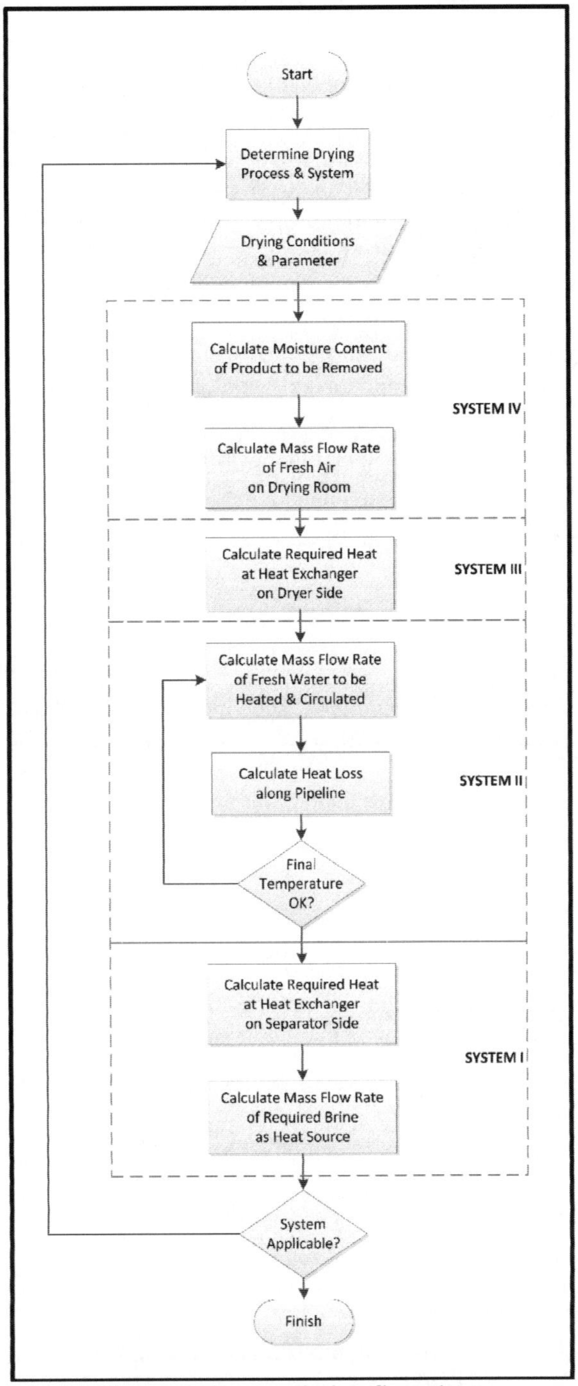

Figure 2. System design flow chart.

To calculate the mass flow rate of brine which is needed to dry the tomatoes, some parameters will be assumed and used as a constant in those four systems:

- the capacity of the dryer is 1 ton per day of tomatoes. This amount is obtained from total of tomato production in West Java for one year. Tomatoes produced in West Java are about 304.687 tons or about 835 ton/day [3]. And then, that amount is divided by the total area of tomato garden, which is about 15 hectares. Actually, the number of harvested tomatoes is more than 1 ton/day.
- the distance and elevation difference between heat source (brine location) and drying location are about 2 kilometers and 110 meters. These values are estimated by observing the topography in West Java.
- fresh water is circulated by using 3 inch carbon steel pipe.
- the drying room temperature is 50^{0}C.
- the ambient temperature around drying room is 18^{0}C.
- the favorable relative humidity of product in system IV is 26% [5].
- the efficiency of heat exchanger in system III is 84% [5].

The time needed to reduce the moisture content to 26% by flowing hot air into the drying chamber is 14 hours. The moisture content to be removed is calculated using Equation (1):

$$M_r = M * (\frac{RH_1 - RH_2}{1 - RH_2}) \tag{1}$$

Where, M is drying capacity (kg)
M_r is moisture to be removed (kg)
RH_1 is initial relative humidity (%)
RH_2 is final relative humidity (%)

Table 1. Moisture content to be removed [5].

Parameter	Symbol	Value	Unit
Drying capacity	M	1000	kg
Drying time	t	14	hours
Initial humidity	RH_1	95.7%	
Final humidity	RH_2	26%	
Moisture of product to be removed	M_R	941.9	kg

Tomatoes drying system occurs through two processes of energy usage, namely heating and drying by the following Equations(2) - (5).

$$Q_{dryer} = Q_{heatup} + Q_{evaporate} \tag{2}$$

$$Q_{heatup} = m_p c_{pp} (T_{fin} - T_{in}) \tag{3}$$

$$Q_{evaporate} = M_r * h_{fg} \tag{4}$$

$$h_{fg} = h_v - h_l \tag{5}$$

Where, Q_{dryer} is total heat required for dryer process (kW)
$Q_{heat\text{-}up}$ is heat required for heat up tomato (kW)
$Q_{evaporate}$ is heat required for decreasing moisture content (kW)
h_{fg} is heat laten of moisture (kJ/kg)
C_p is heat capacity of product (kJ/kg°C)
m_p is mass rate of product (kg/s)
$T_{f,\,in}$ is final temperature (°C)

$T_{f,in}$ is initial temperature (°C)

Based on the Equations (2) – (5), Table 2 shows calculation of demand of total heat:

Table 2. Total heat required for drying process.

Parameter	Symbol	Value	Unit
Tomato heat capacity	C_{pp}	4.6	kJ/kg.°C
Initial temperature of product	$T_{p,in}$	18	°C
Required temperature of product	$T_{p,fin}$	50	°C
Heat latent of moisture	h_{fg}	2420.3	kJ/kg
Required heat for heat up	Q_{hu}	2.92	kW
Required heat for evaporation	Q_{ev}	45.2	kW
Total heat required for drying process	Q_{net}	48.2	kW

The value of moisture content is taken from Psychrometric diagram.The demand of drying air flow rate can be calculated by Equation (6) and (7):

$$M_a = \left(\frac{M_r}{Hr_1 - Hr_2}\right) \tag{6}$$

$$m_a = \frac{M_a}{t} \tag{7}$$

Where,M_a is required mass of drying air (kg)
M_r is moisture to be removed (kg)
H_{r1} is initial humidity ratio
H_{r2} is final humidity ratio
m_a is mass flow rate of drying air (kg)
t is drying time (s)

Based on the Equation(6) and (7), Table (3)shows demand calculation of total heat and mass flow rate of drying air:

Table 3. Mass flow rate of drying air.

Parameter	Symbol	Value	Unit
Atmospheric pressure	P_{atm}	1	atm
Ambient temperature	T_a	18	°C
Relative humidity inside drying room	RH_{dr}	35%	
Drying room temperature	$T_{u,fin}$	50	°C
Inital humidity ratio	$H_{r,1}$	0.005	kg/kg
Final humidity ratio	$H_{r,2}$	0.028	kg/kg
Required mass of dyring air	M_a	40,9	kg
Mass flow rate of drying air	m_a	0.81	kg/s

Based on the calculation of system IV, the value of mass flow rate of drying air which flows to HE in system III can be used to obtain the mass flowrate of fresh water.The type of heat exchanger material selected is brazed plate heat exchanger. It consists of two plates that are combined into a pressed plate

[6]. By using the energy balance of Equation (8), the output temperature and heat exchanger area can be determined:

$$Q_i = \eta \, x \left[m_f \, x \, c_p \, x \, (T_{in} - T_{out})\right]_{heater} = \left[m_f \, x \, c_p \, x \, (T_{out} - T_{in})\right]_{heated} \quad (8)$$

Where, Q_i is heat required for heating (W/m^2)
m_f is mass flow rate of fresh water (kg/s)
C_p is specific heat capacity of fluid ($J/Kg^{o}C$)
T_{in} is fluid temperature in heat exchanger inlet (^{o}C)
T_{out} is fluid temperature in heat exchanger outlet (^{o}C)

LMTD (Logarithmic Mean Temperature Difference) is average temperature difference along the heat exchanger. In counter flow heat exchanger, the inlet temperature difference between hot fluid and cold fluid is always greater than the outlet temperature. So, that its the temperature of hot fluid decreases and cold fluid increases. Therefore, to achieve greater heat recovery, a counterflow design is preferred to that of a drying design. LMTD can be calculated based on the change of temperature between the two types of fluid which flow into the heat exchanger[1]. LMTD can be calculated using Equation (9):

$$LMTD = \frac{(T_{hot\,in} - T_{cold\,out}) - (T_{hot\,out} - T_{cold\,in})}{\ln\left(\frac{T_{hot\,in} - T_{cold\,out}}{T_{hot\,out} - T_{cold\,in}}\right)} \quad (9)$$

The next step is to calculate the heat exchanger area with Equation(10):

$$Q = U.A.LMTD \quad (10)$$

Where, U isoverall heat transfer coefficient ($W/m^2.K$). The typical value of U for water-to-water and water-to-air type heat exchanger are 2000 $W/m^2.K$ and 600 $W/m^2.K$ respectively (Leinhard, 2008)
A is heat transfer area (m^2)
LMTD is Logarithmic Mean Temperature Difference (K)

Table 4 shows the values in system III that are calculated by using the Equations (8) – (10).

Table 4. Calculation of water and air of heat exchanger in system III.

Parameter	Symbol	Value	Unit
Temperature water in	$T_{w,in}$	70	^{o}C
Temperature water out	$T_{w,out}$	50.0	^{o}C
Heat capacity of water	Cp_w	4.197	$kJ/kg^{o}C$
Massa water	Mw	2.8	kg/s
Temperature air in	$T_{a,in}$	18	^{o}C
Temperature air out	$T_{a,out}$	50	^{o}C
Heat capacity of air	Cp_a	1.008	$kJ/kg^{o}C$
Heat exchanger efficiency	η_{HE}	84%	-
Mass flow rate of air	$\dot{m}_a$	0.813	kg/s
Required heat	Q	48.15	kW
Logarithmic Mean Temperature difference	LMTD	25.5	K
Area	A	3.14	m^2

In order to produce an outlet temperature of 70^{0}C when the fresh water is entering HE2, the pipe must be designed by minimum heat loss. Heat transfer which occur along the pipe until entering station of the heat exchanger were dominated by two process, which are conduction and convection (Figure3).

The following data are the required heating water flow rate and type of pipe material and its insulation that will be used to calculate outlet temperature of 70^{0}C as listed on Table 5.

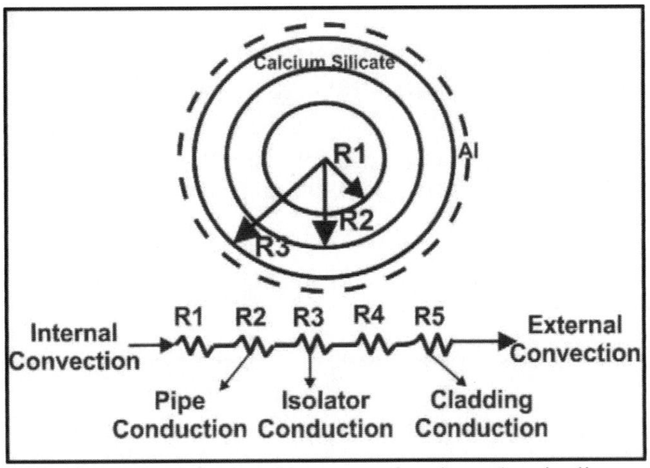

Figure 3. Schematic of heat transfer along the pipeline.

Table 5. Calculation of heat loss along the pipeline (system II).

Parameter	Symbol	Value	Unit
Final fluid temperature	Tfin	70	°C
Ambient temperature	T_{∞}	18	°C
Wind velocity	v_a	2	m/s
Mass flow rate of water	$\dot{m}_w$	2.84	kg/s
Fluid velocity	v_w	1.0	m/s
Cross section area of pipe	A_c	0.003	m^2
Inside pipe diameter	D_i	0.0762	m
Pipe length	L	2000	m
Radius of inside pipe	R_1	0.038	m
Radius of outside pipe	R_2	0.048	m
Radius of outside insulation	R_3	0.058	m
Radius of outside cladding	R_4	0.063	m
Thermal conductivity of pipe	R_p	71.52	W/m.°C
Thermal conductivity of insulation (Calcium Silica)	k_i	0.05	W/m.°C
Thermal conductivity of Al cladding	k_c	227	W/m.°C
Initial fluid temperature	$T_{w,in}$	85.1	°C
Heat Loss Total	Q_{loss}	179.85	kW
Initial fluid temperature	$T_{w,in}$	85.1	°C

Based on the calculation of heat loss in system II and final temperature which maintains to 70^{0}C, outlet temperature and mass flow rate of required brine can be obtained as shown in Tabel 6.

Table 6. Calculation of brine and fresh water of heat exchanger in system I.

Parameter	Symbol	Value	Unit
Temperature brine in	$T_{b,in}$	180	°C
Temperature brine out	$T_{b,out}$	140	°C
Mass flow rate of brine	M_b	7	kg/s
Heat capacity of brine	$C_{p,b}$	4.4	kj/kg °C
Temperature water in	$T_{w,in}$	23	°C
Temperature water out	$T_{w,out}$	85.06	°C
Heat capacity of water	$C_{p,w}$	4.2	kj/kg °C
Heat exchanger efficiency	η_{he}	60%	-
Mass flow rate of water	M_w	2.8	kg/s
Total heat required (for heating fresh water)	Q_w	741.2	kW
Logarithmic Mean Temperature difference	LMTD	105.5	K
Area	A_{HE}	3.5	m^2

5. Drying Room Design

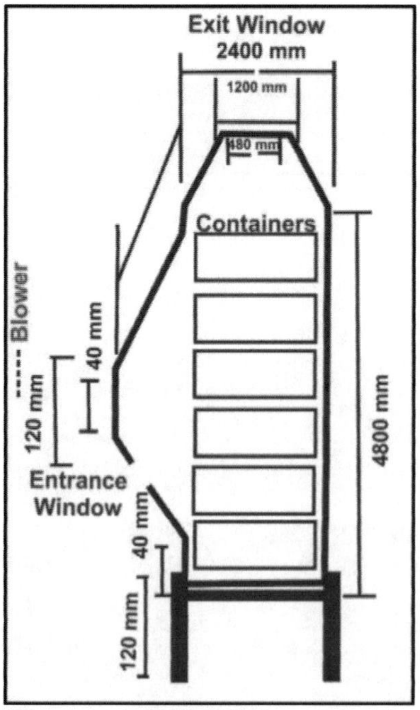

Figure 4. Modified schematic diagram of the geothermal tomato drier system from Ehiem's research[5].

Figure 4 is a design of the drying room for the tomatoes drying application system. Some data used for this design are refered to Ehiem's study. The drying room consists of 6 tomato containers. Calculation of required dryer area is determined by the amount of harvested and dried tomatoes in a day.

The hot air will flow from the blower outside to the top room. This design can produce a final moisture content of tomatoes by 26%. Heat transfer process is divided into two stages. The first stage warms the tomatoes from the initial temperature (around 18^0C) until optimum drying temperature is achieved (about 50^0C). The second stage is the drying stage at constant temperature of 50^0C.

6. Conclusion

Based on this study, it can be concluded that this kind of direct use is technically feasible. Drying 1 ton per day of tomatoes requires 7 kg/s of brine with brine temperature from separator of 180^0C (inlet temperature) and brine temperature from HE1 should be at least 85.1^0C.

The total heat required to transfer the heat from brine to water is about 741.2 kW and from water to air is about 27.9 kW. This energy transfer process is ensued by using the brazed plate heat exchanger. The heat exchanger area for heating the water is 3.14 m^2 and for heating the air is 3.5m^2.

A more detailed study by conducting economic analysisis required to assess the feasibility of tomatoes drying system.

References

[1] Moran, M. J., Shapiro, H. N., Munson, B. R., DeWitt, D. P.,2002, Introduction to Thermal `Systems Engineering: Thermodynamics, Fluid Mechanics, and Heat Transfer,John Wiley & Sons, Inc., United States.

[2] Andritsos, N., Dalampakis, P., and Kolios, N., 2003, Use Of Geothermal Energy For Tomato Drying, GHC Bulletin, March, 9-13.

[3] Badan Pusat Statistik dan Direktorat Jenderal Hortikultura, 2014, Produksi, Luas Panen, dan Produktivitas Sayuran di Indonesia, Jakarta, Indonesia.

[4] Cahyono, 1998, Tomat – Usaha Tani dan Penanganan Pasca Panen, Kanisius, Yogyakarta.

[5] Ehiem, J. C., Irtwange, S.V., and Obetta, S.E., 2009, Design and Development of An Industrial Fruit and Vegetable Dryer, Research Journal of Applied Sciences, Engineering and Technology, United Kingdom..

[6] Rafferty, K. D., 1991, Heat Exchanger, 247-267, in Geothermal Direct Use Engineering and Design Guidebook, 445 p., Oregon Institute of Technology, Oregon.

Assessing the possibility of Enhanced Geothermal System in western Indonesia

R N Hendrawan and W A Draniswari

Geological Engineering, Master Program, Bandung Institute of Technology, Indonesia

rezkinaufan@gmail.com

Enhanced Geothermal System (EGS) is one of geothermal system where abnormally hot masses of low permeability rocks are found at depth. Most of EGS are related to thick and deep sedimentary basin with plutons. EGS focused on drilling at hot crystalline rock and create fractured reservoirs. Indonesia has a complex tectonic setting. There were more than 86 basins in Indonesia. Most of sedimentary basins in Indonesia were tectonically stressed. Moreover, the tectonic setting which is generally a part of the Active Continental Margin also allows the formation of plutons that can be act as hot dry rock. Several hot springs or other geothermal manifestation were found in several areas within the basin and its surrounding area. This study compares structural controls and models in different tectonic setting of main basin in Western Indonesia to get the most suitable potential reservoir for EGS. The characteristic of the South Sumatra Basin, North Sumatra Basin, East Java Basin and Tarakan Basin were evaluated and compared based on its geology and geophysics characteristic. As the result, South Sumatra Basin is the first rank for EGS suitability in Western Indonesia.

1. Introduction

Technology related to natural resources currently began to evolve from conventional methods to unconventional methods. Geothermal technology that applied to the dynamic geothermal system (natural geothermal system) began to move toward the Enhanced Geothermal System (EGS) in sedimentary basin and hot dry rocks.

EGS is a method that aims to create a geothermal reservoir in the basin. EGS can apply to the hot mass with low permeability at depth within thick and deep sedimentary basin. In principle, EGS focused on drilling at hot, crystalline rock, and create fractured reservoirs [3]. The fracture will increase the permeability of the rock so that the fluid can be circulated and bring the heat energy stored in rocks to the surface where electricity can be generated [19].

Studies from Massachusetts Institute of Technology (MIT) [11] showed that in the United States, 100 GWe of cost-competitive capacity could be provided by EGS in the next 50 years. EGS has the potential to be an important source of clean and renewable energy in US because it emits little to no greenhouse gasses. It showed us that EGS is a promising alternative renewable energy. The MIT studies also assume that EGS should target geologic setting where fractures are critically stressed. Economic EGS field usually related to oil-prolific basin because it requires deep and thick sedimentary basin and high heat flow characteristic.

One of EGS production sites was located in the United States. The current geothermal electrical generation capacity in the United States is approximately 3-4 GWe and hence the installed base provides approximately 20.000 GWh of electrical energy [18]. Its heat source was associated with uplift and extension of the Basin and Range, due to the subjection of the East Pacific Rise under South-Western North America. EGS's site in the United States was concentrated in the higher heat flow area of the western region.

Based on those conditions, EGS is very likely to be developed in Indonesia. Its complex tectonic setting and tectonically stressed sedimentary basin is a fine target for EGS preliminary study. In addition,

Indonesian crust relatively had a good heat generation due to thick sediment. This study compares tectonic controls and heat characteristic models of main basin in Western Indonesia to get the most suitable potential reservoir for EGS.

2. Method and Available Data

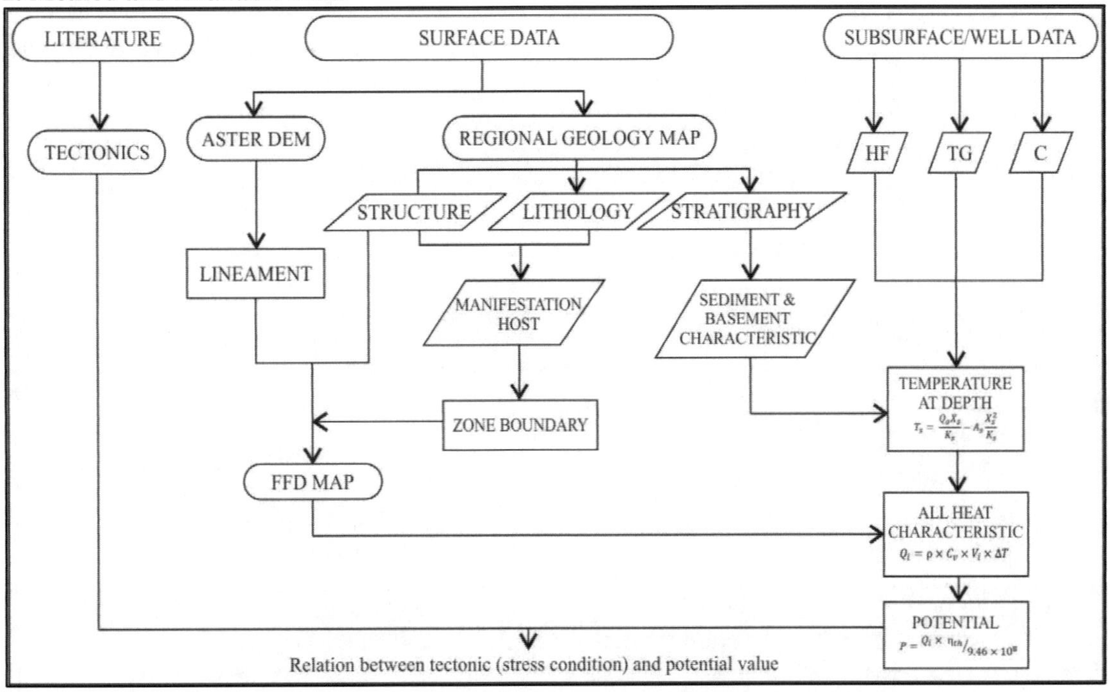

Figure 1. Workflow diagram for EGS Potential Assessment.

EGS potential assessment of Western Indonesia was done based on geological and heat characteristic from surface and subsurface data (figure 1). Literature study and surface-subsurface analysis was done for this assessment [1]. Tectonic, ASTER GDEM, and regional geology data that consists of lithology, stratigraphy and structural data were evaluated in this research. All geological data then combined with well data that contain heat characteristic data to provide potential value and conclusions.

Regional Geology Maps and ASTER GDEM [2] were used to evaluate the surface characteristic of sedimentary basin in Western Indonesia. Regional Geology Maps were provided lithology, structure and stratigraphy data and guided us to determine the manifestation host and both basement and sediment characteristic. ASTER GDEM was used to visualize the structure of landscapes. Its shaded relief maps help to better understand the relationship between geological and geomorphological features. Lineament pattern of structural geology from the regional geology map were combined with lineament pattern from ASTER GDEM to obtain Lineaments Distribution Map. By dividing the cell into 5x5 km, which the same gridding size with assessment cells, and picking the lineament we could create the Lineaments Distribution Map . Later the lineament distribution will compare with the potential value whether it affects the potential value or not.

Subsurface study for assessing EGS focused on its heat characteristic. Analysis was done based on All Heat Flow Data published by Royal Holloway – South East Asia Research Group [1]. could measure Temperature at Depth, Heat Characteristic, and Potential based on Heat Flow (HF or Q), Thermal Gradient (TG) and Thermal Conductivity (C or K) values. The heat characteristic data specially the heat flow data is the starting point to get the thermal energy value and the potential value. Relationship between heat flow, thermal conductivity and thermal gradient can be noted as below:

$$Q = c \times dT/dZ \ [11] \tag{1}$$

The combination of thermal conductivity and thermal gradient data gave more representative data about the heat characteristic. Thermal conductivity is the property of material's ability to conduct heat. It is represents the rate of energy loss (in watts, W) through a piece of material. Thermal conductivity is measured in watts per-Kelvin meter (W/Km). Thermal gradient described the rate of temperature change per-depth. Thermal gradient is measured in Kelvin per meter (K/m).

An estimation of temperature, heat characteristic and potential was done based on equation from Massachusetts Institute of Technology previous research (figure 1). Assessed zones were determined based on manifestation data. Grids composed of 5km x 5km cells were assigned for each zone, and then calculations were done for each square. Temperature at depth are derived by assuming that the heat was produced by sediment contribution and radioactive heat generation (A). There are 1D heat conduction transfer and two system layer (basement and sediment) which the thermal conductivity between the sediment and basement are identical.

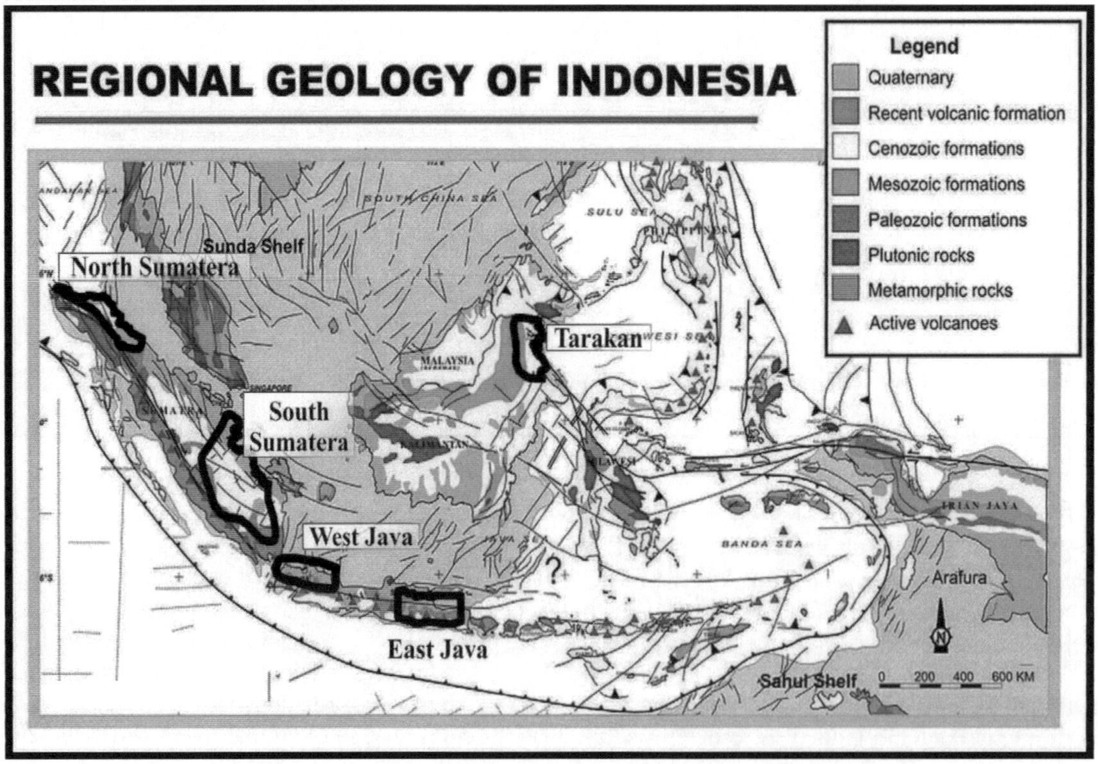

Figure 2. Regional Geology of Indonesia Map [7].

Each zone were divided in to square with an area of 25km^2 and each square contains the thermal conductivity, the heat flow, the thermal gradient, the sediment thickness, the radioactive heat generation, the regional heat flow and the average surface temperature information. Radioactive heat generation (A) value was defined as 1 μW/m^3 [11]. and the average surface temperature was 25C. Temperature at Depth value then used to define the difference of temperature between surface and subsurface (ΔT). It multiplied by the rock density (ρ), heat capacity (c), and volume (v) to get the heat energy (Q_i) value. At real not all of the heat energy at the subsurface could generate the energy through EGS, there are some aspect include technical and non-technical that related to efficiency of potential generation, but for this assessment we proposed the ideal scenario with 100% efficiency. At last, the potential value was calculated from its heat energy by multiplied the heat energy with the efficiency factor (μ) and the constant.

The last part of this study was comparing the potential value both similar and different basement depth of each basin with its tectonic condition and also other variable such as lineament distribution, geological structure and lithology to determine the correlation between them.

3. Western Indonesia Basin

Several hot springs has been studied in Western Indonesia and its surrounding by Center of Geological Resources, Ministry of Energy and Mineral Resources. Hot spring manifestation in Western Indonesia not only related to volcanic activity but also in oil-prolific sedimentary basin. Five basins were selected due to the presence of geothermal manifestations in sedimentary basin or it surrounding (figure 2). There were Central Sumatera Basin, South Sumatera Basin, West Java Basin, East Java Basin, and Tarakan Basin. Basin boundaries were established based on beach line and manifestation distribution. The basin boundary had drawn landward from beach line due to the absence of fault and fracture data in seaward.

3.1. North Sumatera Basin

North Sumatra Basin located in 95.334-99.280 E and 3.110-5.680 N and has an area of 28.370km^2. Geothermal manifestations in North Sumatra Basin were related to volcanic system and concentrated around Dua Volcano and Sibayak Volcano. Hot spring manifestations lies on Quarter volcanic rock units formed a NW-SE trending lineament.

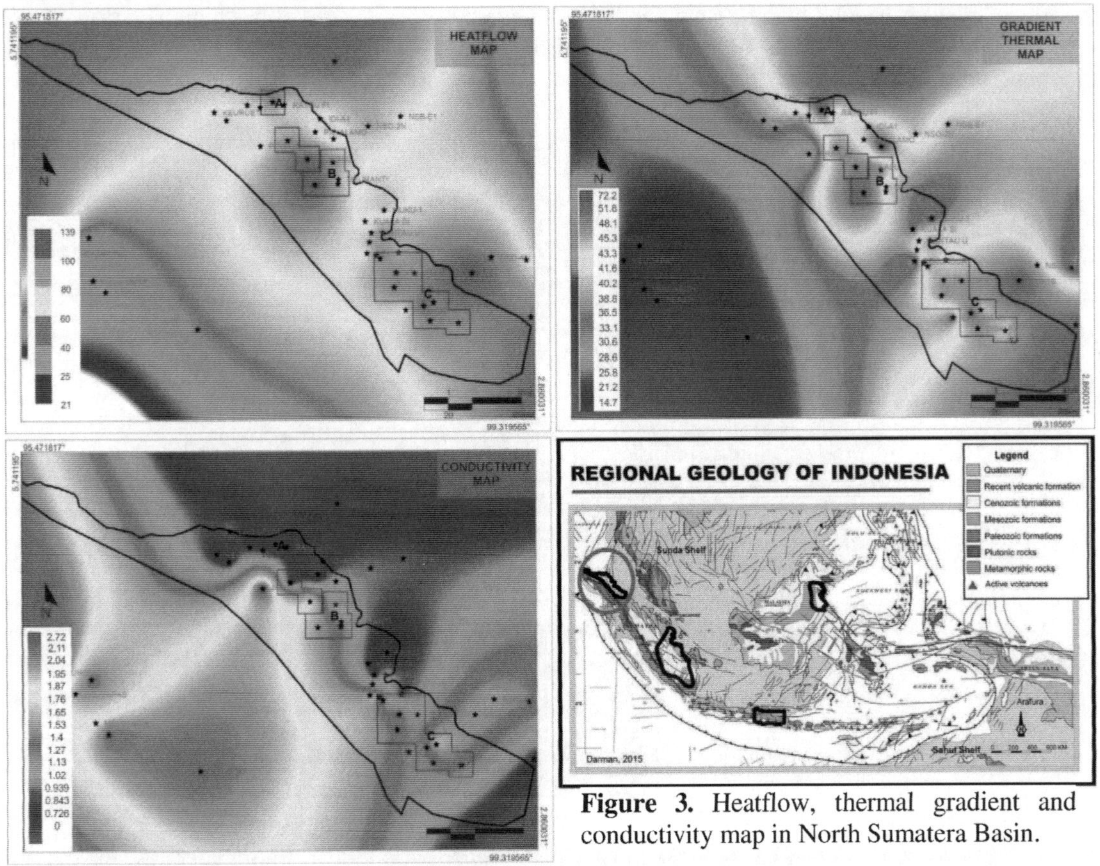

Figure 3. Heatflow, thermal gradient and conductivity map in North Sumatera Basin.

North Sumatra Basin was bordered by Malacca Platform in the east, Asahan Arc in the south, Barisan Mountain in the west and open to the north up to Andaman Islands [13]. During Pre-Tertiary, volcanism activity played major role in North Sumatra Basin. Volcanism activity commonly produced acid intrusive such as granite and granodiorite. Tectonic activity during Pre-Miocene to Miocene mainly produced the N-S structural pattern while at post Miocene the structural pattern was mainly relevant to

Bukit Barisan uplift with NW-SE trending. Later, the main activity was orogeny processed that produce higher heat flow in North Sumatra Basin and its surrounding.

Heat characteristic of North Sumatera Basin were made from 34 wells data [1]. The heat flow within the basin is significantly higher than its surrounding (figure 3). The highest heat flow areas were concentrated near Peulalu and NSO-1S Well. The heat flow value showed lower value to the southern part. The maximum heat flow amount 139 mW/m^2 from S.Pineu Well and the minimum value was 86 mW/m^2 from Ray Fi Well. The minimum temperature gradient was 43 at Rayeu-Fi Well and reached maximum value 62 at Piada-A1 well while thermal conductivity data were ranged from 1.91 W/m (Piada-A1 Well) to 2.38 W/m (Peulalu Well). Thermal gradient in North Sumatra Basin showed opposite trend with its thermal conductivity map (figure 3). Higher thermal gradient values were concentrated in SE area and continued in the center part to North area while higher thermal conductivity map was N-S trending, started from Piada-A1 Well.

Lineaments Distribution Map combined with Geological Structure Map (figure 4) was created to know the stress condition of the research area and help us to delineate the assessment zone for potential calculation in North Sumatra Basin. Based on figure 4 below, the highest fault and fracture density was concentrated in the central part of South Sumatra Basin with the lineament trending NE-SW.

Three zones from North Sumatra Basin were picked for EGS potential calculation based on 3 parameters: thermal gradient, heatflow and fracture density (figure 4). Zone A characterized by a high thermal gradient value and high heat flow value (above 100 mW/m^2). Zone B has high heat flow value, high fracture density and high thermal gradient value and zone C has high values on heat flow and thermal gradient parameters.

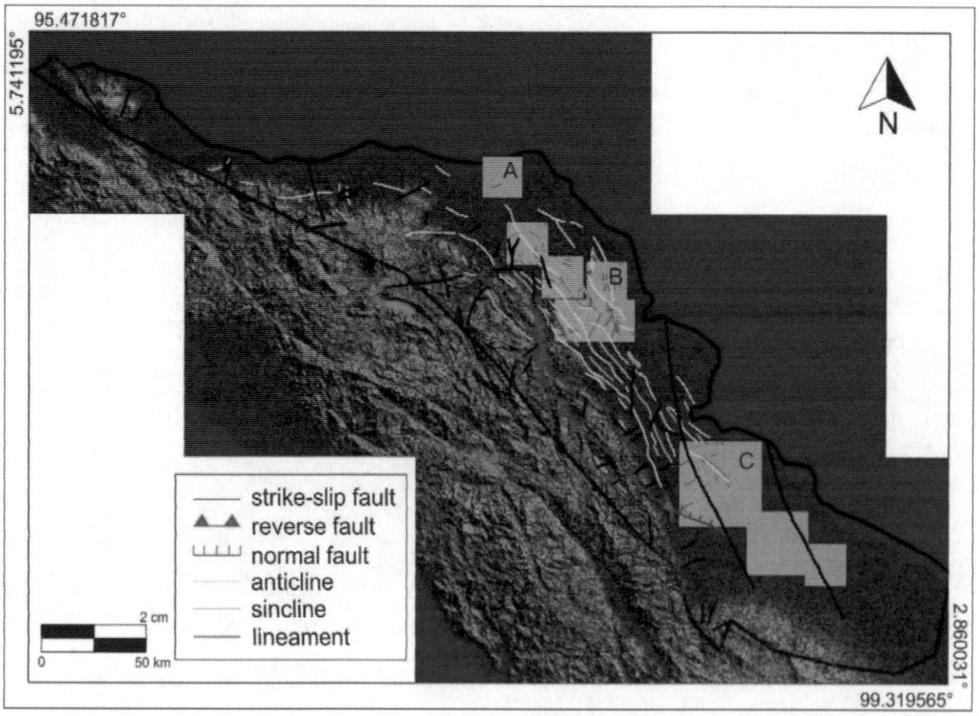

Figure 4. Geological Structure and Lineament Distribution in North Sumatera Basin.

3.2. South Sumatera Basin

South Sumatra Basin ranged from 102,014 – 105,508 E and 4,617 S – 0,517 N with area amount 112.100km^2. Geothermal manifestations in South Sumatra Basin were related to fault and sedimentary rocks. It lies in Air Benakat Formation which consists of claystone intercalated with fine sandstone.

South Sumatra Basin was a back arc basin that located in Bukit Barisan. Regionally, this basin bounded by Barisan Mountain to the Southwest and pre-Tertiary Sunda Shelf in the northeast. South Sumatra Basin influenced by 3 major tectonic episodes, there are : (1) initial extensional periods during Latest Cretaceous to Early Oligocene that produce NE-SW and N-S trending horst block and half graben, (2) tectonic quiescent period during Late Oligocene to Early Miocene that continued with wrench tectonics produced fold in Early to Middle Miocene and (3) strong compressional and uplift period during Pliocene-Pleistocene indicated by structure reactivation that produce NW-SE trending reverse inversion fault and basement uplift.

Heat characteristic map of North Sumatera Basin were made from 49 wells data [1] (figure 5). The heat flow values within the basin are significantly above the average. The highest heat flow areas were concentrated near Arang-AR and Lebong F Well. The heat flow values were lowered to the southern part. The maximum heat flow amount 129 mW/m^2 from Arang AR and Lebong F Well and minimum value was 92 mW/m^2 from Bajubang Well. The minimum temperature gradient was 43 at Malapari Well and reached maximum value 62 at Setiti Well while thermal conductivity data were ranged from 1.97 W/m (Beringin Well) to 2.27 W/m (Sigoyang Well). Higher thermal gradient values were concentrated in North and Northwest area and lowered to the Southern part while higher thermal conductivity map was E-W trending, lowered to the Western part (figure 5).

Geological structure and lineament distribution map (figure 6) was used to evaluate the stress condition of the research area and help us to delineate the assessment zone for potential calculation in South Sumatra Basin. Based on Figure 6 below, the highest fault and fracture density was concentrated in the central and southern part of South Sumatra Basin with the lineament trending NE-SW.

Two zones from South Sumatra Basin were picked for EGS potential calculation (figure 6) based on 3 parameters; thermal gradient, heatflow and fracture density. Zone D characterized by a high thermal gradient value, high fracture density and high heat flow value (above 100 mW/m^2) while zone E has high values on heat flow and thermal gradient parameters.

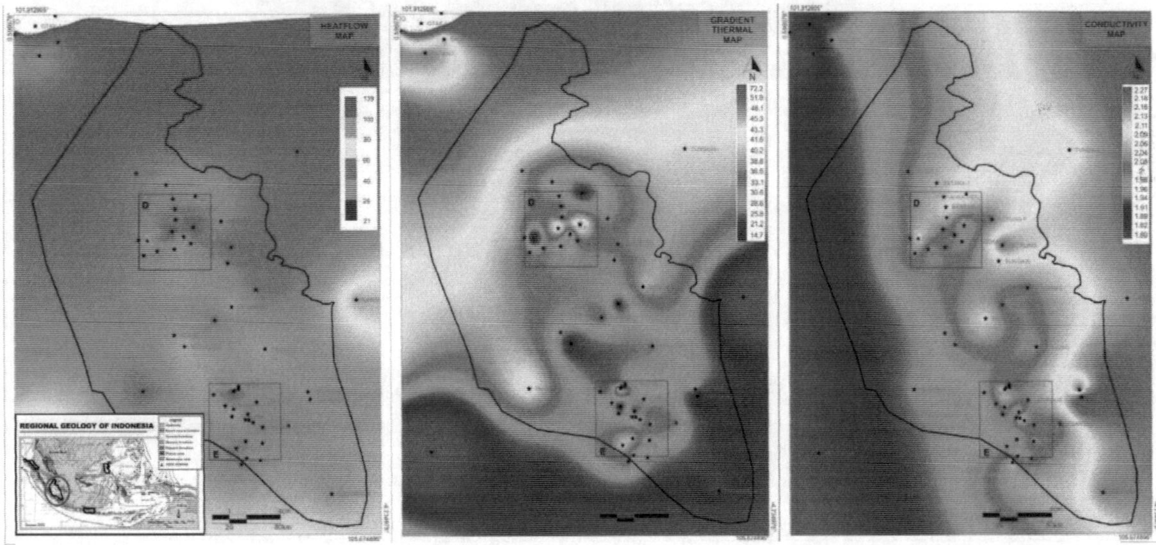

Figure 5. Heatflow, thermal gradient and thermal conductivity map in South Sumatera Basin.

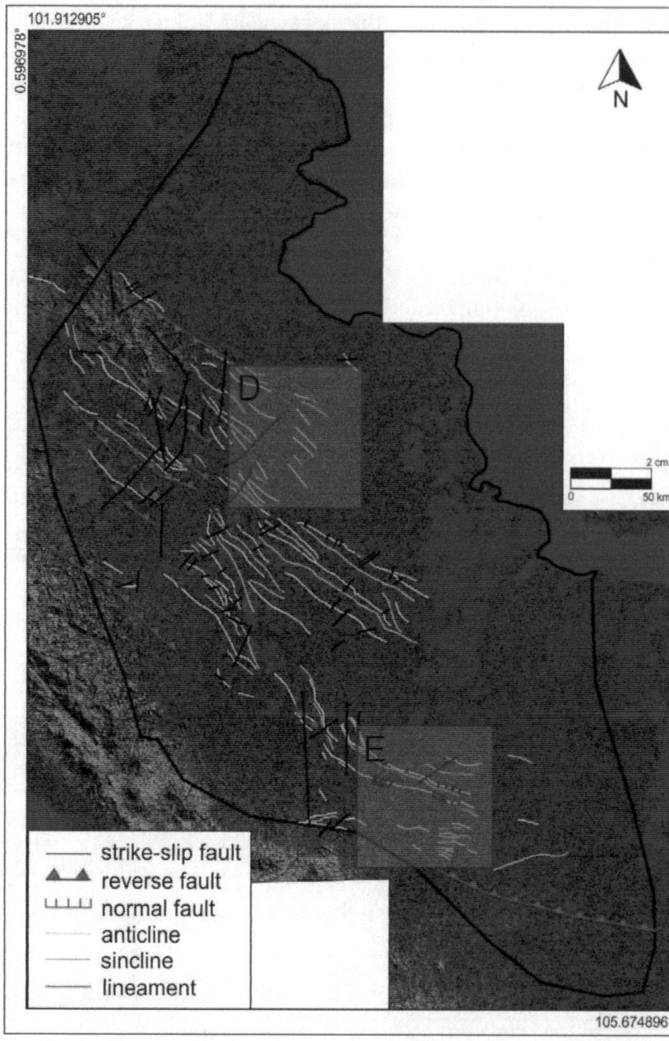

Figure 6. Geological structure and lineament distribution in South Sumatera Basin.

3.3. West Java Basin

West Java Basin ranged from 105.852 – 108.776 E and 7.077 S – 5.938 S with area about 16.420km^2. There is no hot spring manifestation in West Java Basin. The nearest hot spring manifestation located outside of the basin boundary. It were located in Banten around Rawadanau and related to volcanic system. It lies in Quartenary rock formations that consist of Lake Deposits, Collovium, Young Volcanic Rocks, and Lower Banten Tuff.

North West Java Basin was a back arc basin that located between Sunda micro-plate and India-Australia Plate. It was formed by collision of the Eurasian Plate with the India-Australia Plate during Late Cretaceous to Early Eocene. Compressional stress trending North-South were influenced North West Java Basin and caused the formation of regional structure trending East-West. While tensional stress were resulted in North-South block faulting that was responsible for the development of the North West Java Basin and its numerous sub-basin and basement high[13].

Because of the absent of geothermal manifestation and the lack of well data around Rawadanau, we eliminated West Java Basin from potential calculation.

3.4. East Java Basin

East Java Basin located in 111.263 – 114.142 E and 7.990 – 6.610 S and has an area about 38.420 km^2. Hot spring manifestations in southern parts of East Java Basin were related to volcanic activity from Welirang Volcano and Lamongan Volcano while hot springs in northern part were related to sedimentary rocks. It lies in Ngrayong formation – that consist of sandstone, shale, mudstone, siltstone and coal, Pacitan formation – that consist of limestone and dolomitic limestone, Siltstone member of Kujung Formation, and Prupuh Limestone member of Kujung Formation[17].

The North East Java Basin was classified as a back arc basin. There are 2 major trends in North East Java Basin, NE-SW orientation and E-W orientation. During extensional period, half graben and faults associated with this rifting were developed throughout the area in principally NE-SW and E-W orientation. During Early Miocene, deformation began following the extensional period. Tectonic activity dominated by compressional with some wrenching. Fault inversion then caused uplift of the older NE-SE and E-W trending Eocene graben systems [13].

Heat characteristic maps of East Java Basin were made from 17 wells data [1] (figure 7). The heat flow values within the basin were significantly below the average. The highest heat flow areas were concentrated near AR-1, Ledok F1 and JS 13A-1 Well. The heat flow values were lowered to the south-eastern part. The maximum heatflow amount 112 mW/m^2 from AR-1 and its minimum was 21 mW/m^2 from Madiun Well. The minimum temperature gradient was 25 at Madiun Well and reached maximum value 47 at AR-1 well. The higher thermal gradient values were concentrated in North West area and lowered to the South-eastern part. The thermal conductivity values data were ranged from 0.84 W/m (Madiun Well) to 2.37 W/m (AR-1 Well). The highest thermal conductivity areas were concentrated near AR-1, GR-1 and JS 13A-1 Well. The thermal conductivity patterns were lowered to the south (figure 7).

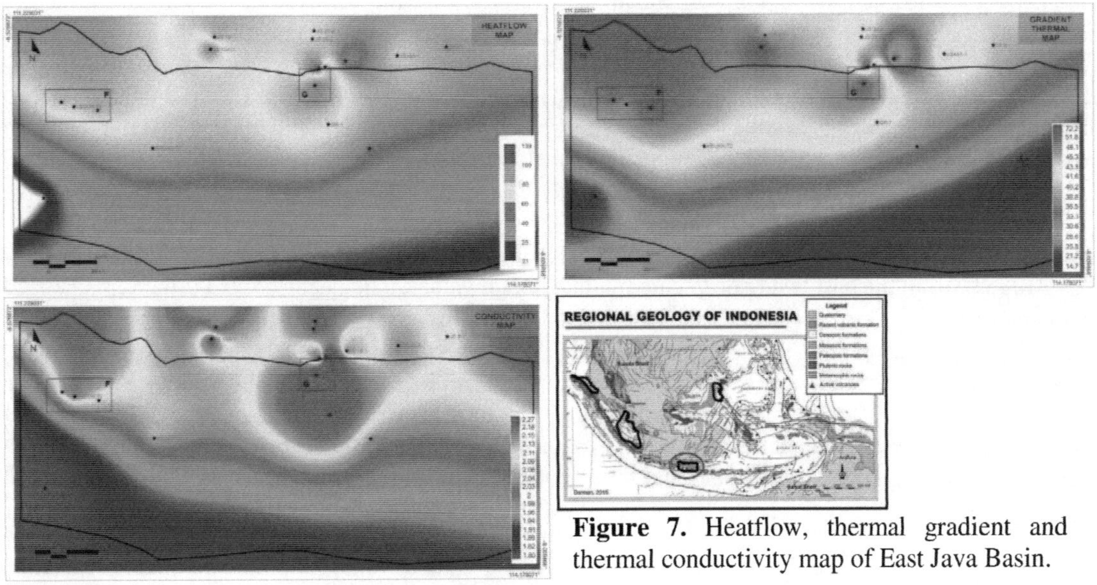

Figure 7. Heatflow, thermal gradient and thermal conductivity map of East Java Basin.

The stress condition of the research area had assessed by Fault and Fracture Density Map combined with Geological Structures and Lineament Map (figure 8). It displayed that the highest fault and fracture density was concentrated in northern and north-western part of East Java Basin with the lineament trending NW-SE. Later, this information used to help us to delineate the assessment zone for potential calculation in East Java Basin.

Two zones from East Java Basin (Zone F and G) (figure 8) were picked for EGS potential calculation based on 4 parameters; thermal gradient, heatflow, thermal conductivity and fracture density. Both of them characterized by a high thermal gradient value, high fracture density, high thermal conductivity, and high heat flow value (above 85 mW/m^2).

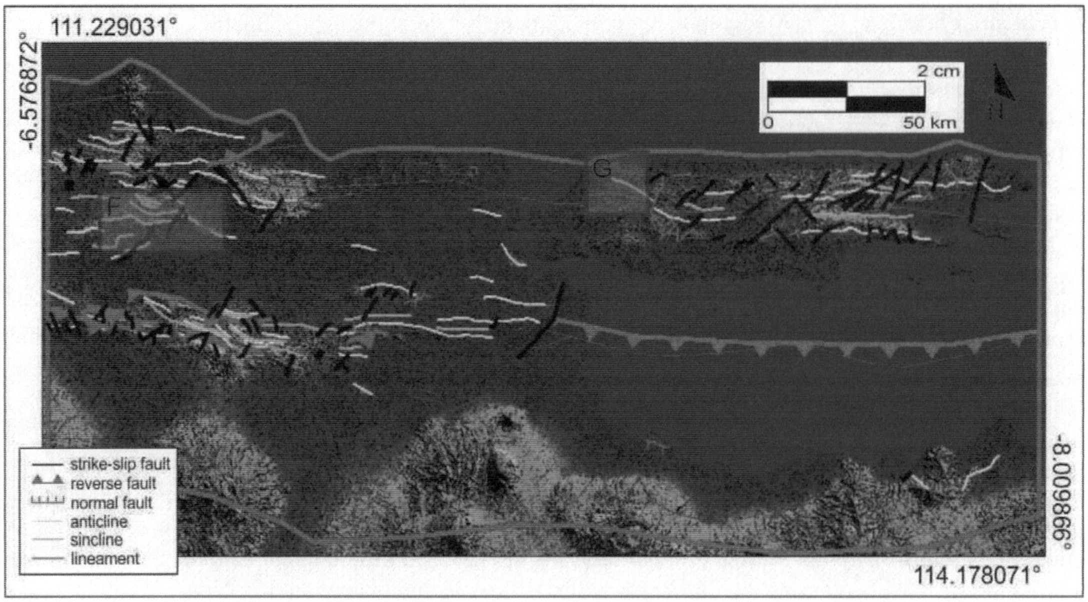

Figure 8. Geological Structures and Lineament of East Java Basin.

3.5. Tarakan Basin

Tarakan Basin is located in 116,555 E – 118.122 E and 1.807 – 4.251 N with area about 31.950km^2. Geothermal manifestations in Tarakan Basin were related to fault and sedimentary rocks. It lies in Embakat Group from Mentarang and Lurah Formation which consists of sandstone [20].

Tectonic activity in Tarakan Basin was initiated extension and subsidence during the Middle to Late Eocene. It produced wrench fault and resulted in the formation of major NW-SE trending and stopped at Early Miocene. During Middle Miocene through Pliocene, Tarakan Basin was in stable tectonic condition but because of basin subsidence and gravity effects, listric faults were produced in this area. Later, during Pliocene-Pleistocene, structure reactivation produced reverse and thrust fault. This deformation reoriented the folding to be NE-SW trending [13].

Heat characteristic maps of Tarakan Basin were made from 11 wells data [1] (figure 9). The heat flow values within the basin were dominated with average values. The heat flow patterns were lowered to the eastern part. The maximum heat flow within the basin was 89 mW/m^2 from Bunyuk Well and minimum value was 61 mW/m^2 from Giru-1 Well. The minimum temperature gradient was 27 at Giru-1 Well and reached maximum value 41 at Linapaca1 Well. Higher thermal gradient values were concentrated in the western part (landward) and lowered to the eastern part (seaward). Thermal conductivity data were ranged from 1.9 W/m (Sesayap Well) to 2.27 W/m (Giru 1 Well). Opposite with thermal gradient pattern, the thermal conductivity values within the basin were lowered to landward (figure 9).

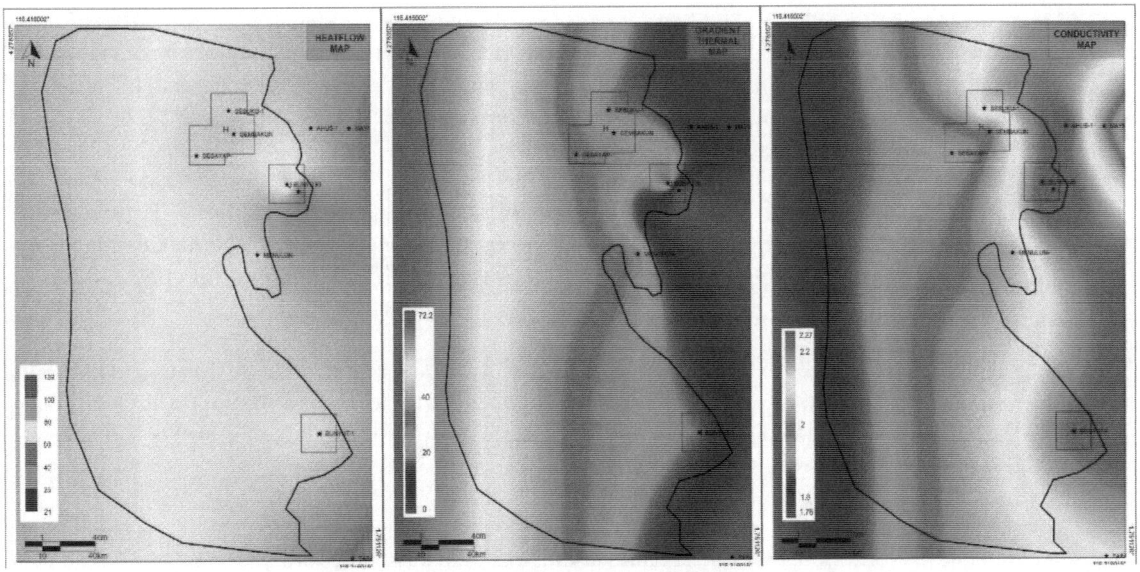

Figure 9. Heatflow, gradient thermal and conductivity map of Tarakan Basin.

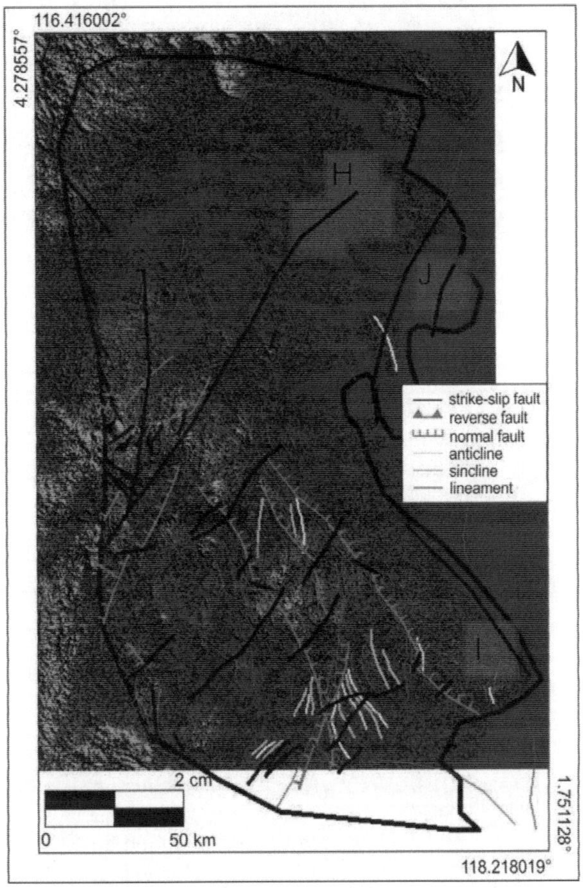

Figure 10. Geological Structures and Lineament Distribution Map of of Tarakan Basin.

Lineament Distribution Map combined with Geological Structures Map was created to know the stress condition of the research area and help us to delineate the assessment zone for potential calculation in Tarakan Basin. Based on figure 10 below, the highest fault and fracture density was concentrated in the south-western part of Tarakan Basin with the lineament trending NE-SW.

Three zones from Tarakan Basin were picked for EGS potential calculation (figure 10) based on 4 parameters; thermal gradient, heat flow, thermal conductivity, and fracture density. Zone H characterized by a high thermal gradient value, low thermal conductivity and high heat flow value (above 75 mW/m^2). Zone I has high heat flow value, high fracture density and high thermal conductivity value while zone J has highest heat flow value, high thermal gradient and high thermal conductivity.

4. Calculation

Calculation was done to determine the potential value of Enhanced Geothermal Systems (EGS) in Western Indonesia basins based on heat flow, thermal conductivity and depth of basement data. There were two calculation of potential. First, we used the similar basement depth for all basins and second, we used different basement depth value in same variable.

4.1. Similar Basement Depth

Basement depth in North Sumatera, South Sumatera, Northeast Java and Tarakan Basin were assumed to have same value. This method was done to find the pattern and relation between EGS potential value and heat characteristic data that consist of heat flow and thermal conductivity.

Table 1. Potential Calculation of Similar Basement Depth in each Basin

(a) Several calculation in South Sumatera Basin

SQUARE	HEAT FLOW	CONDUCTIVITY	POTENTIAL (MW)	LITHOLOGY	REFFERENCES	ZONE
D1	107	1.91	114,19	Alluvium	[10]	A
B1	108	1.91	114,162	Alluvium		
C1	111	1.91	114,08	Alluvium		
C2	112	1.91	114,05	Alluvium		
D4	94	1.97	111,08	Alluvium		
B4	90	1.97	111,08	Alluvium		
A4	88	1.98	110,7	Alluvium		
average		112,85				
C3	103	1.97	110,9	Sandstone	[4]	B
D1	93	1.98	110,6	Alluvium		
D2	96.5	1.98	109,9	Alluvium		
B5	105	2.37	92,08	Mudstone, Sandstone		
F7	127	2.36	91,95	Mudstone, Sandstone		
F6	126	2.37	91,6	Mudstone, Sandstone		
average		100,2				
E3	94	204	107,3	Alluvium	[6]	C
F3	98	2.04	107,6	Alluvium		
G3	100	2.04	107,2	Alluvium		
J10	112	2.3	94,77	Sandstone	[5]	
K9	112	2.2	94,77	Alluvium		
K10	111	2.31	94,32	Alluvium		
average		100,76				

(b) Several calculations in South Sumatera Basin

SQUARE	HEAT FLOW	CONDUCTIVITY	POTENTIAL (MW)	LITHOLOGY	REFFERENCES	ZONE
M9	121	1.99	109,20	Alluvium	[10]	D
N9	121	1.099	109,20	Alluvium		
N8	124	1.99	109,12	Alluvium		
B9	105	2.18	100,10	Claystone, Marl	[14]	
C10	101	2.19	99,74	Claystone, Marl		
B10	102	2.19	99,72	Claystone, Marl		
average		106,12				
D10	125	2	111,70	Tuff	[10]	E
C2	105	2.02	110,65	Sandstone		
I14	110	1.97	110,62	Tuff		
Q2	103	2.1	103,97	Tuff		
F2	124	2.09	103,9	Tuff		
F3	124	2.09	103,9	Tuff		
average		107,78				

(c) Several calculations in East Java Basin

SQUARE	HEAT FLOW	CONDUCTIVITY	POTENTIAL (MW)	LITHOLOGY	REFFERENCES	ZONE
B4	77	1.8	122,10	Breccia, Sandstone	[12]	F
C4	90	1.87	117,14	Sandy Limestone		
B3	82.5	1.88	116,75	andy Limestone		
E2	91	2.12	103,30	andy Limestone		
F1	94.5	2.12	103,21	Breccia		
E1	94	2.13	102,74	Sandy Limestone		
average		108,50				
C1	67	2.05	107,49	Alluvium	[16]	G
B1	74	2.05	107,30	Alluvium		
A1	78.5	2.08	105,63	Alluvium		
D3	99.5	2.34	93,40	Claystone, Sandstone		
C4	102.2	2.35	92,92	Sandstone		
C3	106.7	2.36	92,43	Claystone, Sandstone		
average		97,34				

(d) Several calculations in Tarakan Basin

SQUARE	HEAT FLOW	CONDUCTIVITY	POTENTIAL (MW)	LITHOLOGY	REFFERENCES	ZONE
A6	76	1.9	115,70	Alluvium	[15]	H
A7	76	1.9	115,70	Alluvium		
B6	76	1.9	115,70	Alluvium		
F2	76	2.11	104,20	Alluvium		
E1	76.5	2.11	104,20	Alluvium		
F1	76	2.12	103,70	Alluvium		

average		110,76				
A1	82.5	2.14	102,56	Alluvium	[15]	I
A4	73	2.15	102,33	Alluvium		
A3	78	2.15	102,20	Alluvium		
C3	76	2.22	99,02	Alluvium		
D4	62.5	2.25	98,04	Claystone, Sandstone		
D3	64	2.25	98,00	Claystone, Sandstone		
D2	76	2.28	96,43	Alluvium		
average		100,15				
A1	79	2.15	102,17	Alluvium	[43]	J
A4	78.5	2.16	101,71	Alluvium		
A3	79	2.16	101,70	Alluvium		
B3	79.5	2.18	100,75	Alluvium		
D3	76.5	2.19	100,37	Claystone, Sandstone		
C3	78	2.19	100,33	Claystone, Sandstone		
average		101,08				

Based on calculation (Table 1), the maximum potential value was 122,10 MW from square B4 zone F in Northeast Java Basin while the minimum value was 91,58 MW from zone B in North Sumatra Basin. Nevertheless, the maximum average potential value was showed by zone A from North Sumatra Basin and zone H from Tarakan Basin. Generally, the comparison of average potential values at each basin showed that South Sumatera Basin has the greatest potential value. Total average potential value for South Sumatra Basin is 106,95 MW.

4.2. Different Basement Depth

Different tectonic setting would produce different structural product and different basement depth. Basement depth variable for this calculation was obtained from literature study and regional approach.

Each square and zone showed relatively large interval value. The maximum value was 163,4 MW from square C6 with depth of basement 4500m while the minimum value was showed by square F12 which has basement depth about 1000m and potential value 1,33 MW. Compared with other basins, Tarakan Basin showed high average potential value.

Table 2. Potential Calculation of Different Basement Depth

(a) Several calculations in North Sumatera Basin

SQUARE	DEPTH	POTENTIAL	LITHOLOGY	REFFERENCES	ZONE
C3	4000	113,89	Alluvium	[10]	A
D3	4000	112,62	Alluvium		
E4	4000	111,25	Alluvium		
E5	2000	13,51	Alluvium		
B5	2000	13,24	Alluvium		
A5	1000	1,582	Alluvium		
Average		52,059,			
C1	5000	214,73	Alluvium	[4]	B
D1	4500	157,82	Alluvium		
D2	4500	156,90	Alluvium		
K6	4000	106,48	Limestone		
F10	1500	4,66	Limestone		

SQUARE	DEPTH	POTENTIAL	LITHOLOGY	REFFERENCES	ZONE
G12	1000	1,34	Mudstone, Siltstone		
F12	1000	1,33	Marl		
Average		67,415			
B6	5000	194,85.1	Sandstone	[6]	C
A5	5000	194,13	Limestone		
D3	4500	152,33	Alluvium		
B3	4000	104,02	Sandstone		
K10	3000	39,41	Alluvium		
P15	2000	12,82	Tuff	[5]	
O15	2000	12,77	Tuff		
Average		62,34			

(b) Several calculations in South Sumatera Basin

SQUARE	DEPTH	POTENTIAL	LITHOLOGY	REFFERENCES	ZONE
M9	4000	109,27	Alluvium	[10]	D
N8	4000	109,120	Alluvium		
E15	4000	108,54	Tuffaceous sandstone		
I4	3000	43,10	Tuffaceous sandstone	[14]	
Q1	3000	43,10	Clay, Silt		
O1	3000	43,080	Basalt, Andesite		
Average		55,73			
D10	3500	74,8.1	Tuff	[10]	E
H12	3500	73,5	Tuff		
B6	3500	73,30	Sandstone		
D6	3500	72,16	Alluvium		
J8	3000	43,99	Tuffaceous sandstone		
P1	3000	43,89	Clay, silt		
P2	3000	43,89	Tuffaceous sandstone		
Average		54,68			

(c) Several calculations in East Java Basin

SQUARE	DEPTH	POTENTIAL	LITHOLOGY	REFFERENCES	ZONE
B4	4000	122,10	Claystone, Limestone	[12]	F
C4	4000	117,144	Marl		
B3	4000	116,74	Marl		
E4	4000	111,41	Limestone, claystone		
F2	4000	104,26	Marl		
D1	4000	104,20	Marl		
G1	4000	103,67	Marl		
Average		108,50			
C1	4000	107,48	Alluvium	[16]	G
B1	4000	107,29	Alluvium		
A1	4000	105,62	Alluvium		

D3	4000	93,38	Claystone, Mudstone		
C4	4000	92,92	Marl, Limestone		
C3	4000	92,42	Claystone, Mudstone		
Average		97,33			

(d) Several calculations in East Java Basin

SQUARE	DEPTH	POTENTIAL	LITHOLOGY	REFFERENCES	ZONE
C6	4500	163,40	Alluvium	[15]	H
C5	4500	162,57	Alluvium		
D6	4500	162,57	Alluvium		
G3	3500	70,61	Alluvium		
E2	3500	69,60	Alluvium		
F2	3500	69,60	Alluvium		
Average		116,94			
A1	4000	102,56	Alluvium	[15]	I
A4	4000	102,33	Claystone, sandstone		
A3	4000	102,20	Claystone, sandstone		
C1	3500	66,92	Alluvium		
D1	3500	66,68	Alluvium		
D2	3500	64,41	Alluvium		
Average		94,00			
A1	4000	102,17	Alluvium	[15]	J
A4	4000	101,71	Alluvium		
A3	4000	101,70	Alluvium		
B3	4000	100,75	Alluvium		
D3	4000	100,37	Claystone, sandstone		
C3	4000	100,33	Claystone, sandstone		
Average		101,08			

Table 3. Summary of Potential Calculation

Similar Basement	**North Sumatera**			**South Sumatera**		**East Java**			**Tarakan**	
Statistics	**Zone A**	**Zone B**	**Zone C**	**Zone D**	**Zone E**	**Zone F**	**Zone G**	**Zone H**	**Zone I**	**Zone J**
Max	114,19	110,9	107,3	109,2	111,7	122,1	107,49	115,7	102,56	102,17
Min	110,69	91,6	94,32	99,72	103,9	102,74	92,43	103,7	96,43	100,33
Average	112,85	100,2	100,76	106,12	107,78	108,5	97,34	110,76	100,15	101,08
Total Area (m^2)	500	2000	2550	7225	7225	800	400	800	400	400
Area above 100MW	500	925	1524.9	7175.1475	7225	800	94.12	800	225	400
Percentage area above 100 MW	100.00%	46.25%	59.80%	99.31%	100.00%	100.00%	23.53%	100.00%	56.25%	100.00%

Different Basement	North Sumatera			South Sumatera		East Java		Tarakan		
Statistics	**Zone A**	**Zone B**	**Zone C**	**Zone D**	**Zone E**	**Zone F**	**Zone G**	**Zone H**	**Zone I**	**Zone J**
Max	113,89	214,73	194,85	109,27	74,8	122,1	107,48	163,4	102,56	102,17
Min	1,582	1,33	12,77	43,08	43,89	103,67	92,42	69,6	64,41	100,33
Average	52,059	67,415	62,34	55,73	54,68	108,5	97,33	116,94	101,08	101,08
Total Area (m2)	500	2000	2550	7225	7225	800	400	800	400	400
Area Above 100MW	75	50	475.065	924.8	0	800	94.12	557.6	200	400
Percentage area above 100 MW	15.00%	2.50%	18.63%	12.80%	0.00%	100.00%	23.53%	69.70%	50.00%	100.00%

5. Discussion

Potential value using similar basement depth and different basement depth had contrast number. Examples several comparison square data with diverse variable in table 4.

From the table below, basement depth was the most important variable to establish potential value, mostly every 1 meter could change potential value around 0,03 MW. Different depth basement number due to tectonic activity at surrounding area that produced both of basement high or deepest blocks like basin etc.

Table 4. Potential comparison between similar and different basement calculation in several location

SQUARE	HEAT FLOW	BASEMENT DEPTH	THERMAL CONDUCTIVITY	POTENTIAL (MW)	FFD	LITHOLOGY	ZONE
D1	93	4500 4000	1.98	157, 110,5	0	Alluvium	B
K10	111	3000 4000	2.31	39,41 94,32	0	Alluvium	C
D10	12.5	3500 4000	2	74,8 111,7	0.256	Sandy tuff, tuffaceous sandstone	E
D2	76	3500 4000	2.28	64,5 96,42	0.167	Alluvium	I

Furthermore, thermal conductivity also had an impact on the significant potential changes. The calculation tables showed that maximum values tend to had small thermal conductivity, and vice versa. In contrast, variation of heat flow did not display significant difference rather than thermal conductivity.

Lineament Distribution Map, Geological Structure (both fold and fault) and lithology were geological aspect for EGS assessment. There is no direct effect of lineament distribution to EGS potential value. Lineament distribution only explained surface structural abundant as well as structure and lithology that just showed the top surface condition even though sometimes could helped to decided sediment thickness, uplift and subsidence on subsurface analogue.

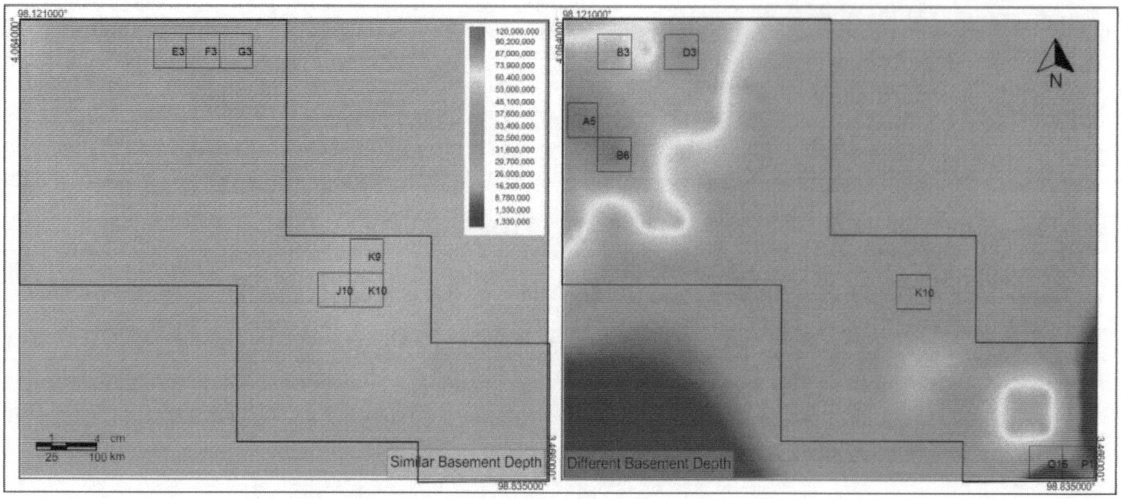

Figure 10. EGS Potential comparison in Zone C.

Based on the analysis performed, potential distribution maps from zone C, D, and H with similar and different basement depth parameters were compared to guide the interpretation of what factor affected the EGS potential value in North Sumatra Basin, South Sumatra Basin, and Tarakan Basin. Red color displayed high value of potential while yellow and green showed average potential value. The blue color indicated smaller potential number than average value.

Based on figure 11 there were different locations where the potential reach highest values. When basement depth was similar for each location, the highest potential value was concentrated in the northern part of this zone otherwise if we used different basement depth for each location, we got that the most potential area were located in the north-west part. The location for minimum potential values did not show the difference location, it was concentrated around K10.

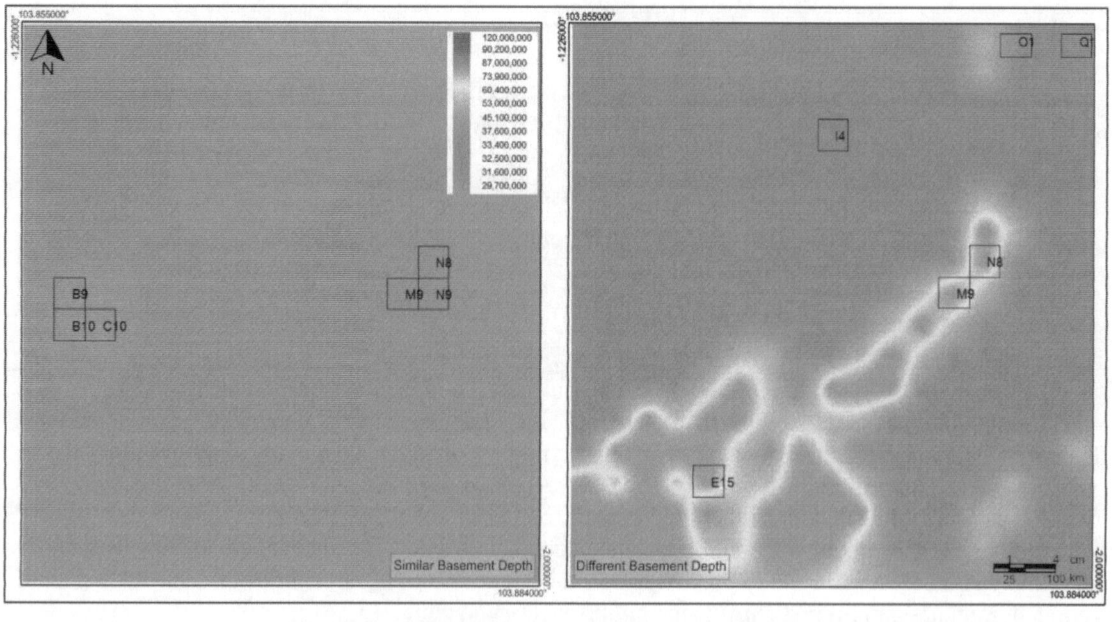

Figure 11. EGS Comparison in Zone D.

Comparison in figure 12 between similar and different D2 basement depth displayed that the highest potential values concentrated at same location around N8 and M9 whereas the minimum potential values were showed by different location. Calculation result with the similar basement depth gave square B9, B10, and C10 the lowest potential value but calculation result with the different basement depth showed that the lowest potential values were located in north-east part of zone D, on square I4, O1, and Q1. Overall, zone D had higher potential in middle to southwest. We analyzed that differential of potential value in zone D also relatively influenced by tectonics same as North Sumatera Basin.

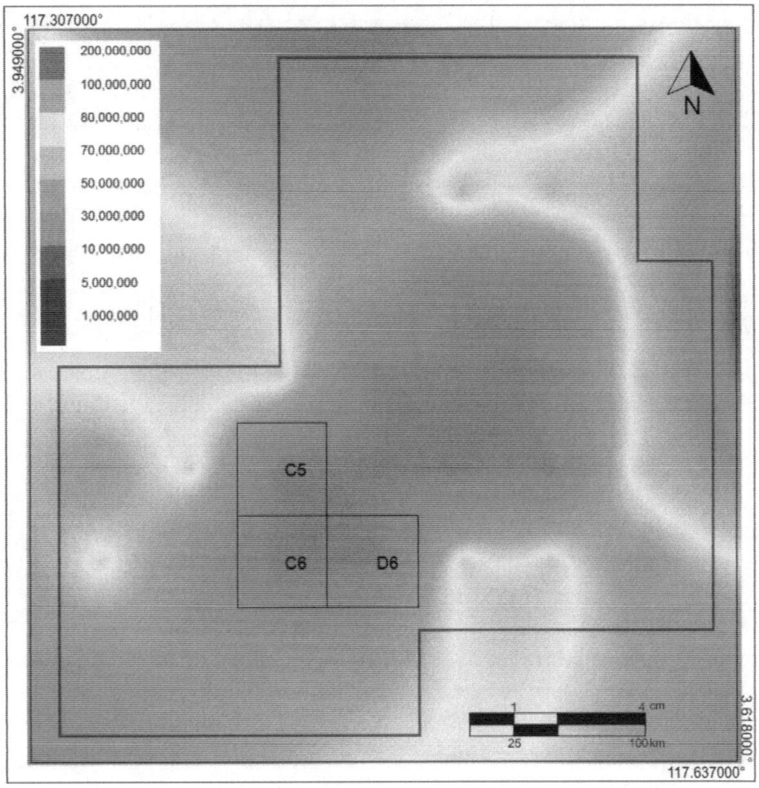

Figure 12. EGS Potential in Zone H.

Tarakan Basin had a uniform number of depths. Figure 13 showed square C5, C6, D6 bvc and surrounding area was highest potential value in several western Indonesia basins, amount more than 162,5 MW. Nevertheless, Tarakan Basin had a small potential zone, no more than 250km^2 of total area.

Based on analysis in the previous paragraph, we interpreted that South Sumatera Basin had a biggest potential area amount more than 650km^2 total area from zone D and E after correlation between similar and different basement depth calculation.

6. Conclusion

South Sumatra Basin was a biggest EGS potential area in Western Indonesia Basin amount 928.4m^2 even the high value existed in North Sumatera Basin. Basement depth is the most influential aspect of the EGS potential. Higher potential value of EGS relatively existed on extensional part from tectonics activity, not in contraction part, in case Tarakan Basin is formed by Makassar Strait rifting and other basins as the product of subduction between Australian plate and Eurasian plate. Basin which had non-volcanic geothermal manifestations could be developed by EGS method in thickest part of basin (high basement depth). Geological aspects such as lineament distribution, geological structure and lithology could not be the main parameters for assign EGS potential area. Measurement of basement depth value accurately using gravity or similar methods and conduct assessment with similar data variable and method would give better value of EGS potential assessment.

7. References

[1] All Heat Flow Data (Open Access) Royal Holloway South East Asia Research Group (http://searg.rhul.ac.uk/current-research/heat-flow/, accessed Desember 23rd, 2015)
[2] ASTER GDEM v2 Hillshade-Gray Global Elevation (Open Access) J Space System (SpatialOnDemand.com, accessed November 28th, 2015)
[3] Beardsmore G, Rybach L, Blackwell D and Baron C 2010 A Protocol for Estimating and Mapping Global EGS Potential *Australian Geothermal Conference 2010*
[4] Cameron N R, Djunuddin A, Ghazali S A, Harahap H, Keats W, Kartawa W, Miswar, Ngabito H, Rock N M S and Whandoyo R 1981 *Peta Geologi Lembar Langsa Sumatra Skala 1:250.000* (Bandung: Pusat Penelitian dan Pengembangan Geologi).
[5] Cameron N R *et al* 1983 *Peta Geologi Lembar Takengon, Sumatra, Skala 1:250.000* (Bandung: Pusat Penelitian dan Pengembangan Geologi).
[6] Cameron, N R *et al* 1982 *Peta Geologi Lembar Medan, Sumatra, Skala 1:250.000* (Bandung: Pusat Penelitian dan Pengembangan Geologi).
[7] Darman H, Sidi H F 2000 *An Outline of the geology of* Indonesia (Jakarta: Indonesian Association of Geologists).
[8] Gafoer S, Obrie T and Purnomo J 1986 *Peta Geologi Lembar Lahat, Sumatra, Skala 1:250.000* (Bandung: Pusat Penelitian dan Pengembangan Geologi).
[9] Keats W *et al* 1981 *Peta Geologi Lembar Lhokseumawe, Sumatra, Skala 1:250.000* (Bandung: Pusat Penelitian dan Pengembangan Geologi).
[10] Mangga S A, Santosa S, Hermanto B, 1993 *Peta Geologi Lembar Jambi, Sumatera, Skala 1:250.000* (Bandung: Pusat Penelitian dan Pengembangan Geologi).
[11] Massachusetts Institute of Technology 2006 *The Future of Geothermal Energy—Impact of enhanced geothermal systems (EGS) on the United States in the 21st century* (United States: MIT Press)
[12] Pringgoprawiro H and Sukido 1992 *Peta Geologi Lembar Bojonegoro, Jawa Timur, Skala 1:250.000* Pusat Penelitian dan Pengembangan Geologi Bandung Indonesia.
[13] Patra Nusa Data 2006 *Indonesia Basin Summaries (IBS)* (Jakarta: PT. Patra Nusa Data)
[14] Simandjuntak T O, Budhitrisna T, Surono, Gafoer S and Amin T C 1994 *Peta Geologi Lembar Muarabungo, Sumatra, Skala 1:250.000* (Bandung: Pusat Penelitian dan Pengembangan Geologi).
[15] Situmorang R I and Burhan G 1995 *Peta Geologi Lembar Tanjung Redeb, Kalimantan, Skala 1:250.000* (Bandung: Pusat Penelitian dan Pengembangan Geologi).
[16] Sukardi 1992 *Peta Geologi Lembar Surabaya & Sapulu, Skala 1:100.000* (Bandung: Pusat Penelitian dan Pengembangan Geologi).
[17] Susilohadi 1995 *Late Tertiary and Quartenary Geology of the East Java Basin, Indonesia* (Australia: University of Wollongong).
[18] U. S. Department of Energy 2008 An Evaluation of Enhanced Geothermal Systems Technology *Global Technologies Program. U. S. Department of Energy*
[19] U. S. Department of Energy 2012 What is an Enhanced Geothermal System (EGS)? *http://energy.gov/eere/geothermal/enhanced-geothermal-systems-0* (accessed December 23, 2015).
[20] Wight A W R, Hare L H. Reynolds J R, 1992 *Tarakan Basin, NE Kalimantan, Indonesia : a century of exploration and future potential* (Kuala Lumpur: Geological Society of Malaysia)

Subsidence: Causes, Effects, and Mitigations in Geothermal Field

Akta Sektiawan[1], Ganung Adi Prasetyo[2], Dida Patera Adli[3] and Ethis Yuantoro[3]

[1] Petroleum Engineering Department, Institut Teknologi Bandung, Bandung, Indonesia
[2] Geophysical Engineering Department, Institut Teknologi Bandung, Bandung, Indonesia
[3] Geological Engineering Department, Institut Teknologi Bandung, Bandung, Indonesia

E-mail: [1]akta.sektiawan@gmail.com

Abstract
Subsidence is the motion of the ground surface as it moves down relatively. It can occur in a wide range of area. It is an impact of production of large mass and volume of saturation from the reservoir. It doesn't happen especially in geothermal fields only, but also in oil and gas industry. Large fluid volume production leads to the decrease of pore pressure inside reservoir. This decline disturbs the pressure stability and overburden pressure compress the pores. It results the drop in ground surface. The decrease in ground surface level induces a devastating effect in the construction of some facilities, such as building, pipeline, canal, and river. It may interrupt the balance in the ecosystems nearby. Good management and several survey methods (such as levelling and gravity) will reduce the risk of subsidence and the other effects related to it. This discussion output can be used as a guide for minimizing subsidence impact in the geothermal field in general.

1. Introduction
Energy demand increases rapidly over the past year and will be predicted to escalate for the following years. Therefore, people try to harness energy from nature, or renowned as a renewable energy, because fossil fuel is limited for only couple of years.

The geothermal energy has been known as a renewable energy in Indonesia and researcher has found a large number of field which has potential geothermal manifestation and resources. Indonesian government plans to develop many of those fields in the near future. However, we have to develop it in a good management because despite it is a sustainable energy, bad reservoir management will cause a problem in the future, such as land subsidence. Subsidence happens because of excessive withdrawal of fluid and rock compaction. We will discuss about the cause, the impact, and how the subsidence mitigated generally.

2. Definition
The land subsidence is the movement land on the reference datum that causes of differences variable [1]. The land subsidence is caused by loading on the surface, the lose of the fluid when it exploited excessively, the quake that wreck the land structure, the instability of the land related by a process, etc.

There are two kinds of subsidence. There are endogenic subsidence and exogenic subsidence. The endogenic subsidence related with the force from the earth such as tectonic movement, folding and fault. The exogenic subsidence cause by the human activities such as exploitation the hydrocarbon, taking groundwater and changing on land composition.

Theory that support on this paper refers to Terzaghis Principle,1925 [2]. It will be explained with the following figure :

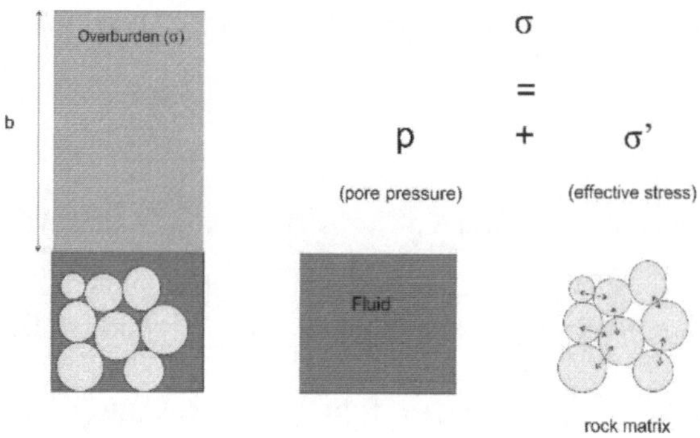

Figure 1. Overburden pressure concept [2].

Over burden is the total pressure of mass rock including pore pressure and effective stress. Pore pressure is pore related to cavity from the grain of the rock. Effective stress is related to grain rock contact. Compaction is related to decrease the pore pressure that causes the thickness of the rock reduce then affect the land subsidence. However, the failure compaction probably can be occurred. It is called disequilibrium compaction.

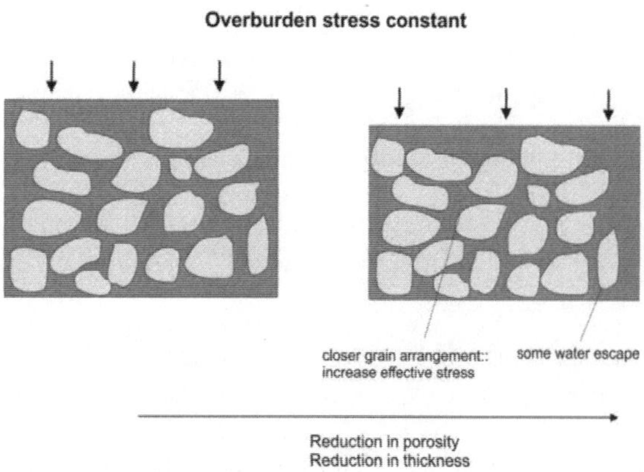

Figure 3. Compaction [2].

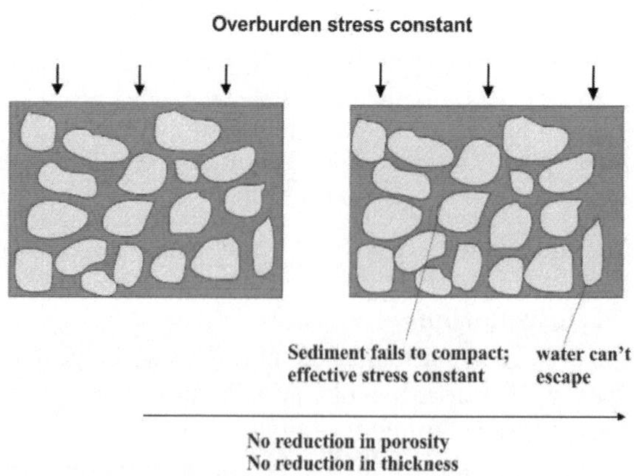

Figure 2. Disequilibrium Compaction [2].

3. Methods for identifying subsidence

In the paper that has been written, it uses two methods for identifying the subsidence in the geothermal field. The methods are levelling and gravity. The levelling has been used to know the size of the subsidence without knowing the inside of subsurface. The dimension of the subsidence is measured in length unit (m, feet). Then the gravity method is used to know the size of subsidence from the mass flow. This method knows where the brine (fluid) moving along with the mass flow. The size of the subsidence can be represented by the value of gravitation field that is measured in miliGal (mGal).

3.1. Levelling method

Levelling method is the one way to determine the elevation coverage the area. Several procedure must to be followed and would be explained in this section. The first is to set up the instrument within 100 meters (110 yards) of a point of known or assumed elevation. A rod or staff is held vertical on that point and the instrument is used manually or automatically to read the rod scale. This gives the height of the instrument above the starting (backsight) point and allows the height of the instrument (H.I.) above the datum to be computed.

The rod is then held on an unknown point and a reading is taken in the same manner, allowing the elevation of the new (foresight) point to be computed. The procedure is repeated until the destination point is reached. It is an usual practice to perform either a complete loop back to the starting point or close the traverse on a second point whose elevation is already known. The closure check guards against blunders in the operation, and allows residual error to be distributed in the most likely manner among the stations.

Some instruments provide three crosshairs which allow stadia measurement of the foresight and backsight distances. These also allow use of the average of the three readings (3-wire leveling) as a check against blunders and for averaging out the error of interpolation between marks on the rod scale.

The two main types of levelling are single-levelling as already described, and double-levelling (Double-rodding). In double-levelling, a surveyor takes two foresights and two backsights and makes sure the difference between the foresights and the difference between the backsights are equal, thereby reducing the amount of error. Double-levelling costs twice as much as single-levelling.

The curvature of the earth means that a line of sight that is horizontal at the instrument will be higher and higher above a spheroid at greater distances. The effect may be significant for some work at distances under 100 meters.

The line of sight is horizontal at the instrument, but is not a straight line because of refraction in the air. The change of air density with elevation causes the line of sight to bend toward the earth.The combined correction for refraction and curvature is approximately:

$$\Delta h_{meters} = 0.067\, D_{km}^{2} \tag{1}$$

$$\Delta h_{feet} = 0.021\, \left(\frac{D_{ft}}{1000}\right)^{2} \tag{2}$$

For precise work these effects need to be calculated and corrections applied. For most work it is sufficient to keep the foresight and backsight distances approximately equal so that the refraction and curvature effects cancel out. Refraction is generally the greatest source of error in leveling. For short level lines the effects of temperature and pressure are generally insignificant, but the effect of the temperature gradient dT/dh can lead to errors.

Assuming error-free measurements, if the Earth's gravity field were completely regular and gravity constant, leveling loops would always close precisely:

$$\sum_{i=0}^{n} \Delta h_i = 0 \tag{3}$$

around a loop. In the real gravity field of the Earth, this happens only approximately; on small loops typical of engineering projects, the loop closure is negligible, but on larger loops covering regions or continents it is not.

Instead of height differences, *geopotential differences* do close around loops:

$$\sum_{i=0}^{n} \Delta h_i g_i \tag{4}$$

where g_i stands for gravity at the leveling interval *i*. For precise leveling networks on a national scale, the latter formula should always be used.

$$\Delta W_i = \Delta h_i g_i \tag{5}$$

Equation (5) should be used in all computations, producing geopotential values W_i for the benchmarks of the network.

3.2. Gravity method

This method is based on the gravitational force. Gravitational force has been generated from two particles or more that have a number of mass then the particles should be have a pull force to each other [3]. The force can be represented by the following equation:

$$F=G\frac{m_1 m_2}{r^2} \tag{6}$$

where F is gravity force (N), G is Newton gravity constant, m_1 is first mass (kg), m_2 is second mass (kg), r is distance between first and second mass (m).

3.2.1. Correction. There are several correction that must be done to have an anomaly gravitation value. Instrument reading correction, which adjusts device reading by converting into miligal unit.

$$\text{Read (mGal)} = ((\text{Read (scale)-Interval}) \text{ x Counter Reading}) + \text{Value in mGal} \tag{7}$$

Tidal correction, which eliminates the gravitational influence of sun and moon that changes with latitude position and time

$$g_{st}=g_s+t \tag{8}$$

where g_{st} is tidal corrected gravity, $g_s$is gravity reading, and t = tidal corrected value.

Drift correction, which eliminates fatigue effect of the device that makes different gravity reading in the same measurement station at different time.

$$D_n=\frac{g_{st(n)}-g_{st(1)}}{T_N-T_1}(T_n-T_1) \tag{9}$$

where D_n is Drift in n-th station, $g_{st(n)}$is tidal corrected gravity at station-n, $g_{st(1)}$is tidal corrected gravity at station-1, T_N is last station measurement, T_n is initial station time measurement, andT_1 = n-station time measurement.

Lattitude correction, which corrects latitude position because of the shape of earth that makes different gravity measurement

$$g(\emptyset) = 978031.8(1+0.005304 \sin^2\emptyset+0.0000059 \sin^2 2\emptyset) \tag{10}$$

where $g(\emptyset)$ is latitude corrected gravity ($\emptyset$ = angle in radian).

Free air correction, which compensates point of measurement with datum (mean sea level).

$$g_{FA}= -0.3086 \text{ x } h \tag{11}$$

where g_{FA} is free air corrected gravity and h is elevation (m).

Bouguer correction, which compensates rock mass that is neglected when making free air correction

$$g_B= 0.04193 \text{ x } \rho \text{ x } h \tag{12}$$

where g_B is bouguer corrected gravity, ρ is density $\left(\frac{\text{gr}}{\text{cc}}\right)$, and h is elevation (m).

Terrain correction, which is done to get gravity reading in the surface without topographic impact around

$$TC\ (mGal) = 0.04191\ \frac{\rho}{n}[r_2-r_1+(r_1{}^2+z^2)^{0.5}-(r_2^2+z^2)^{0.5}] \tag{13}$$

where ρ is density $\left(\frac{gr}{cc}\right)$, r_1 is compartment inner radius (m), r_2 is compartment outer radius (m), and n is length of zone.

After several corrections, gravity's data become a data that called by Complete Bouguer Anomaly. Complete Bouguer Anomaly is measured in Gal unit. Complete Bouguer Anomaly should be process to get anomaly density variation as spatial function with kriging method and creating depth anomaly density section (then to be modelled the geolgy's strata and structure). Anomaly density variation (as figure 10 and 11) is the the one of final result after creating depth anomaly density section.

4. Subsidence in several geothermal field

4.1 Deformation analysis on macro and micro scale in oil field

Macro scale deformation are caused by pore pressure decrease and overburden pressure. Some studies describe the amount of subsidences [4]. They have correlated pressure drop to subsidence and time to subsidence on non-fractured reservoir. The datas are indicated below:

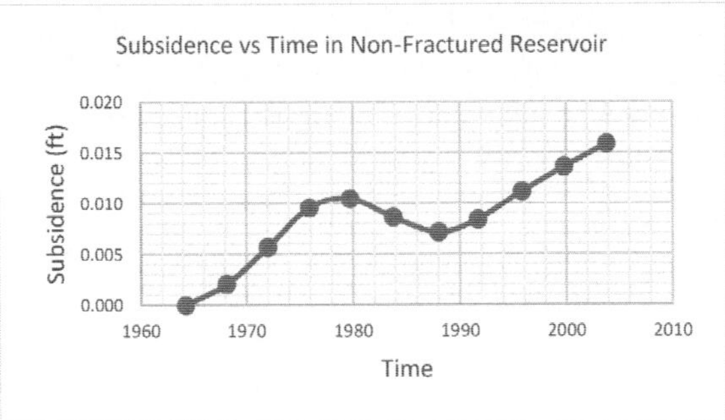

Figure 5. Subsidence variances during time on Doroud Field (Non-fractured reservoir) [4].

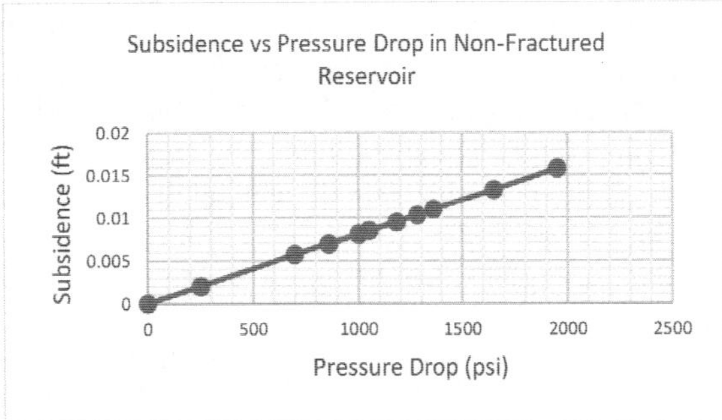

Figure 4. Subsidence variances vs. pressure drop on Doroud Field (Non-fractured reservoir) [4].

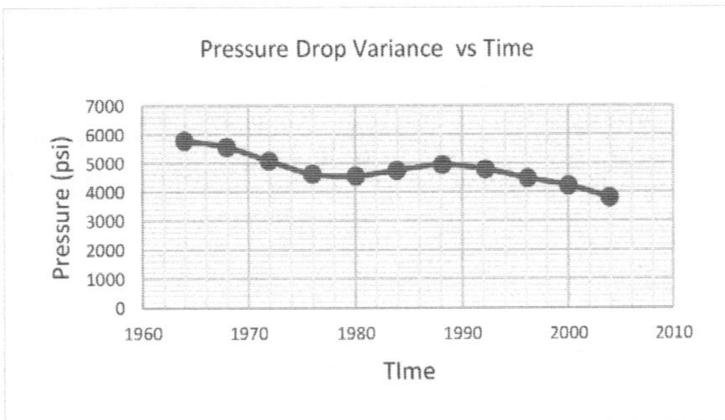

Figure 6. Pressure drop variances during time on Doroud Field [4].

Micro scale deformation is caused by fractured rock. The fractured rock properties are consist by aperture, orientation and fracture density. The same as before, they have generated the relationship between pressure drop, subsidence and time. The data are indicated below:

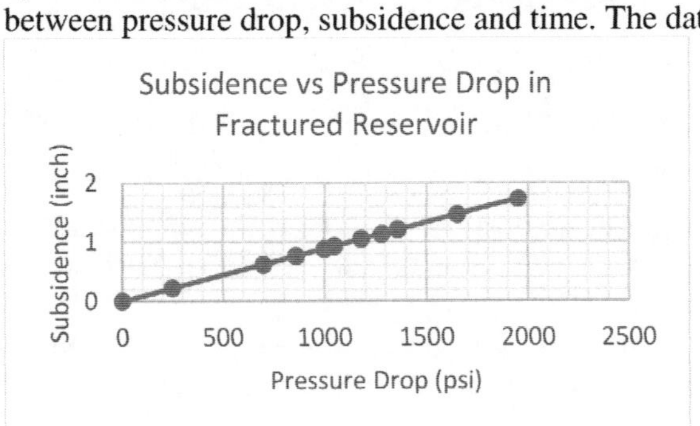

Figure 7. Subsidence vs pressure drop variances on Doroud Field (Fractured reservoir) [4].

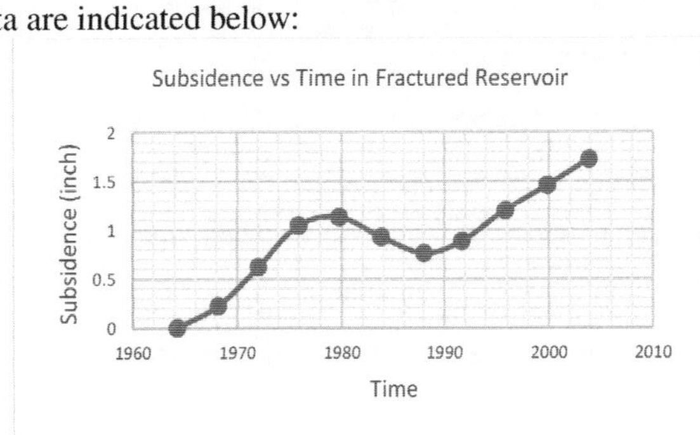

Figure 8. Subsidence variances during time on Doroud Field (Fractured reservoir) [4].

4.2 Levelling and gravity measurement in geothermal field

For example, we look closer to Reykjanes Peninsula, which is located in Iceland [5]. There are 3 high temperature geothermal fields, Svartsengi, Eldvorp, and Reykjanes and only Svartsengi which has been put into operation since 1975. The elevation and gravity has been monitored since 1976. Svartsengi geothermal field started operation in 1976 and is now producing 45 MWe electricity. The productive area as outlined by 7 productive wells is in 1 km^2. The average production in 1998 was 226 kg/s and 28 kg/s mass is reinjected into reservoir.

Some 220 benchmarks have been used in the outer part of Reykjanes peninsula for motoring subsidence after production in Svartsengi commenced in 1976. Every benchmarks was not used in every measurement since the initial strategy was to determine some part of the network or so. This led to the difficulty in comparing the result from one survey to another survey, therefore since 1992 all benchmarks was measured and the survey only repeated in longer period of time than before. The reference point is the northernmost point on SE-NW profile. Figure below shows us the outcome of survey at benchmark SN-H2, which indicated that it is the maximum subsidence measured during 1976-1999.

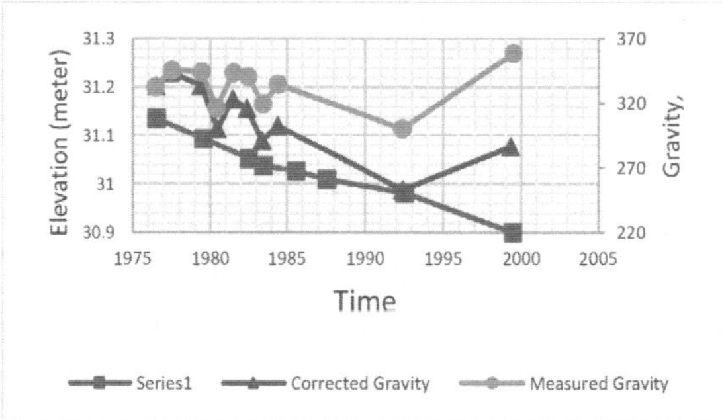

Figure 9. Elevation and gravity variation with time at benchmark SN-H2, Svartsengi geothermal field *[5].*

The following figure shows about the value of variation gravity. The minimal value is shown on red color to white. The red color to white color could explain why the geothermal field run into subsidence. The subsidence is closely related to the injection water and taking production on reservoir fluids. The little time-lapse gravity variation will be shown by the following figure :

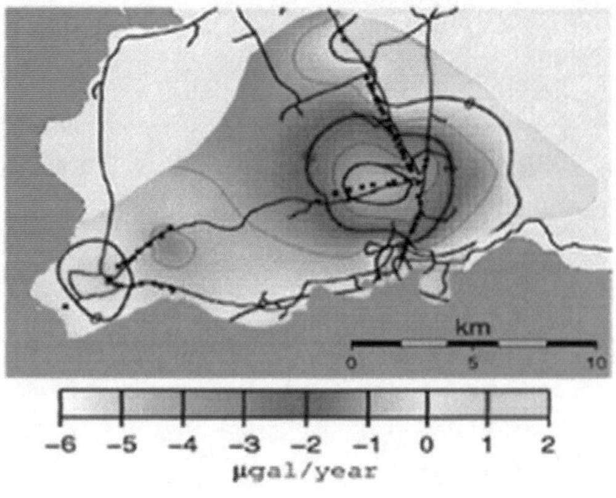

Figure 11. Mean gravity variation 1975-1999 *[5].*

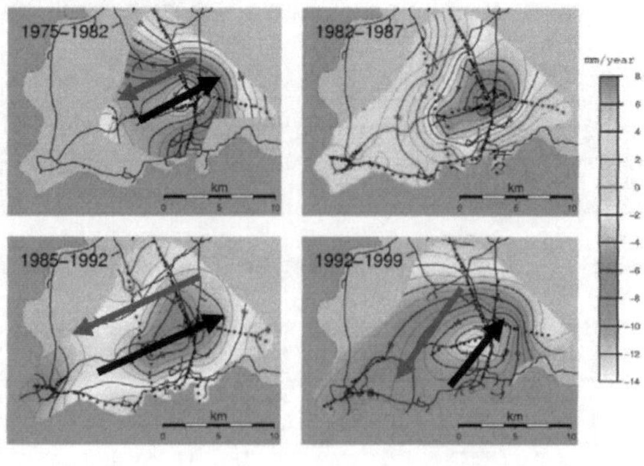

Figure 10. Subsidence rate at the outer part of Reykjanes peninsula for four times interval *[5].*

Figure 11. shows 4 periods of monitoring in 1975 to 1999. The first monitoring is in 1975-1982, the figure shows that the subsidence is high from the sign of red-white color. The trend of the over subsidence is indicated by the sign of black arrow. The cause of highly subsidence is strongly related by the water withdrawal on geothermal well. The second monitoring in 1982 to 1987, the figure shows that the subsidence happened around the well production. The enclosed contour represented the subsidence over the area has diminished to become more stable than. This condition related by reducing of the production reservoir fluid and injecting more water. The third monitoring is in 1985-1992, the figure shows that the subsidence spreads to a larger area, which is shown by the green color contour from the second monitoring changes to the yellow-red contour that has more negative value. The forth monitoring in 1992-1999 shows that the subsidence is high from the sign of red-white color. The trend of the over subsidence is similar to the direction of the first monitoring.

Another field example case is the Wairakei field, NZ [6]. The area produce heat with maximum fluid withdrawal of 2300 kg/s which started the production in 1950. The subsidence is started in 1960 after 10 years production (Figure 12). Benchmark A97 in the eastern borefield, and at P128 in the centre of the subsidence bowl. The rate of the subsidence has increase until in maximum poin in 1970s (Figure 13). But after that the subsidence rate has decrease because the situation in the aquifer already stable.

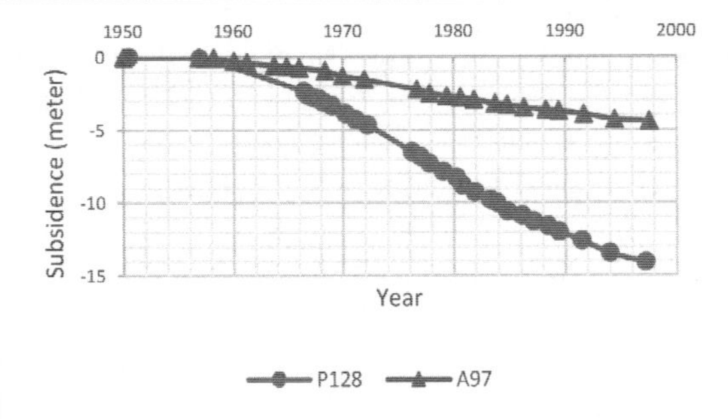

Figure 13. Subsidence and time diagram in Wairakei [6].

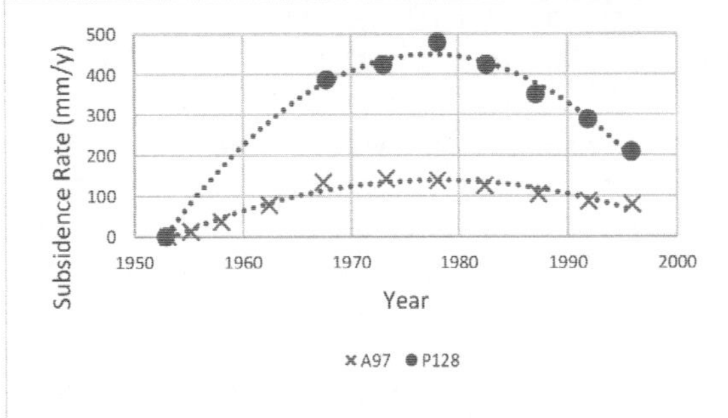

Figure 12. Subsidence rate and time diagram in Wairakei [6].

Mass and heat flow rate went up to the peak in 1960s. When the subsidence began to take place, mass and heat flow rate decreased gradually to be more stable than before. As can be seen from figure 12 and 13 above, rate of subsidence is significant because the pore pressure decreases overtime and there was no fluid injection to the reservoir. Injection process was undertaken in 1993 and Figure 14 below tells that after injection, the reservoir pressure remains constant.

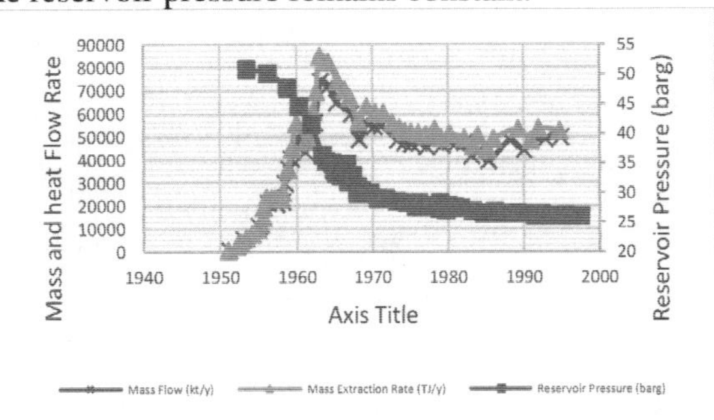

Figure 14. Mass and Heat flow rate and reservoir pressure vs time in Wairakei [6].

5. Impact of subsidence in the surface

5.1 Flooding

Subsidence probably doesn't cause any damage and can only be observed by using precise survey. However, slight subsidence which occurs in flat area or near a water body can cause specific environment effect nearby. The effect impacts dramatically when it happens close to sea or lake. The water from them can flow to the region which is lower and overwhelms the area. The remedy action should be prepared to restore the area, such as pumping water out of the swamp, constructing sewers, repairing the well, and building embankment which cost lots of money [7].

5.2 Structural effect

Building. Building structure usually provide maximum limits in deformation, which is based on total load applies in the beam or floors movement. Building component such as doors or windows,will alter because of excessive sags and the function will reduce. The effect of subsidence in building can be felt if the subsiding rates is different in respect to the building. The floors doesn't remain in a plane that leads to cracking of components or leakage in drainage. But, uniform subsiding rate will not be noticed because of no or slight damage [7].

Surface facilities. The effects of subsidence is most noticeable on long structure like pipelines, drainages, roads and powerlines. Pipelines are impacted by compression and tension forces due to land depression, therefore such a treatment needs to be undertaken. It is necessary to realign the pipelines by removing some sections of the pipe or replacing the stretched part of the pipe.

An example of the effect in drainage is the concrete at Wairakei [7], separated water drainage channel that has been damaged by compression in the edge of subsidence bowl. Repairing action has been undertaken and it is mitigated by placing sliding joints where the drainage descends to the Wairakei stream.

There are two high-voltage transmission lines which cross over maximum subsiding area in Wairakei. The tilting of some of the stake is extreme that is necessary to undertake a remediation as the lines are over-tension. It causes the insulators to rotate more than operational limits. Movement of the pylon at the top is about 1 meter for 24 meters high pylon [7].

Roads experience a moderate damage caused by subsidence. Near steam main overbridge, there has been a cracking in the pavement of State Highway 1 near Wairakei [7]. The lateral movement is up to 40mm.year, with subsidence of 100 mm/year in steam main bridge. Pavement cracking is still taking place proximate to the steam main bridge. A narrow depression across the highway, parallel to the bridge abutments is found a damage. Subsurface cracking may cause the failure. The direction of cracking is toward the center of subsidence and about perpendicular to the highway.

6. Mitigation

As shown in the field example of Wairakei that maintain a constant after injection since 1993, injection strategy has proved to be the best practice to overcome and reduce subsidence rate. An injection management should be adapted and configured based on the field condition [8]. The goal is to control and restrict excessive induced stress and strain changes without losing the long-term benefits of reservoir development for sustainable energy extraction.

7. Conclusion

Subsidence happens because of pore pressure decreases, so the effective stress goes up. The physical effect related to it is pore volume drops and compression takes place in the rocks.

Monitoring activity for alleviating subsidence is undertaken by leveling and gravity measurement. Both measurement should be performed regularly in a period of time. The result then is analyzed to predict the subsidence pattern over the measured area. It is known that fluid withdrawal is strongly related to subsidence. Blue arrow as shown in figure 11 indicates fluid movement and black arrow shows subsidence magnitude from the smallest to the biggest number. A benchmark point (SN-H2) which is located in the northern part of Svartsengi Field has the biggest subsidence level as indicated in figure 9.

The impact of subsidence which had been observed in some geothermal field in New Zealand are flooding in an area near water body and structural deformation in building which is located in the subsidence-influenced area as well as pipeline and transmission grid failure.

Monitoring activity for alleviating subsidence is undertaken by leveling and gravity measurement. Both measurement should be performed regularly in a period of time. The result then is analyzed to predict the subsidence pattern over the measured area. It is known that fluid withdrawal is strongly related to subsidence. Blue arrow as shown in figure 11 indicates fluid movement and black arrow shows subsidence magnitude from the smallest to the biggest number. A benchmark point (SN-H2) which is located in the northern part of Svartsengi Field has the biggest subsidence level as indicated in figure 9.

Injection management also should be undertaken that is based on specific needs of the field, because it has been proved that it could maintain reservoir pressure, as shown in Wairakei field in figure 14.

References

[1] Marfai MA, King L 2006 Impact of the Escalated Tidal Inundation due to Land Subsidence in a Coastal Environment *Geophys Res Abstract* **8**

[2] Osipov VI 2015 *Physicochemical Theory of Effective Stress in Soils* (Cham: Springer International Publishing)

[3] Telford WM and Geldart LP and Sheriff RE 1990 *Applied Geophysics* (Cambridge: Cambridge University Press) pp 6-61

[4] Atashbari V and Ahmadi M 2007 *SPE Production and Operation Symp.* 31 March - 3 April Oklahoma City

[5] Eysteinsson H 2000 *Proc World Geothermal Congress* 28 May - 10 June Kyushu pp 559-564

[6] Allis RG 2000 *Geothermics* **29** Issues 4-5 pp 455-478

[7] Bloomer A 2001 Effects of Geothermal Induced Subsidence *Proc. 23rd NZ Geothermal Workshop*

[8] Chris BJ 2005 Advances in Enviromental Management of Geothermal Developments. *Proc. World Geothermal Congress* 24-29 April Antalya

Feasibility of Geothermal Energy Extraction from Non-Activated Petroleum Wells in Arun Field

M Syarifudin[1], F Octavius[1], K Maurice[1]

[1] Department of Petroleum Engineering, Institut Teknologi Bandung, Bandung 40132, Indonesia.

E-mail: msyarifudin22@yahoo.com

Abstract. The big obstacle to develop geothermal is frequently came from the economical viewpoint which mostly contributed by the drilling cost. However, it potentially be tackled by converting the existing decommissioned petroleum well to be converted for geothermal purposes. In Arun Field, Aceh, there are 188 wells and 62% of them are inactive (2013). The major obstacle is that the outlet water temperature from this conversion setup will not as high as the temperature that come out from the conventional geothermal well, since it will only range from 60 to 180°C depending on several key parameters such as the values of ground temperature, geothermal gradient in current location, the flow inside of the tubes, and type of the tubes (the effect from these parameters are studied). It will just be considered as low to medium temperature, according to geothermal well classification. Several adjustments has to be made such as putting out pipes inside the well that have been used to lift the oil/gas and replacing them with a curly long coil tubing which act as a heat exchanger. It will convert the cold water from the surface to be indirectly heated by the hot rock at the bottom of the well in a closed loop system. In order to make power production, the binary cycle system is used so that the low to medium temperature fluid is able to generate electricity. Based on this study, producing geothermal energy for direct use and electricity generation in Arun Field is technically possible. In this study case, we conclude that 2900 kW of electricity could be generated. While for-direct utility, a lot of local industries in Northern Sumatera could get the benefits from this innovation.

1. Introduction

The suitable well characteristics and abundance of ex-petroleum wells in several fields are why they are potentially a beneficial source of geothermal energy. Sumatera is a part of Pacific Ocean's "Ring of Fire" and scattered with both active and dormant volcanoes. Estimates of the magnitude of the available Indonesian geothermal resource vary greatly. The 2007 West Japan Engineering Consultants study estimated the exploitable potential across 50 fields at 9,000 MW. In 2011, MEMR revised the country's geothermal potential to 29,215 MW from 27,000 MW a decade earlier, indeed, the 27,000 MW figure is cited in many World Bank reports, and appears to be the basis for claims that Indonesia possesses 40% of the world's geothermal resources.

However, no matter what report that we use, Indonesia still has abundant geothermal resources that can help meet the country's rising electricity demand and increase the electrification rates. Indonesia's estimate of conventional geothermal resource base is generally considered to be among the largest in the world. This is why Indonesia has the potential to use geothermal resources not only to generate electricity but it can also be used to utilize the direct use and most of this potential is located in the Island of Sumatera.

Indonesia has a large number of geothermal fields and most ambitious plans for accelerated geothermal power development in the world: 4,000 MW by 2014-2015 and 9,500 MW by 2025. Figure 1 is a map of the country showing the most well-known geothermal sites.

However, there is a perception among some geothermal stakeholders and sponsors outside Indonesia that the resource related risk in geothermal development in Indonesia is substantial. Yet, we find no statistical basis for this perception, even though there are some 100 known geothermal fields and prospects in the country, and nearly 300 deep wells. This ambitious plan will require strong government support to materialize. Any shortfall in the expansion of geothermal power generation capacity will most likely be met by additional coal-fired power plants. Over the past decade, the government has intensified its efforts to scale-up and speed-up geothermal power development.

Despite government efforts, progress in the last few years was slow. The perception that the Indonesian geothermal program has stalled is widespread, and exists among many stakeholders. From 2010–2013, only 135 MW was added, and the best estimation suggests that by the end of 2016, no more than an additional 190 MW is likely. No power purchase agreements (PPAs) were signed under the 2012 FIT (feed-in tariff).

One of the main problem to develop geothermal is usually came from the economical viewpoint. The cost will mostly came from the drilling cost to make the well, which responsible for a big chunk of the project cost. Over the last decade, geothermal drilling costs have increased at unprecedented rates, which cannot be explained merely by any increase in depths or general inflation. The driving force has been the demand for drilling in the oil and gas sector, which has been reactivated by the increase in global oil prices and (in the US) by drilling for shale oil and gas (fracking).

Figure 2 is a histogram of the drilling cost per MW of successful wells in Indonesia, based on the correlation in GeothermEx (2004). It shows that the drilling cost per MW well capacity in Indonesia ranges from US$100,000 to nearly $2,000,000 the most probable range being $300,000 to $400,000 per MW.

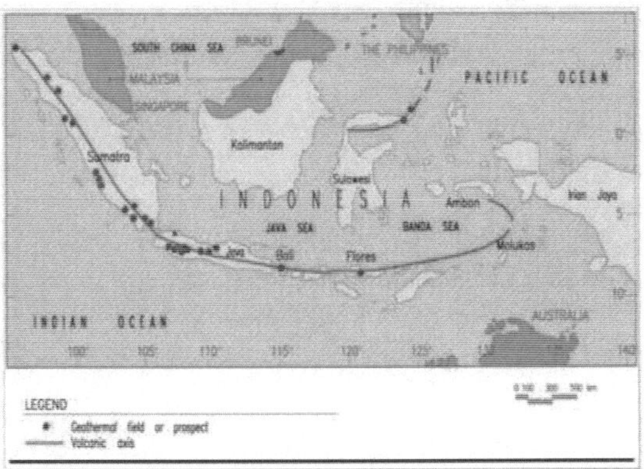

Figure 1. Selected Indonesian geothermal areas (Sanyal S K et al 2011)

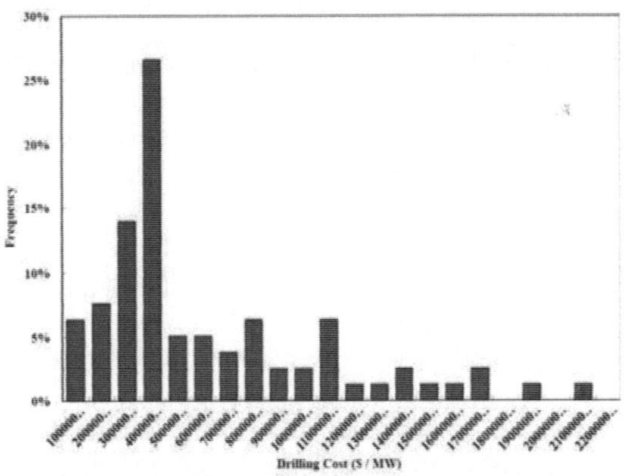

Figure 2. Successful wells drilling cost per MW, Indonesian fields (Sanyal S K et al 2011)

When a petroleum reservoir is depleted beyond an economically feasible point, the wells will be abandoned, decommissioned, and reclaimed. These wells are sometimes referred to as abandoned or inactive wells. Abandoned petroleum wells are an enduring liability to oil/gas companies, as these companies are responsible for the possible environmental contaminations and litigations. Indonesia also has many these abandoned wells (around 13,097).

Compared to a conventional geothermal system, repurposing an abandoned petroleum well for a geothermal project can reduce the capital costs of the project by up to 50% (Bu and Ma 2012). It can also make a constructive use for an enduring liability. The capital costs for a geothermal project can be further diminished by employing a closed loop configuration (i.e. closed circuit of pipes), as it can be retrofitted to a single borehole. Alternatively, an open loop configuration requires at least two boreholes to operate, namely an injection well and an extraction well. Additionally, a closed geothermal loop does not require a water management system, compared to a compulsory water management system for open loop configurations. Closed loop configurations will also need less pumping power compared to an open loop configuration which requires overcoming gravitational head. Moreover, closed loop configurations can use a non-aqueous fluid to increase the rate of heat exchange with the earth.

The present study aims to do an early feasibility and engineering design study in optimizing the utilization of decommissioned wells to generate electricity and direct use in Arun Gas Field. The heat extraction on the outlet fluid temperature are determined. Total electricity generation potential are estimated using comparison method. Potential for direct use utilizations around this field are also investigated.

2. Conversion Issues

Unique measures have to be done in solving several technical constraints in converting a petroleum well into geothermal well.

Firstly, the reservoir area in a petroleum well will contain toxic and dangerous substances such as brine, oil, gas or even hydrogen sulfide. It will be a disaster if those fluids rise into the surface or contaminate the fresh water zone. Before conversion can be run, all connection between the well and formation has to be terminated. One way to do it is using cement plug. We must ensure that these plug is properly installed and will not leak or fail in the future.

Secondly, another issue in this conversion is the temperature. Most petroleum wells do not located in a high earth temperature gradient area. Thus their formation temperature is not as high as what available in the conventional geothermal wells, even in the bottom of those wells. Technologies such as binary cycle will be needed in order to utilize these wells for electricity generation.

3. Result

3.1. Indirect use

The best geothermal configuration for this conversion case is closed loop binary cycle which can be seen in Figure 3.

The fluid temperature that are expected from those converted wells will not be what we call as high temperature for electricity generation use. For this reason, the most prominent way of utilizing them is by using the binary cycle technology. Binary cycle is able to produce electricity using low temperature outlet fluid because it uses organic fluid (e.g. isobutene, isopentane or ammonia) in its system which has a very low boiling point compared to fresh water, thus making it possible to spin the turbine even with low temperature fluid.

Closed loop well configuration is used in term of increasing the efficiency. This means that it will neither produce nor inject any fluid into or from the formation. It helps to reduce impacts for the environment. Also, if compared to the conventional geothermal open loop, this closed loop will not let the fluid inside its pipe to interact with the environment so it minimize the possibility of contamination into the freshwater zone. Second reason is that it will prevent production related problems to happen such as scaling, pressure depletion or subsidence.

Another move is to improve the heat harvested from the formation by using spiral type heat exchanger in the bottom of the well. This type has proven to be the best configuration when compared with W-type, 2U-type or U-type heat exchanger, in case of efficiency. Average heat exchange rates for 100 hours of U, W 2U, and spiral type GHEs were 35.71 W/m, 40.76 W/m, 39.03 W/m, 76.80 W/m respectively (Lee, Seung-Rae et.al., 2013).

For its conductive grout around the spiral, silica sand is selected because it has a good conductivity and common setup for ground heat exchanger.

An analytical solution can be used, however some assumptions are needed. The first assumption is that because the spiral is located in the bottom of the well (ex-reservoir zone), we used constant average reservoir temperature instead of depleting depth-dependent temperature. The second assumption is that there is no heat loss in this system because insulators are used to keep the precious heat and improve efficiency. The third assumption is the heat flow from the reservoir is equal to the heat given to pipe wall as well as fluid.

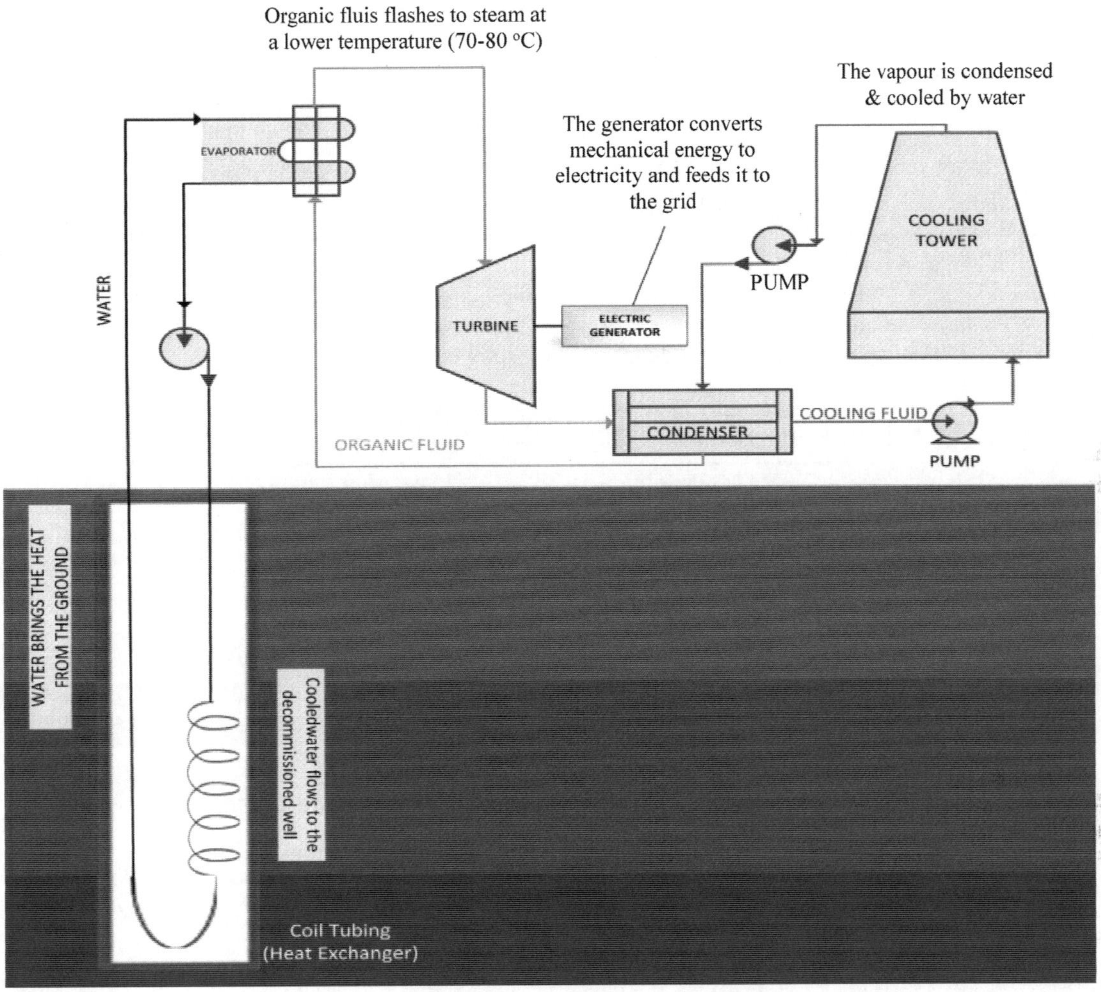

Figure 3. Process flow schematic for closed loop binary cycle.

Heat flow from the pipe wall in a length (dx) is expressed by (1):

$$\dot{q}\pi D dx = h\pi D(T_0 - T)dx \tag{1}$$

The heat given to the fluid is given by (2):

$$\rho u_m c_p \frac{\pi D^2}{4} dT = \dot{m} c_p dT \tag{2}$$

By setting both of the equation equal and integrating them from the start of the heat exchanger, we get (3):

$$\int_{T_1}^{T_2} \frac{dT}{T_0 - T} = \int_0^x \frac{4h}{\rho u_m c_p D} dx \tag{3}$$

The result of integration for distribution along the heat exchanger pipe is (4):

$$ln\left(\frac{T_0-T_2}{T_0-T_1}\right) = -\frac{4hx}{\rho u_m c_p D} \tag{4}$$

$$\frac{T_0-T_2}{T_0-T_1} = e^{-\frac{\pi h D L}{\dot{m} c_p}} \tag{5}$$

Where c_p is water specific heat (kJ/kg.K), D is heat exchanger pipe inside diameter (inch), T_0 is heat exchanger pipe temperature (°C), T_1 is water temperature at heat exchanger inlet (°C), T_2 is water temperature at heat exchanger outlet (°C), h is heat transfer coefficient (W/m^2.K), L = heat exchanger pipe length (m), $\dot{m}$ is mass flow inside the heat exchanger (kg/s), u_m is overall heat transfer coefficient (W/m^2.K), and ρ is fluid density (kg/m^3).

A good simulation for this is by doing it for the case in Arun Field, Aceh. Arun is a very mature gas field in Northern Sumatera that is currently on its closing phase. Their field profile can be seen in Table 1. The properties of spiral coil tube heat exchanger are shown on Table 2.

Table 1. Profile of Arun Field. (Soeparjadi R A 1982)

Profile	Spesification
Producing formation	Limestone
Mean Depth, m	3,048
Productive area, acre	21,450
Average net pay, m	153.3
Average pressure, psi	7100
Average temperature, °C	178
2013 producing well	72
2013 non-producing well	116

Table 2. Example specification of coil tube heat exchanger used in this case.

Design	Spesification
Heat exchanger diameter, m	0.152
Helical height, m	0.5
Helical length, m	0.692
Total helical	306.6
Total heat exchanger pipe length, m	212.248
OD, inch	1.25
ID, inch	1.1
Average water c_p, kJ/kg.K	4.2
Silica sand h, W/m^2.K	1.4

There is also a rule of thumb of maximum flow rate constraint inside a 1¼-inch coil tubing which is approximately 1 barrel/min (Ackert, D. et al., 1989).

T1 or the heat exchanger inlet temperature is the independent variable, while T2 is the dependent variable. Figure 4 shows the result of equation (5) adjusted with the condition and setup in Arun Field. We get the relationship between the inlet and outlet temperature of the spiral heat exchanger as:

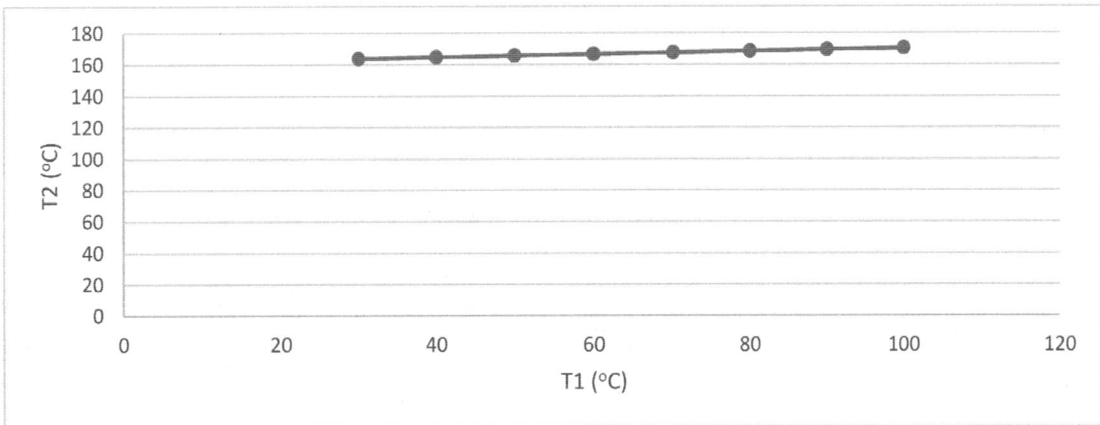

Figure 4. Outlet temperature vs inlet temperature of spiral heat exchanger in Arun Field.

Table 3. Binary cycle plants around the world (Franco A and Villani M, 2009)

Plant	T2 (°C)	Working Fluid	Brine Consumption	Gross Capacity (kWe)
Nigorikawa, Japan	140	R114	50 kg/s	1000
Otake, Japan	130	Isobutane	14.7 kg/s	1000
Husavik, Iceland	124	NH_3-H_2O	90.1 kg/s	1700
Nagqu, China	110	Isopentane	69 kg/s	1000
Altheim, Austria	106	C_5F_{12}	86 kg/s	1000
Wabuska, USA	104	Isopentane	60.025 kg/s	1750
Chena, USA	74	R134s	23.16 kg/s	400
Kuthya-Simav, Turkey	145	R124	121.8 kg/s	2900

Which means that the average outlet fluid from a single conversed well in Arun is 2.65 kg/s with temperature of 170°C. If all the non-produced wells in Arun field are utilized, this conversion might produce about 296.96 kg/s of 170°C fluid. For comparison of the other binary cycle plants in the world that can be seen in Table 3.

It is clear from our result that Arun's fluid production is greater and hotter than the biggest producer on Table 3, which is in Kuthya-Simav, Turkey. So it is realistic if we expect that the wells conversion in Arun is able to generate electricity more than 2900 kW.

3.2. Direct use

Geothermal also can be used as direct use usually for ground temperature normally below 180°C or in a case where the temperature is not hot enough to produce sufficient electricity. There are many examples of direct use and each of them has its own optimum temperature. One of the references about direct use activities that can be done at a certain temperature range is known as Lindal Diagram (Gudmundsson, et all, 1985*)*. Based on this diagram, the direct use activities for the region around Arun gas field can be projected.

Basically, Geothermal is used to replace conventional energy like fossil fuel (oil or gas), wood, etc. Benefits we can get using geothermal are:

1. Less cost than most of other energies. According to Green Mountain Geothermal in 2014, Geothermal price is around 10$/MMBtu, while wood is 14.65$/MMBtu, Natural Gas 18.28$/MMBtu, and Fuel Oil $34.88/MMBtu.

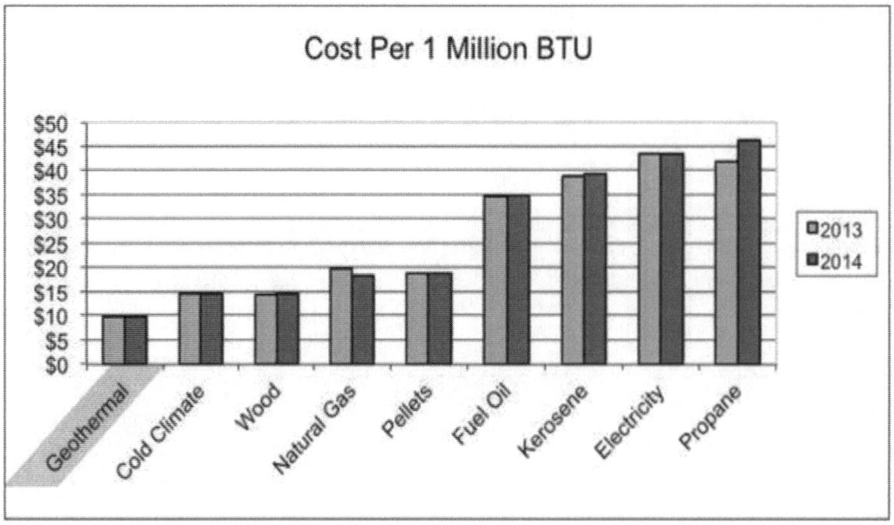

Figure 5. Energy cost per 1 Million BTU (Green Mountain Geothermal, 2014)

2. High efficiency.
3. Less pollution. The emission of geothermal mainly only Carbon Dioxide and it is usually treated to reduce the concentration so it can be released to the air safely. While the other fuel also release dangerous gas like Carbon Monoxide, Sulfide Gas, etc.

Through a quick research, various activities around Arun Field will be able to gain benefits from the well conversion. Uses of geothermal in this area can be categorized into drying, heating, and cooling process.

3.2.1. Drying. In drying process, geothermal may able to replaces fossil fuel like oil and gas or other fuel, including wood. Temperature for this purpose usually around 95°C – 121°C. Benefits we can get using geothermal are:

1. Less cost compared to oil and gas
2. More efficiency compared to wood
3. Less pollution

Examples of industries in North Sumatera that have been existed and need drying process in their business are:

1. Lumber industry: To dry lumbers, after been cut, so it can continue its next processing steps.
2. Pulp & paper industry: Pulp drying process.
3. Desalination: To dry the seawater in producing salt.

3.2.2. Heating. This utilization is similar to drying, but in heating the temperature needed is not as high as in drying, only about 38°C – 66°C. Just as in drying, several benefits can be acquired from using geothermal in heating, such as lesser cost and pollution.
Industries in Sumatera that have heating process include:

1. Biogas: To maintain optimum temperature in biogas processing.
2. Aqua cultural: To be flowed into fish ponds in order to help stimulating growth of the fish.
3. Recreational: To be used for recreational aqua activities.

3.2.3. Cooling. Beside for heating or drying, geothermal also can be used for cooling.

1. Building Cooling and Water Heating
2. Refrigeration

3. Soft Drink Carbonization: Lower temperature give better result because more gas CO_2 can be dissolved. Optimum temperature is around: 2 – 5°C.

Building cooling and water heating also refrigeration can be used not only for industry. These functions are common needs for everyone. But, if industry use these functions, the other benefit is it can reduce the operating cost. For example, a pulp and paper industry may not require cooling process in pulp and paper processing. But the function of geothermal in cooling can be used to cool the building or replace the air conditioner. With this, the operating cost for these things also can be cut so in overall, lower operating cost. As long as heating or cooling are needed in industries, there are wide possibilities that geothermal from this petroleum well to be useful.

4. Conclusion

This study used several assumptions such as neglecting heat loss, steady state heat source, uniform reservoir condition, sustain production, and single phase fluid flow. Thus, a more comprehensive and complex studies have to be done in assessing the next project feasibility. Economic assessment must also be made to know the worthiness of the project.

1. Petroleum well conversion in Arun Field is technically feasible due to its abundance of decommissioned wells and relatively high reservoir temperature.
2. To tackle some constraints in indirect utilization, special configuration is needed which involves heat exchanger, closed loop cycle and binary cycle generation.
3. Conversion in Arun Field is able to produce 2.65 kg/s, 170°C fluid, thus able to generate more than 2900 kW of electricity.
4. Using outlet temperature fluid around 170°C, several direct applications that can be used for industries around Arun Field are drying, heating, and refrigeration.

References

[1] Franco A and Villani M 2009 *Optimal Design of Binary Cycle Power Plants for Water-Dominated, Medium-Temperature Geothermal Fields* Geothermics 38 p 28
[2] Ackert D et al 1989 The Coiled Tubing Revolution *Schlumberger Oilfield Review* vol 1 Number 3 pp 4-16
[3] Soeparjadi R A 1982 *Geology of the Arun Gas Field* SPE-10486-PA
[4] Ghoreishi-Madiesh S A et al 2014 *Geothermal Energy Extraction from Decommissioned Petroleum Wells*
[5] Lee S R et al 2013 *Evaluation of Heat Exchange Rate for Different Types of Ground Heat Exchangers* (Alaska)
[6] Sanyal S K et al 2011 *Geothermal Resource Risk in Indonesia a Statistical Inquiry*
[7] West Japan Engineering Consultants 2007 Master Plan Study for Geothermal Development in the Republic of Indonesia
[8] Green Mountain Geothermal 2015 *Heating Fuel Cost Comparison per MMBTUs* (Vermont)
[9] Gudmundsson Jon S, Freeston Derek H, Lienau Paul J 1985 *Geothermal Resources Council Transactions* vol 9 Part I: The Lindal Diagram

Decline Curve Analysis for Production Forecast and Optimization of Liquid-Dominated Geothermal Reservoir

I Hidayat

Petroleum Engineering Department, Institut Teknologi Bandung (ITB), Jl. Ganesha No.10, Bandung 40132 - Indonesia

Email : ikihidayah@gmail.com

Abstract. Power projects in the geothermal field has a long span of about 30 years. The power supply should be maintained at a certain value across a range of time. A geothermal field, however, has the characteristics of natural production decline with time. In a geothermal field, development of decline curve model of steam production is important for forecasting production decline in the future. This study was developed using decline curve by production data along 3 years liquid-dominated geothermal reservoir in Ulubelu field. Decline curve in geothermal field based on decline curve in petroleum industry. The decline curve was correlated by reservoir management in geothermal. The purposes of this study to get best match model decline curve and forecasting production in the future. Based on decline curve analysis by production data in Ulubelu field, the result model decline curve is exponential model. From the model, we can get the value of decline rate in the field is 9.4 %/year. Then, the formula of forecasting steam flow used exponent decline to forecast in the future. By using separated system cycle in Ulubelu field, the minimal steam flowrate towards turbine was 502018.4 ton/month. Based on formula of forecasting production and minimal steam flowrate, we can get the time make up wells to maintain steam supply for stability in generator power capacity.

Keywords: decline production, curve, liquid- dominated

1. Introduction

Indonesia is a country with a plenty resource of geothermal energy which is also as the biggest one in the world. Geothermal energy potential resources which is reach the value of 28,910 MW (Indonesia Energy Outlook, 2014). The mostly reservoir type in the world is water dominated. Especially in Indonesia, water dominated system is majority in geothermal system about 80% (Saptadji, 2015).

1.1 Production Management

A geothermal field that are developing and start to production, management of production and injection operation is important thing to be done. This is a multidisciplinary tasks involving production monitoring, geochemistry, reservoir engineer, and simulation to get cumulative field data in production and injection that measured in subsurface and surface. One of tasks is to maintain sufficient supply of steam to electrical generator.

Electrical power project in geothermal field has time range about 30 years. All this information must be brought together into a coherent model that can be used to predict future production, injection capacity, and to identify any areas where remedial action must be planned in order to maintain the production plant at full capacity. These actions range from scheduling from simple well cleanout to remove calcite scaling to moving the injection well field to avoid returns of cool fluids between injection and production wells.

Some data will be available frequently: wellhead pressure (WHP), valve settings, separated-steam and brine flows, total flow (mass and enthalpy) from separator stations, injection well flows and temperatures, and so on, together with reservoir monitoring data that mainly comprise pressures in observation wells. Other data may be available only intermittently, such as mass flow rate, enthalpy, and chemistry of individual wells.

In production wells, if the production WHP and enthalpy remain constant, the mass flow changes in response to changes in reservoir pressure. This is cause of production decline, initially rapid and then flattening out as time goes on. A change in trend indicates some change in the well that may require further investigation. The model production decline can identify some problem in the field with look the trends. A detail of the model provides a better predictor of performance and give explicit ability to model effect of particular changes, such as deposition or cooling of one feed zone.

1.2 Decline Curve

Production model is obtained from decline curve analysis. Initially decline curve analysis is used in petroleum industry and then used in geothermal field because the same principal. The decline model discusses a model or trend-fitting of well performance in order to establish a simple model to project future performance. Such trend analysis provide the best projections for short term extrapolations, provide that condition do not change. They are not reliable as models of long term reservoir performance or capacity because large-scale reservoir processes are not reflected in the fitting process. In geothermal field, there are two kind of decline curve, are exponential model and harmonic decline. Exponential model is a form decline where the flow decreases by constant fraction or percentage per year. The other forms, harmonic decline plot rate versus time. The plot used logarithmic scale in time and rate production used linear scale generate linear plot.

In this paper, the decline model just used surface production data. The production data used steam flow rate towards turbine versus time along three years in Ulubelu field. For optimizing the productions, developing model could easier to check condition of production in the current and the future. The example, the big trouble in Geyser field is declining flow rate about 50% in 8 and 12 years (Ripperda, 1987). The installed generation capacity in this field peaked at about 2000 MWe in 1989. But present generation capacity of about 850 MWe will have declined to about 700 MWe over the next two decades (Sanyal and Enedy, 2012). The other problem happened in water dominated geothermal field (Dieng Field). In Dieng geothermal Field, the geothermal plant contain a lot of Silica in fluids, making scales. As a consequence, at present, it only generates 40MW of electricity compared to the expected 60MW output (ECFA, 2006). One of the ways to identify the problem is decline curve analysis.

2. Methodology

Based on decline curve analysis, develop the model is used field production data in plot to time. We required production data especially fluids flow rate versus time. In this case, the production data is taken along 3 years production versus time. The result of graphic then used to determine model of decline curve.

Based on literature study, decline curve in geothermal field is basically from decline curve petroleum industry. The most using model in geothermal field are harmonic model and exponential model depend to decline rate. Then, understanding about production management principal especially in liquid-dominated reservoir geothermal field.

Selection of the model is making match actual data by plotting the model. Then, comparing of both model that matching with actual data. The best matching model with actual data is chosen. There are two way to determine best model, the first one is qualitative and the second is quantitative way. The qualitative way is to find the best model by look the most linear graphic. But the quantitative way is to find the best model by using R-square value. The bigger R–square value is the best model. So that, both way is consider to determine the best model decline curve in liquid-dominated geothermal.

The best model is used to analysis production problems and forecasting for the future production. Then, we calculate the minimum steam flowrate by separated steam cycle and the assumptions corresponding liquid-dominated geothermal in Ulubelu field. The important thing in future production is determine how much make up wells for maintaining the production rate along 30 years project corresponding to demand power capacity in State Electricity Company.

3. Result and Analysis

3.1 Decline Curve Model

Production in geothermal field is very important to observe, to maintain steam flowrate in generate electricity to be produced. Without production stage, revenue from geothermal energy project is not running. The consequence, production optimizing on geothermal development is very crucial stage to maintain power capacity. Naturally, production in oil and gas will decline corresponding to time as well as geothermal fluids.

The purposes of decline curve analysis is to test decline curve model that used in petroleum industry in geothermal production data and to examine and develop new analysis. Based on production data from control room in liquid-dominated geothermal, we could obtain production decline curve model. But, the data contain scatter. According to Zais (2008), reservoir that correlate with scatter data is cause of rainfall, recharge, earthquake, and subsidence. While, production correlate with scatter data cause of change production schedule, bad well completion, workovers, bad calibration, and bad data collection way.

The problems in production data is scatter data. The scatter data is very difficult to be avoided. But the method to decrease scatter data such as (Zais, 2008):

a. Average production data.
b. Least square.
c. Decreasing known effect and trends.
d. Using knowledge and experience.

In this paper, the decline curve model is specific to steam flowrate toward turbine only. The following below the steps to obtain decline curve model:

a. Gather production data especially steam flowrate from 2013-2015.
b. Make a plot steam flowrate to time.
c. Selection the data for decreasing scatter.
d. Find decline curve model with steam flowrate plot in semi log and time in plot log. If the plot approach a linear graphic, the model is exponential model
e. Try another plot with steam flowrate and time in log plot. If the plot approach a linear graphic, the model is harmonic model.

The following flowchart of developing decline curve model in Ulubelu field:

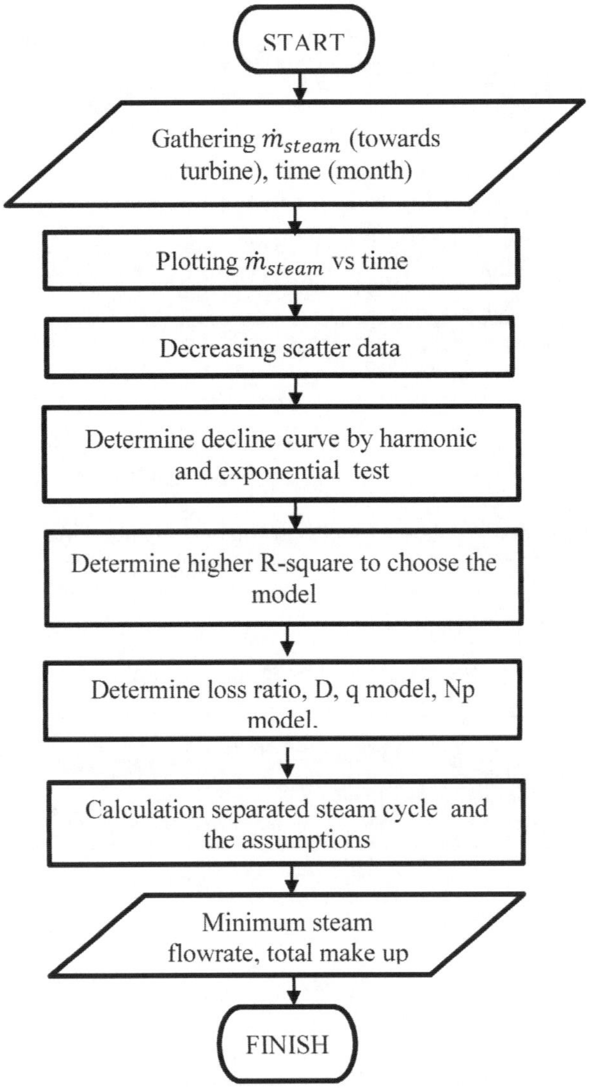

Figure 1. Decline Curve Model Flow Chart

Steam flowrate (ton/month) data toward turbine was made each month from 2013-2015. The flowrate is measured from a field by combine two path steam pipe from separator. The following is the result of steam flowrate graphic :

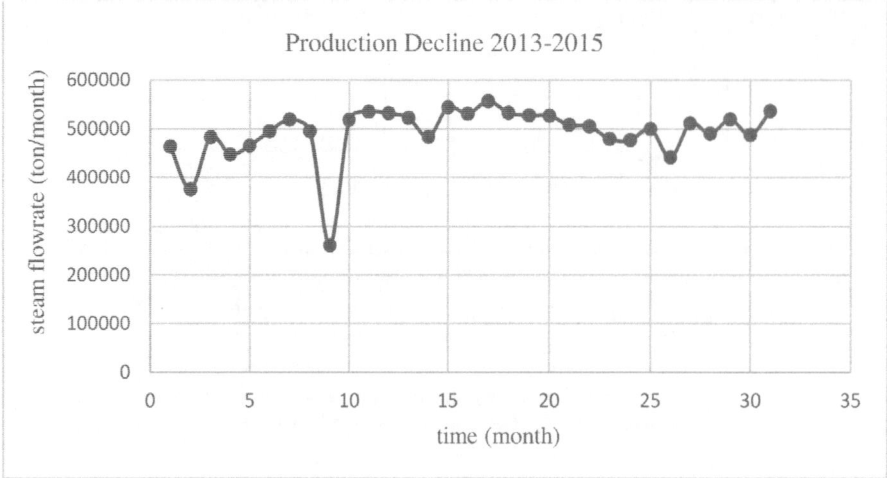

Figure 2. Production Decline 2013-2015

Base from figure 1, production data showed scatter so that need smoothing data by averaging the data. The averaging data is divided to 3 intervals. So, the following results of steam flowrate graphic by averaging all data.

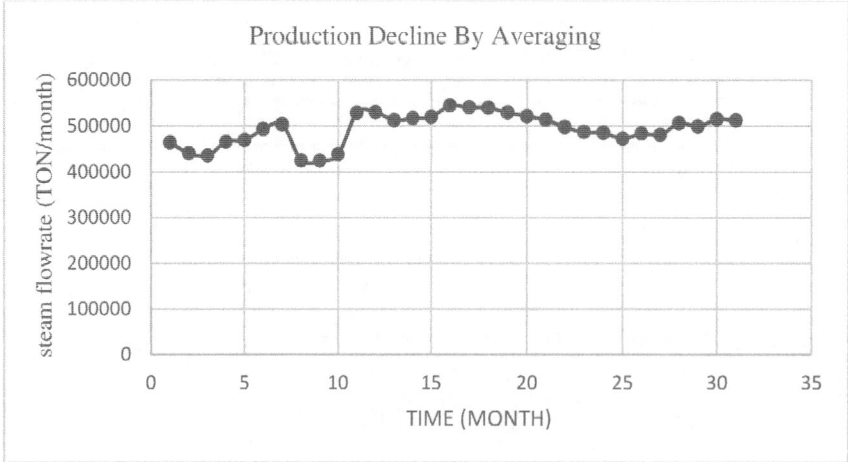

Figure 3. Averaging production decline

After averaging the production data, then the data need to be election. The data is chosen by assuming constant reservoir factor, field condition still is not shut, and no additions new well. The following graphic production flow and injection flow:

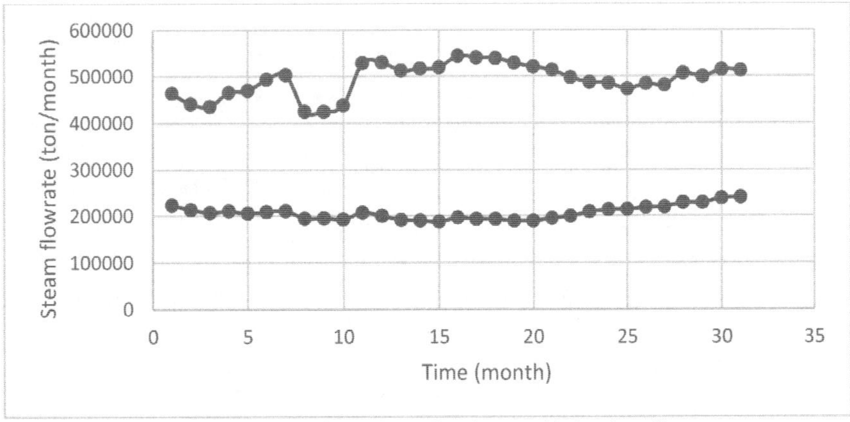

Figure 4. Relation of production and injection flow

Base of figure 3, field condition is no shut, and no addition, the selection data for curve model is 16th to 25th month, the following result of is obtained:

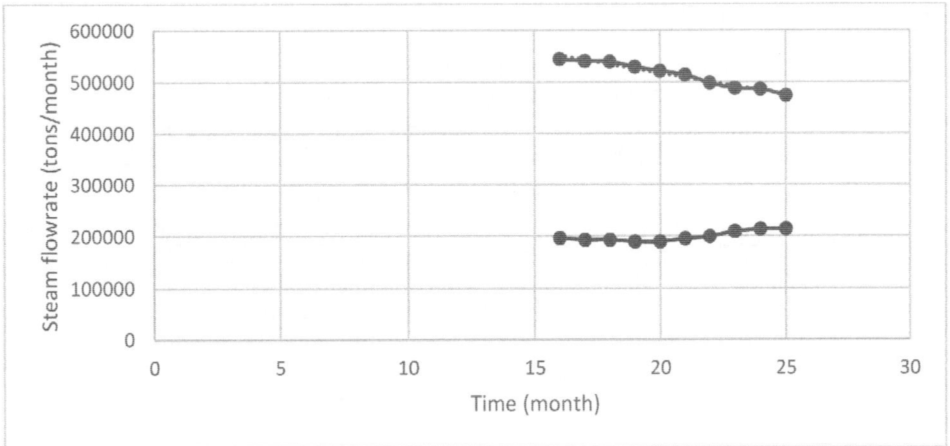

Figure 5. The selection data for curve model

The graphic will be used for testing decline curve model. For determining decline curve model, the following test was conducted for determining suitable model between exponential and harmonic model:

- Production decline figure. 4 is used for testing
- Test harmonic decline by making figure 5 to be axis x and y on log plot.
- Test exponential model by making figure 6 to be axis x on semi log plot and axis y on log plot.
- Determine the graphic that produce linear plot by qualitative and quantitative by R-square value.
- The graphic that produce higher R-square value linear will be production decline model.

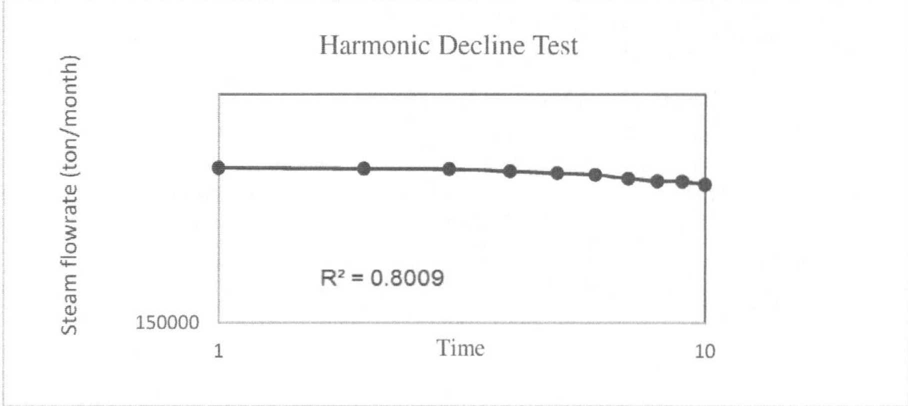

Figure 6. Log plot harmonic model test

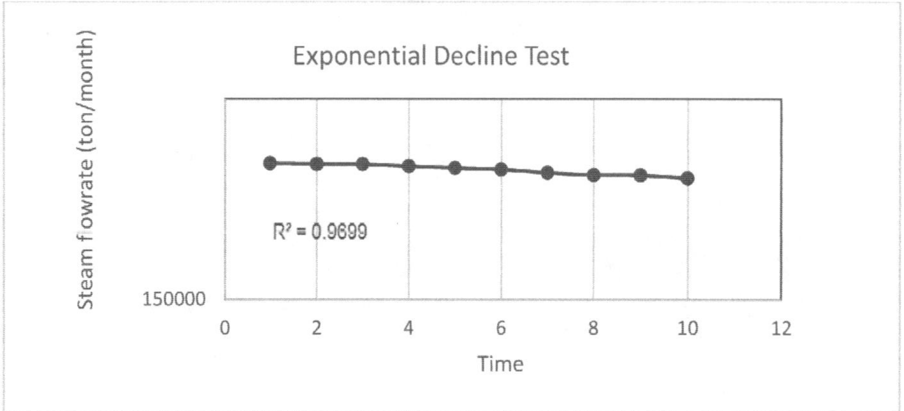

Figure 7. Semi log plot exponential testing

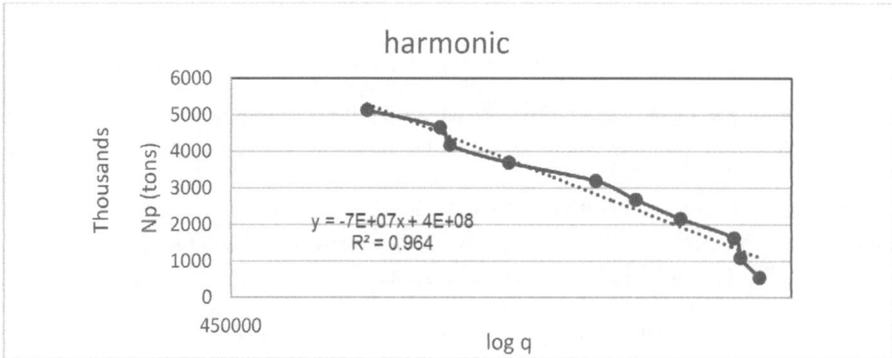

Figure 8. Test harmonic decline model by quantitatif

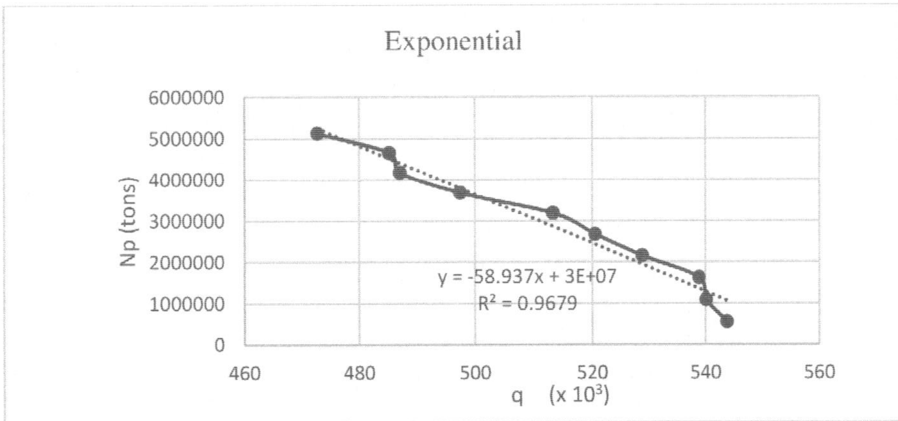

Figure 9. Test exponential decline model by quantitatif

Test decline curve for exponential model by semi log plot but decline curve for harmonic model used log plot. By trending test in excel, R square value exponential model was be higher than harmonic model. It means exponential model better accurately in linear. So that, model from production data in a geothermal field water dominated system is exponential model.

Then, we will calculate decline rate by firstly calculate loss ratio. Loss ratio is calculated by Arps formula:

$$Loss\ Ratio = \frac{q}{\Delta q/\Delta t} \approx \frac{q}{\Delta q} \tag{1}$$

Table 1. Loss ratio production data

Month	Steam flow (tons/month)	Average Steam flow **(tons/month)**	Loss Ratio
16	530943.3	543836.5	
17	556713.9	540181.7	147.8
18	532887.9	538901.9	421.1
19	527104.0	528852.3	52.6
20	526564.9	520580.5	62.9
21	508072.6	513278.9	70.3
22	505199.2	497470.6	31.5
23	479140.2	487066.9	46.8
24	476861.4	485291.9	273.4
25	499874.2	472688.4	37.5
Average Loss Ratio			127.1

Correlation between loss ratio and decline rate are inversely proportional, where

$$Loss\ Ratio = \frac{1}{D} \tag{2}$$

$$D = \frac{1}{Loss\ Ratio} = \frac{1}{127.1055} \times 100\% = 0.7867\ \%/\text{month}$$

Finally, decline rate value from production decline in a Geothermal water dominated system is 0.7867 %/month. This decline rate is effective decline rate. But the type of decline is divided by two are effective decline rate (De) and nominal decline rate (D). If we can see the characteristic of D and De, it prefers used D because for D value relatively easy to change time units by multiple or divide with time conversion factor (Asep, 2004).

The following is the correlation between De and D:

$$D = -\ln(1 - De) \tag{3}$$

$$D = -\ln(1 - De) = 0.0521 = \frac{0.7899\%}{month} = 9.4\ \%/year$$

So that, decline rate value is 9.4 $\%/year$. This decline rate value is called nominal decline rate. The value will be used for forecasting to the future production.

3.2 Prediction steam flow and cumulative production in the future

Base of decline rate value, steam flow production can be predicted by exponential decline curve

$$q = q_i\, e^{-Dt} \tag{4}$$

$$q = 543836.5167e^{-0.007899t}$$

Where, t is time in month and q is steam flow in ton/month. By using equation (4) above, then we can predict steam flow in the future. Here q_i is value steam flow at April 2014. Equation 4 can be assumed that no add wells, no recharge, and no rainfall that can change reservoir factor. While for predicting steam production to the future is:

$$Np = \frac{q_i(1-e^{-0.007899t})}{0.007899} \tag{5}$$

3.3 Calculate minimum steam flow by separated steam cycle

The important thing in geothermal field is to maintain steam flowrate in certain power capacity. A geothermal field must produce 2x55 MWe. It means that we must calculate the minimum steam flow rate to produce the value of electricity to State Electricity Company. To calculate minimum steam flowrate, we used separated system cycle because the field is water dominated (two phase). The assumption to calculate minimum steam flow by separated steam cycle is

a) Turbine efficiency : 85%
b) Reservoir temperature : 250°C, single phase liquid
c) Make up wells produce : 10 MWe/well

Based on calculated separated steam cycle system in appendix I, the minimum steam flow rate required to produce power capacity of 110 MWe is 502018.4 ton/month.

3.4 Planning make up wells

The minimum steam flow rate to supply 110 MWe need to be maintained. When the steam flow rate decline to the minimum, a make-up well need to be drilled. Make up well will be done to maintain steam flow along 30 years. The below graphic decline steam flow before and after make up well.

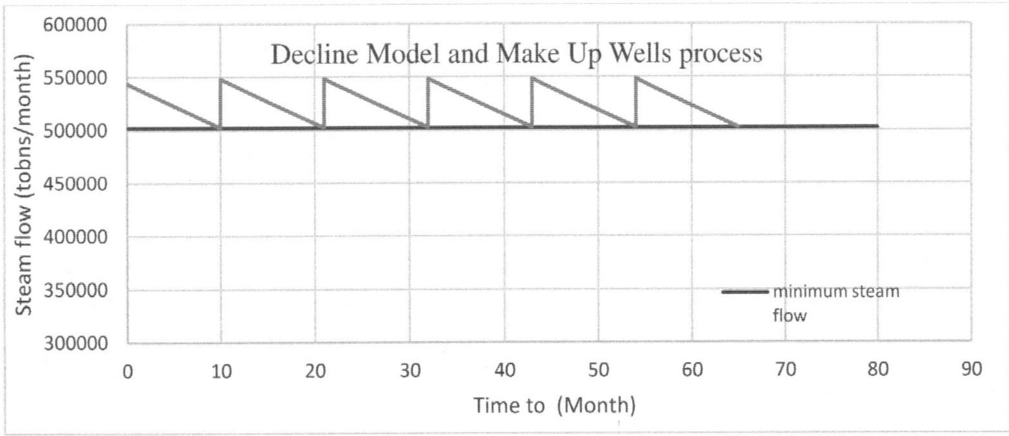

Figure 10. Decline model and make up wells

From figure 7, we can analyze that one make up well can maintain steam flow along 11 month average by assuming the well produce 10 MWe/well and decline rate is constant. It means that the geothermal project must drill 32 make up wells totally to maintain steam flow along 30 years production.

4. Conclusion

Decline curve analyses is most important to predict production performance in the future and maintain production to supply 110 MWe along 30 years project. From the results of this study it can be concluded of the following:

a) Decline curve model in a geothermal field water dominated is exponential decline model with decline rate 9.4 %/year.
b) Prediction for forecasting decline curve in the future based exponential decline get formula :
   - Steam flow forecasting :
   $$q = 543836.5167e^{-0.007899t}$$
   - Steam production forecasting : $Np = \frac{q_i(1-e^{-0.007899t})}{0.007899}$
c) The minimum steam flow to supply power electricity 110 MWe is 502018.4 ton/month.
d) 32 make up wells will be needed to maintain production along 30 years.

5. Recommendation

This paper recommended for further research as follows:

1. Decline curve necessary fluid flowrate data from each well in a field
2. Normalized pressure for each well before construct
3. Decline production data must be matched to decline curve simulation reservoir

6. References

[1] Grant, Malcolm. 2011. *Geothermal Reservoir Engineering: Second Edition*. USA, Elsevier Inc, 249-252
[2] Ripperda, M., Bodvarsson, G.S., 1987. *Decline Curve Analysis of Production Data From The Geysers Geothermal Field.* Berkeley, CA 94720.
[3] Yuniar, D.M., Hastuti, P., Silaban, M., 2015. *Ulubelu, First Year Reservoir Monitoring.* Melbourne, World Geothermal Congress.

[4] Aditya, Hermawan. 2013. *Production Decline Analysis Of Dry Steam Wells In Wayang Windu Field.* California, Workshop on Geothermal Reservoir Engineering Stanford University
[5] Permadi, A.K., 2004. *Diktat Teknik Reservoir I.* Bandung, Departemen Teknik Perminyakan ITB.

7. Appendix
The following below, a summary table of calculating separated steam cycle system:

Table 2. Calculating separated steam cycle system

Power Capacity	**110.00**	**MWe**
Pressure at Well Head	10.00	bar
h5 (at condenser presure)	2097.47	kj/kg
h4 (at turbine)	2765.64	kj/kg
hf@7.5bar	709.38	kj/kg
hg@7.5bar	2765.64	kj/kg
hf@10bar	762.68	kj/kg
hg@10bar	2777.12	kj/kg
fluid enthalpy at well head	1085.69	kj/kg
steam fraction at well head	0.16	
steam fraction at separator	0.18	
steam entrophy to turbine (S4)	6.68	kj/kg K
Sf@condenser	0.61	kj/kg K
Sg@condenser	8.21	kj/kg K
steam fraction at condenser	0.80	
hf5 (at condenser pressure)	178.68	kj/kg
hg5 (at condenser pressure)	2578.30	kj/kg
hfg5 (at condenser pressure)	2399.62	kj/kg
Steam flow minimum	193.68	kg/s
Steam flow minimum	502018.40	ton/month
Addition make up a well	17.61	kg/s
Addition make up a well	45638.03	ton/month

Characterizing Geothermal Surface Manifestation Based on Multivariate Geostatistics of Ground Measurements Data

Ishaq[1], Mohamad Nur Heriawan[1], Asep Saepuloh[2]

[1]Faculty of Mining and Petroleum Engineering, Bandung Institute of Technology (ITB), Jl. Ganesha No. 10, Bandung, West Java, Indonesia 40132.
[2]Faculty of Earth Sciences and Technology, Bandung Institute of Technology (ITB), Jl. Ganesha No. 10, Bandung, West Java, Indonesia 40132

Email: ishaqitb14@yahoo.co.id, heriawan@mining.itb.ac.id, saepuloh@gmail.com

Abstract. Mt. Wayang Windu is one of geothermal field located in West Java, Indonesia. The characterization of steam spots at surface manifestation zones based on the soil physical measurements of the area is presented in this study. The multivariate geostatistical methods incorporating the soil physical parameter data were used to characterize the zonation of geothermal surface manifestations. The purpose of this study is to evaluate the performance of spatial estimation method of multivariate geostatistics using Ordinary Cokriging (COK) to characterize the physical properties of geothermal surface manifestations at Mt. Wayang Windu. The COK method was selected because this method is favorable when the secondary variables has more number than the primary variables. There are four soil physical parameters used as the basis of COK method, i.e. Electrical Conductivity, Susceptibility, pH, and Temperature. The parameters were measured directly at and around geothermal surface manifestations including hot springs, fumaroles, and craters. Each location of surface manifestations was measured about 30 points with 30 × 30 m grids. The measurement results were analyzed by descriptive statistics to identify at the nature of data. The correlation among variables was analyzed using linear regression. When the correlation coefficient among variables is higher, the estimation results is expected to have better Linear Coregionalization Model (LCM). LCM was used to analyze the spatial correlation of each variable based on their variogram and cross-variogram model. In oder to evaluate the performance of multivariate geostatistical using COK method, a Root Mean Square Error (RMSE) was performed. Estimation result using COK method is well applicable for characterizing the surface physics parameters of radar images data.

Keywords: surface manifestations, Linear Model Coregionalization, multivariate geostatistics, cokriging, Root Mean Square Error

1. Introduction

Mt. Wayang Windu Geothermal Field (WWGF) is an area that has geothermal activity and now in production. The surface manifestation zones at WWGF can be characterized based on the soil physical parameters. The multivariate geostatistical methods incorporating the soil physical parameter as primary data with remote sensing images as secondary data were used to characterize the zoning manifestations. Ordinary Cokriging (COK) method has advantage to provide more accurate estimation and knowable relationships of spatial correlation among variables.

IOP Conf. Series: Earth and Environmental Science **42** (2016) 012025 doi:10.1088/1755-1315/42/1/012025

The use of remote sensing images data as secondary variable has its own advantages as it can cover a wide area, so the number of secondary data is much more than primary one. When the secondary variables are much more than the primary ones, COK method generally produces more advantages [1].

The purpose of this study is to determine the nature characteristics of physical parameters i.e. electric conductivity, susceptibility, temperature, and pH of surface manifestation around WWGF using roughness data of the radar images as secondary data.

2. Study Area

WWGF took the name after Mt. Wayang and Mt. Windu, the two small lava domes without eruption history [2]. The field is located at Pangalengan town, approximately 40 km to the south of Bandung city, the capital of West Java Province, Indonesia (see Figure 1). The elevation of Wayang Windu is between 1500 – 2100 m above sea level [3].

Wayang Windu study area includes the Wayang Windu geothermal systems associated with the dominant steam fluid caps. WWGF associated with Mt. Malabar, large volcano-type stratovolcano composed of lava, breccia, lava and dacitic tuffs, and andesite.

Several thermal manifestations including fumaroles, hot springs, mud pools and altered soil are located along the slopes of the mountain range. Fumaroles which are located between Mt. Wayang and Mt. Windu show temperatures between 93°C to 96°C, and slightly superheated if we compared to the normal water boiling point at altitudes of 2000 m. Hot springs which are located at altitudes between 1495 m and 1985 m reach temperatures from 41°C to 88°C. The water discharge of hot springs is bicarbonate type, except for the hot springs located near fumaroles discharged acid sulphate water.

The lithology at WWGF consists of four main units, which were subdivided by Alzwar et al [4] into: the older Malabar-Bedil andesite unit, the Kancana lavas unit, the Malabar-Tilu volcanic unit, and the younger Wayang volcanic unit. According to Bogie et al, the geothermal system at Wayang Windu is classified as transitional between vapour and liquid dominated systems [5].

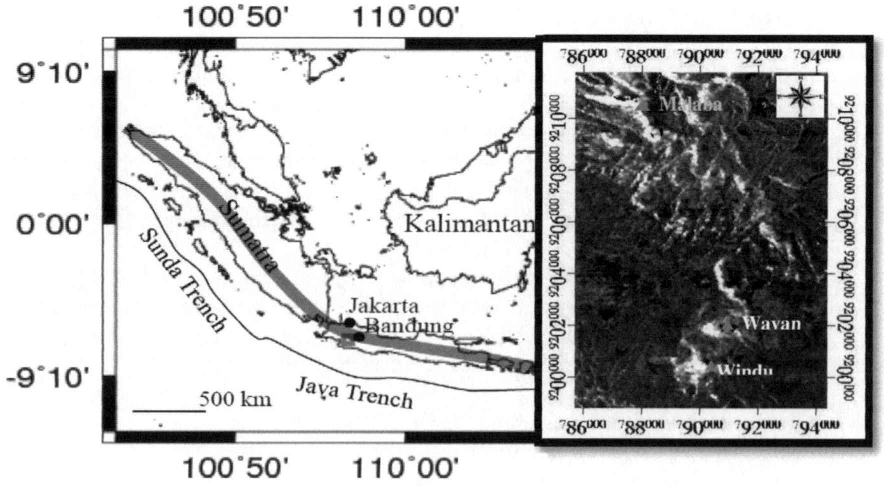

Figure 1. The location of study area at WWGF in West Java, Indonesia shows the lane of geothermal prospecting in gray arc segments [2]. The inset showing local of study area presented by surface roughness model using HV and VH polarized mode.

3. Material and Method

Ordinary Cokriging (COK) is an estimation technique that ensures the value of variable estimated in a point in space, on the basis of neighbouring values of one or several other variables. The COK estimate is a linear combination of both primary and secondary data values [6].

COK therefore is applicable specifically for different but correlated variables, non-collocated variables of different quality that are sampled at different density across a domain. The variogram and cross-variogram can be used in COK system only provided that the requisite constraints on COK weights are met. The cross-variogram can be transformed into a corresponding cross-covariance. The cross-covariance can be modelled using linear coregionalization model to be used in COK system. Since a single secondary variable (*Y*) is considered, the COK estimator of primary variable is:

$$Z^*_{OCK}(\boldsymbol{u}) = \sum_{\alpha=1}^{n} \lambda_\alpha(\boldsymbol{u})\, Z(\boldsymbol{u}_\alpha) + \sum_{\beta=1}^{r} \lambda_\beta^r(\boldsymbol{u})\, Y(\boldsymbol{u}_\beta^r) \quad (1)$$

where λ_α represents the weights applied to *n* primary data (*Z*), λ_β^r represents the weights applied to *r* secondary data (*Y*), and the bold **u** identifies a vector [7].

The model of spatial structure is represented by semi-variogram, γ(h). It represents the average variance between observations data separated by distance *h*, calculated by:

$$\gamma(h) = \frac{1}{2N(h)} \sum_i^n \{Z(x_i) - Z(x_i + h)\}^2 \quad (2)$$

with $Z(x_i)$ is the measurement at location x_i, $Z(x_i + h)$ is the measurement at location $x_i + h$, $\gamma(h)$ is the variogram for distance vector (=lag) h between measurements $Z(x_i)$ and $Z(x_i + h)$, $N(h)$ is the number of couples measurements separated by distance h.

Cross-variogram is used to obtain the covariance between two variables, i.e. ground truth and roughness data. Cross-variogram values are calculated directly from the samples data (co-variable) using the formula:

$$\hat{\gamma}_{AB}(h) = \frac{1}{2N(h)} \sum_i^n \sum_j^m \{Z_A(x_i) - Z_A(x_j)\}\{Z_B(x_i) - Z_B(x_j)\} \quad (3)$$

where $\hat{\gamma}_{AB}(h)$ is the cross-variogram value at distance h [8].

The primary data is derived from field data (ground truth) around the location of geothermal manifestations such as craters and hot springs. Variables of field data are Electric Conductivity, Susceptibility, pH, and Temperature. The fourth parameter is used as reference to extrapolate the whole study area.

Secondary data is derived from the scene backscattering intensity data of the phased array type L-band Synthetic Aperture Radar (PALSAR) on board the Advanced Land Observing Satellite (ALOS), consists of four polarization types known as: HH, HV, VH, VV. ALOS PALSAR is an active microwave sensor using wavelength of 23.6 cm, thus the sensor is not affected by the presence of clouds, and observations is time independent. The backscattering signal σ^0 is function of roughness materials at surface h_o, incident angle θ_i, and scattering matrix element α can be calculated as follows:

$$\sigma^0 = (4k^4 h_0^2 cos^4\theta_i)|\alpha|^2\omega \quad (4)$$

where ω is a spectrum density of surface topography [9].

The initial surface roughness model was used as basis sensitivity identification to the surface roughness at the field. The model could be calculated for each polarization type using the following equation [10]:

$$h_0(\eta\xi) = \lambda \sqrt{-\frac{1}{60} ln\left(1 - \frac{10^{\left(0.1 \times \sigma^0_{\eta\xi}\right)}}{0.04 \cos \theta_i}\right)} \quad (5)$$

where $h_0(\eta\xi)$ is the surface roughness model based on polarized mode, $\eta\xi$ is the polarized mode either in H and V, $\sigma^0_{\eta\xi}$ is backscattering coefficient on polarized mode, and θ_i is the local incidence angle.

The fourth extracted data polarization roughness (h_o) will be analyzed statistically using scatterplot for each ground truth data to see the correlation among variables as the summary in shown in Table 1. The medium to strong correlation between two variables is selected for primary and secondary data in multivariate estimation.

Table 1. The correlation between primary (*ground truth*) and secondary (*satellite images*) variables in each manifestation zones.

Zone 1

	Suscep.	pH	Temp.	EC
HH	-0.29	-0.52	0.35	0.57
HV	-0.28	-0.59	0.44	0.32
VH	-0.28	-0.51	0.35	0.23
VV	-0.35	-0.50	0.35	0.56

Zone 3

	Suscep.	pH	Temp.	EC
HH	-0.11	-0.05	-0.38	-0.56
HV	-0.13	-0.20	-0.43	-0.53
VH	-0.14	-0.26	-0.42	-0.52
VV	-0.17	-0.16	-0.30	-0.42

Zone 2

	Suscep.	pH	Temp.	EC
HH	0.29	0.10	0.28	0.11
HV	0.39	0.31	0.27	0.10
VH	0.42	0.36	0.22	0.02
VV	0.29	0.36	0.19	-0.09

Zone 4

	Suscep.	pH	Temp.	EC
HH	-0.36	-0.52	0.25	0.28
HV	-0.35	-0.54	0.17	0.18
VH	-0.37	-0.57	0.21	0.21
VV	-0.26	-0.39	0.13	0.08

Note: the grey colour shows medium to strong correlation used for multivariate estimation.

Ground truth data measurement is not spread evenly across the surface, but the pattern is clustering based on the location of geothermal manifestations. The distance between two data point is about 30 m referred to the pixels size on radar images. Therefore one data point represents one data ground truth of roughness in the image. Prior to estimation, the data is grouped by spaces to reduce the error when data is analysed together.

4. Results and Discussion

During the first stage of exploratory data analysis, the data distribution is described using descriptive statistics in each zone with the summary is shown in Table 2.

Table 2. Statistical summary of sample data (ground truth and roughness) for Zones 1 to 4.

Zone 1	Mean	Median	SD	Min	Max	RMSE
EC	1.36	0.06	4.71	0.01	20.00	5.87
pH	6.51	6.8	0.89	4.05	7.48	0.96
Suscep.	4.06	4.03	2.68	0.02	15.95	2.92
Temp.	23.14	15.70	20.72	13.80	93	25.06
HH	6.84	5.37	3.50	1.95	18.33	-
HV	7.03	6.49	2.54	1.95	15.13	-
VH	8.35	7.69	3.17	1.95	18.45	-
VV	6.05	4.66	2.90	1.95	15.27	-

Zone 2	Mean	Median	SD	Min	Max	RMSE
EC	0.14	0.11	0.11	0.02	0.44	0.13
pH	6.92	6.94	0.36	6.28	7.92	0.39
Suscep.	3.53	3.57	1.12	1.26	5.59	1.03
Temp.	18.61	19.3	2.40	14.00	22.40	2.90
HH	5.09	5.06	0.55	3.85	6.66	-
HV	6.20	6.16	0.81	4.42	8.10	-
VH	7.34	7.19	1.10	5.19	10.28	-
VV	4.69	4.61	0.50	3.76	6.59	-

Zone 3	Mean	Median	SD	Min	Max	RMSE
EC	0.21	0.09	0.26	0.01	0.75	0.24
pH	7.08	7.02	0.37	6.29	7.76	0.36
Suscep.	3.15	2.99	2.09	0.40	11.26	1.93
Temp.	21.78	20.00	6.95	15.80	45.70	6.15
HH	5.56	5.26	1.08	3.99	8.60	-
HV	6.95	6.52	1.55	4.79	11.23	-
VH	8.28	7.67	1.98	5.11	13.94	-
VV	5.06	4.85	0.92	3.56	8.00	-

Zone 4	Mean	Median	SD	Min	Max	RMSE
EC	3.52	0.13	6.91	0.01	20.00	8.11
pH	5.67	59.00	1.49	3.20	7.23	1.93
Suscep.	2.56	1.73	2.74	-0.003	10.53	3.61
Temp.	28.06	19.90	21.92	14.40	93.30	23.40
HH	6.49	5.34	2.31	3.72	12.74	-
HV	8.45	7.50	3.05	4.75	15.10	-
VH	10.15	8.75	3.77	5.62	19.10	-
VV	6.31	5.02	2.66	3.57	13.71	-

EC: Electric Conductivity (ms); Temperature (°C); HH, HV, VH, VV: Roughness extracted from radar data (cm).

The result of estimation using COK method is shown in Figure 4. In Zone 1, we can see that all physical parameters reveal the heterogeneous occurred in NS - SW direction. It can be concluded that there is a good correlation among parameters within the Zone 1. In S - W direction, the estimated temperature is high; the ground truth shows that this area is Kawah Burung, the crater valley, while the temperature is decreasing toward the N - S direction. When we compared with the electric conductivity (Figure 4a; Zone 1), it shows the same colour pattern means that it is correlated to the temperature. But the pH (Figure 4b; Zone 1) and susceptibility (Figure 4d; Zone 1) has a reversed pattern and it can be concluded that both of parameters were inversely correlated. The same pattern also occurs in Zone 4.

The parameters from estimation results are more difficult identified at manifestations of Zone 2. Colour pattern almost has no resemblance in this zone. The correlation among physical parameters within the Zone 2 has low correlation.

The characteristic of physical parameters from the estimation in Zone 3 is unique. The pattern of colour for all estimates almost shows similar results. This indicates that all parameters linearly correlated. This phenomenon did not occur in the Zones 1 and 4. Theoretically, the increasing temperature would reduce the magnitude of magnetization (susceptibility) material. Using ground truth data, Zone 3 around the springs Cibolang shows temperature about 67.7°C and pH 6.75. The pH should be low because of the present of acidic nature of water. The pH has possibility to be high due to the influence of surface water. Further study related to this phenomenon is indispensable. Zones 1 and 4 have similar physical properties which are located in the fracture zone that related directly to the geothermal reservoir.

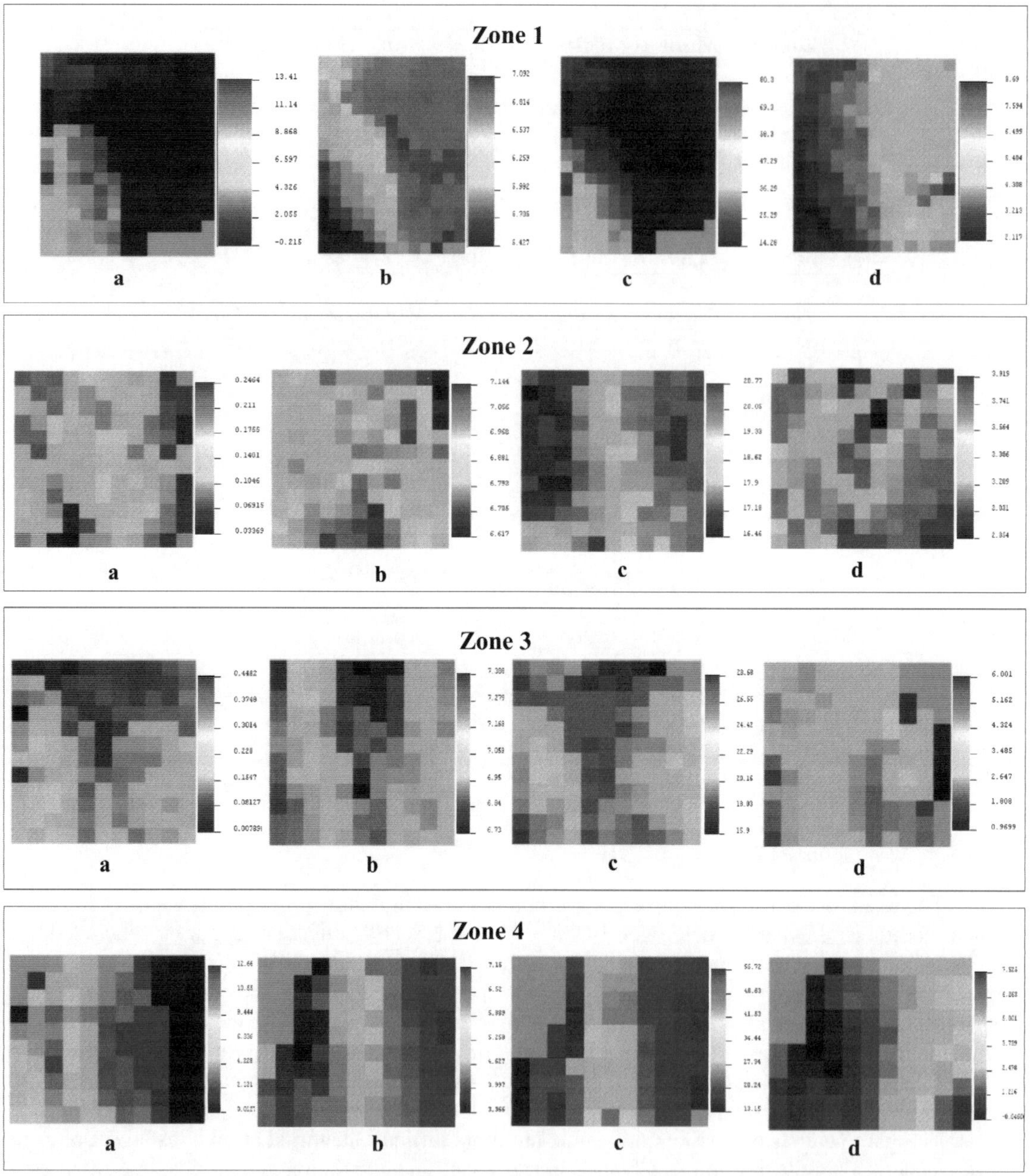

Figure 4. Physical surface characteristics of: (a) EC, (b) pH, (c) Temperature, and (d) Susceptibility in Zones 1 to 4.

The Root Mean Square Error (RMSE) of the estimation results is similar to the standard deviation of data as shown in Table 2. The use of COK method is appropriate to characterize the physical parameters of surface manifestation within geothermal area. Validation using scatterplot for the estimation of measurement data is also performed, but the correlation is less good. The correlation between measurements data and radar image data is also less dominant.

IOP Conf. Series: Earth and Environmental Science **42** (2016) 012025 doi:10.1088/1755-1315/42/1/012025

5. Conclusion

The characteristics of natural physical parameters could be estimated using Cokriging method (COK) which showed respectable estimation. In general, the characteristic of Temperature had positive linear correlation with the Electric Conductivity (EC), and has negative correlation with pH and Susceptibility. The unique characteristics occurred in Zone 3 where all parameters had positive linear correlation, and we need further study to know well about this phenomenon. Zones 1 and 4 have similar physical properties and they are located in fractures zone which is directly related to the geothermal reservoir.

Acknowledgment

The authors sincerely thank to the "BAGUS" SATREPS project for providing field equipment, LPPM-ITB for survey accommodation supports, and Star Energy for collaborating to this research.

References

[1] Wackernagel H.: *Multivariate Geostatistics: An Introduction with Applications*. Springer-Verlag Berlin Heidelberg, New York, 1995.

[2] Hochstein M.P. and Sudarman S.: History of Geothermal Exploration in Indonesia from 1970 to 2000. *Geothermics* 37, 2008, 220-266.

[3] Purnanto M H and Purwakusumah A.: Fifteen Years (Mid-Life) of Wayang Windu. *Proceedings World Geothermal Congress 2015*. Australia: Melbourne.

[4] Alzwar M., Akbar N. and Bachri S.: Geological Map of the Garut and Pamengpeuk Quadrangle, Jawa. *2nd edition, Geological, Research and Development Center*. Indonesia, 2004.

[5] Bogie I., Kusumah Y.I. and Wisnandary, M.C.: Overview of the Wayang Windu Geothermal Field, West Java, Indonesia. *Geothermics. vol. 37,* 2008, pp. 347-365.

[6] Yalçin E.: Cokriging and its effect on the estimation precision. *The Journal of The South African Institute of Mining and Metallurgy. Volume 105,* April 2005.

[7] Minnitt R.C.A. and Deutsch C.V.: Cokriging for optimal mineral resource estimates in mining operations. *The Journal of The Southern African Institute of Mining and Metallurgy. Volume: 114*. March 2014.

[8] Meilianda E, Huhn K, Alfian D and Bartholoma A.: Application of Multivariate Geostatistics to Investigate the Surface Sediment Distribution of the High-Energy and Shallow Sandy Spiekeroog Shelf at the German Bight, Southern North Sea. *Scientific Research: Open Journal of Marine Science,* 2, 2012, 103-118.

[9] Saepuloh A, Urai M, Sumintadireja P and Suryantini: Spatial Priority Assessment of Geothermal Potentials Using Multi-Sensor Remote Sensing Data And Applications. *Proceddings, 1st ITB Geothermal Workshop 2012*. Institut Teknologi Bandung, Bandung, Indonesia. March 6-8, 2012.

[10] Saepuloh, A., Koike, K., Urai, M., and Sri Sumantyo, J.T.: Identifying surface materials on an active *volcano* by deriving dielectric permittivity from polarimetric SAR data, *IEEE Geoscience and Remote Sensing Letters (GRSL),* **12**(8), 2015, 1620-1624.

Occurrence of Sarawet hotspring : what affects the outflow?

W A Draniswari and R N Hendrawan

Geological Engineering, Master Program, Bandung Institute of Technology, Indonesia

draniswari@gmail.com

Geothermal systems located in volcanic arc are generally influenced by volcanic activities. But at the complex geological setting like Sulawesi the geothermal systems probably is not only influenced by volcanic activities but also structural geology features. Sarawet hot spring in Minahasa, North Sulawesi, Indonesia is a volcanic and fault influenced manifestations. Sarawet hot spring is located in north coast Minahasa. Our study shows that the surface geological-structural analysis in combination with the geochemistry and the heat flow investigation is the essential tools for geothermal systems or manifestation characterization. Tectonics, crust characteristic, heat flow, and surface structural could affect the manifestations and its geothermal systems. Geochemically the water is bicarbonate, peripheral-immature water, and characterized by a meteoric water-line trend. The composition of these thermal springs was controlled by a secondary process during ascent. Based on hydrologic gradient, heat flow characteristic and subsurface structural analysis, the mechanism of Sarawet hot spring is a lateral outflow that affected by dynamically maintained fractures system.

1. Introduction

Minahasa is located in North Arm of Sulawesi in Indonesia. It was one of the major geothermal resources in Indonesia. The hot spring was found within several volcanic and sedimentary rock formations. Sarawet hot spring was located in Sarawet, near Likupang. It was related to sedimentary rock formation, Breccia and Sandstone (Tps)[7]. Different with the other hot spring, the Sarawet hot spring was located in the margin of volcanic system boundary. Thus, the mechanism of how this hot spring and permeability zone was formed is still obscure [10]. This study used surface geological and structural analysis in combination with the geochemistry and the heat flow investigation as the essential tools for geothermal manifestation characterization.

2. Methods

Hot Spring Control System could be determined by several variables such as geochemistry of hot spring manifestation, regional geology and ASTER DEM (figure 1). Water sample was from Sarawet hotspring used for geochemistry analysis and the result was represented into 3 diagrams with different parameters.

The Cl-SO_4-HCO_3 ternary diagram was used to classify geothermal fluids on the basis of the major anion concentrations [8]. It helps to discern immature unstable waters and gives an initial indication of mixing relationships or geographic groupings [6].

The Na-K-Mg triangular diagram was used to classify waters into fully equilibrated, partially equilibrated and immature waters. It can be used to predict the equilibrium temperature. The Na-K-Mg triangular diagram shows attainment of the water-rock equilibrium if the data point plots on the full equilibrium line, or suggests a field of immature water below the "immature water curve" which indicates an initial dissolution of minerals before equilibrium reaction sets in [8]. Lithium was used as a tracer, because it is the alkali metal least affected by secondary processes for initial deep rock dissolution and as a reference for evaluating the possible origin of two important 'conservative' constituents of geothermal waters, Cl and B. The Cl/B ratio is often used to indicate a common reservoir source for the waters [9]. All of diagrams decide Hot Spring Characteristic after analysis.

Heat Characteristic consists of heat flow, thermal gradient, and conductivity data. All of them shows thermal condition from research area. Heat flow is the transfer of thermal energy from one body to another or a transfer of temperature. The conveyance of heat through the crust is primarily related to rock type and structure [12]. Thermal gradient is defined as a vector that is dependent on temperature distributed in three dimensions (x, y, and z axes). Knowing three-dimensional temperature distributions within the crust is ideal for determining the true vector of the maximum thermal gradient [2]. Thermal conductivity is a measure of the ability of heat to flow through a particular material, and is a function of temperature. Thus, higher thermal conductivity values for a particular lithology indicate a higher allowance for heat flow [4]. Both hot spring and heat characteristic associated with lithology and geological structure, so it can assist in determine the structural control.

ASTER DEM used for establish lineament to create Fault and Fracture Density (FFD) Map. Both of FFD Map and geological structure map help authors to assign structural control, in addition to Heat and Hot Spring Characteristic. Furthermore, structural control either dynamic or kinematic as results can be state hot spring outflow system related with fault occurrence.

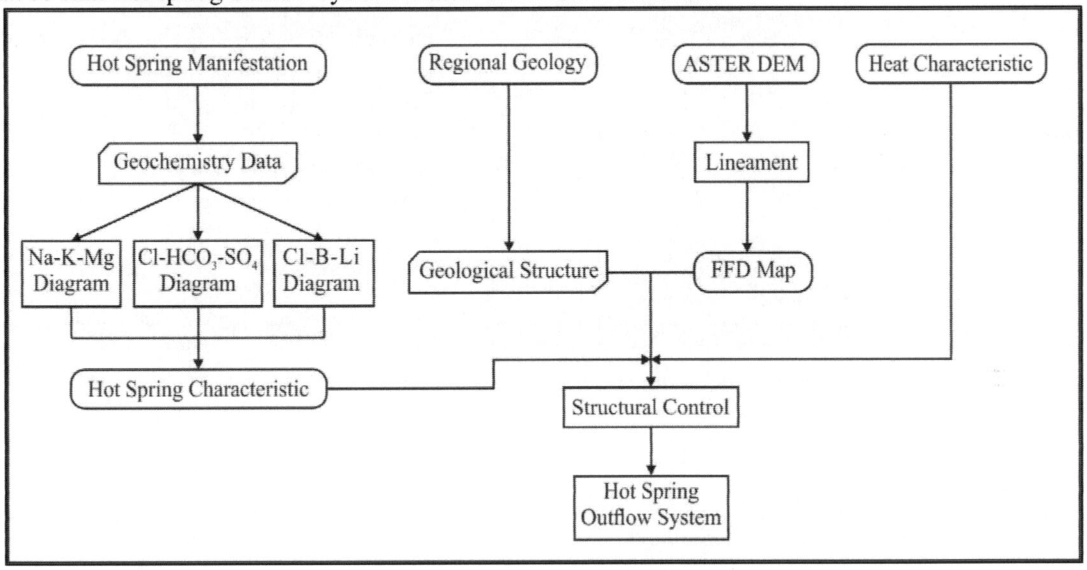

Figure 1. Workflow diagram of structural control identification in outflow.

3. Geological Setting

Minahasa Area consist of Tertiary volcanic and sediment rocks overlain by Quartenary and recent volcanic rocks [7]. The oldest rock formation found in North Minahasa ia Volcanic Rock (Tmw) that consist of breccia, lava and tuff, sandstone, claystone and mudstone which found as intercalation and Breccia and Sandstone (Tps) Formation, consist of breccia and conglomerate intercalated with fine to coarse sandstone, siltstone and claystone. The Quartenary rocks formation of the research area consist of Tondano Tuff (Qtv), Young Volcanic Rocks (Qv), Coral Reef Limestone (Ql), Lacustrine and Fluvial Deposits (Qs), and the youngest is Alluvium (Qal). Qtv consist of coarse volcaniclastic with andesitic composition while the Qv consist of lava, bomb, lapilli and ash with andesit basalt composition that make up the young strato volcano. Ql is distributed in the north coast area and consist of coral reef limestone that mostly found between low and high tide. Not far from the north coast, Qs was distributed as sandstone, siltstone, conglomerate, and marl. And the recent one, Qal, is consist of boulders, cobbles, pebbles, sand and mud

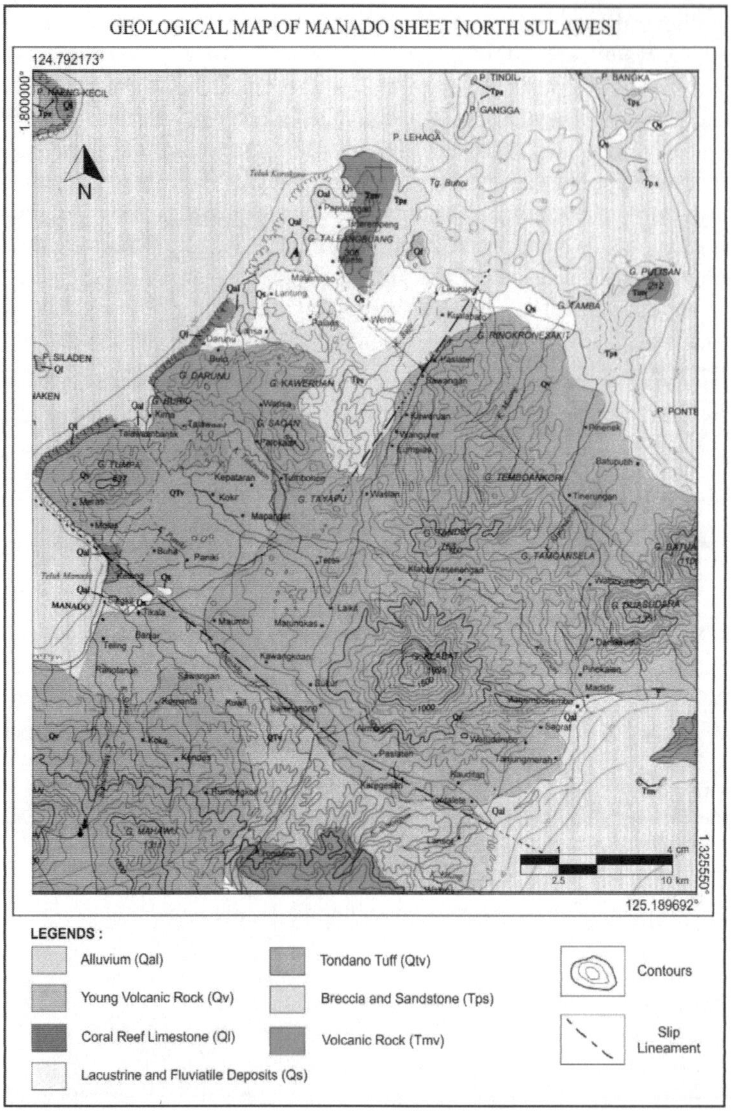

Figure 2. Regional Geology Map of Minahasa showing the distribution of sediment and volcanic rocks and the strike slip of the research area. [7].

The hot springs were not found only within the volcanic area but also on it surrounding areas. In addition, the manifestations were found in the vicinity of a fault that may have leaked the geothermal fluid onto the surface.

Minahasa comprises some fault patterns which are major strike slip fault trending NE-SW, NW-SE and normal fault trending N-S. The fault was created by the movement of Celebes sea-plate. Other geological structure in Minahasa mainly controlled by two major strike-slip fault that cut the North Arms Sulawesi in two places, between Amurang – Malompar in the south and Manado – Kena in the north.

4. Geochemistry

4.1 Cl-HCO3-SO4 Diagram

Based on terniary diagram plotting of the major anions Cl^-, HCO^{3-} and SO_4^{2-} as seen in figure 3 below, Sarawet hot spring composed of bicarbonate water which located in peripheral - surface water zone with dominant concentration of HCO^{3-}. It is interpreted that Sarawet hot spring was associated with CO^{2-} rich fluids that flowing through permeable zone and then condenses in the shallow aquifer. Sarawet hot water is not classified as reservoir water but it was a product of near surface mechanism due to the mixing of the reservoir water with groundwater or surface water.

4.2 Na-K-Mg Diagram

Na-K-Mg diagram (Figure 4) presented the condition of the geothermal fluids. Sarawet hot spring showed high concentration of Mg and plotted as immature water. Immature water zone and near √Mg peak point suggest that there were high influence of groundwater in the systems. This is accordance with Cl^-, HCO^{3-} and SO_4^{2-} plotting diagram that proved the interaction with groundwater or surface water.

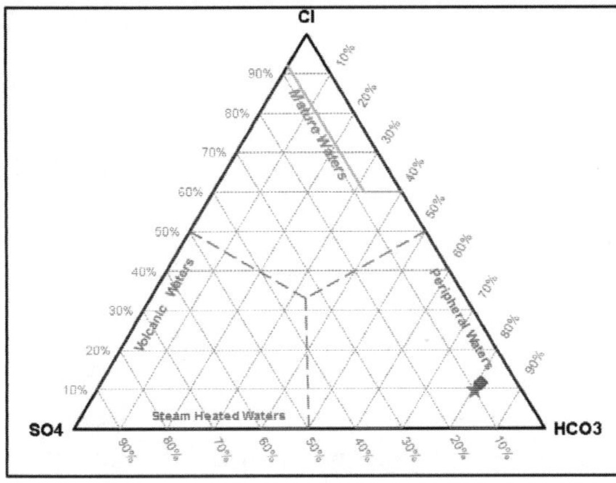

Figure 3. Cl-HCO_3-SO_4 Diagram.

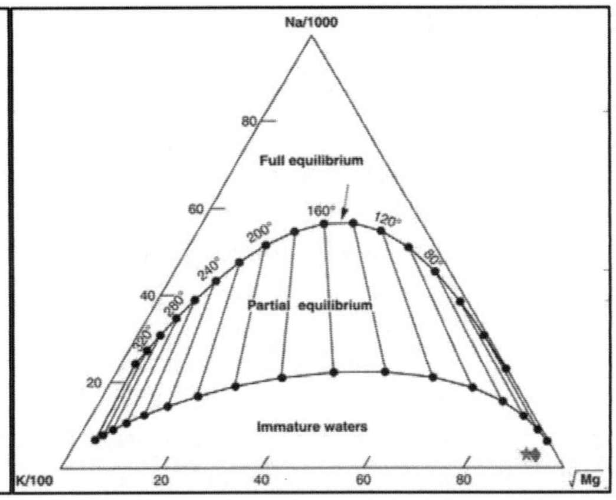

Figure 4. Na-K-Mg Diagram.

4.3 Cl-B-Li Diagram

Figure 5 showed the relatively high Cl content than Li and B at Sarawet hot water. Its indicated the influence of magmatic activity to the hot spring manifestation. The low value of B indicated that the

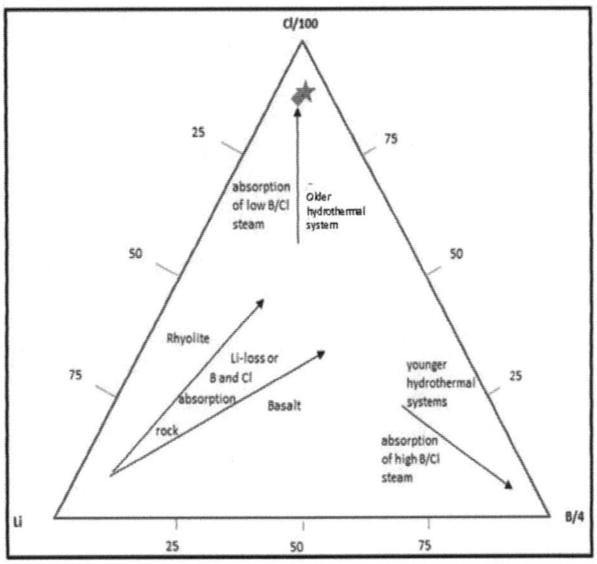

Figure 5. Cl-B-Li Diagram.

IOP Conf. Series: Earth and Environmental Science **42** (2016) 012026 doi:10.1088/1755-1315/42/1/012026

reservoir rocks was volcanic rocks, even the manifestation is located in sedimentary formation.

4.4 Isotopes
A previous isotope study from Pusat Sumber Daya Geologi [10] showed that water samples from Sarawet Hot spring experienced meteoric water enrichment. This is indicated by the results of data that located on the meteoric Water Line area. It displayed that the hot spring is just surface water.

5. Structural Geology
The location, shape and size of upwelling zones near the surface reflect their structural control. Fluid flows in active tectonic areas are mainly controlled by fault and fracture. The fault patterns in Minahasa were a result of Celebes Sea Plate movement from the north and Tomini Microplate from the south. Two stresses from North to South caused the north arm of Sulawesi moved eastward and collide with westward movement of Moluccas oceanic plates. Structurally, the Minahasa fault pattern is majorly strike-slip fault trend in NE-SW, NW-SE and normal faults trending N-S [11]. It is considered that structural pattern around Sarawet hot spring were caused by North-South stress. Fracture and fault analysis done by Digital Elevation Model (DEM) and Regional Geology Map showed that Sarawet hot spring was related to high fault and fracture density area. The most intensive fault and fracture area in research area is placed near Sarawet and as displayed in Fault and Fracture Density (FFD) Map below (Figure 6).

Based on hot spring location with respect to fault, there are two primary mechanism of forming and sustaining fracture permeability in the surface, dynamically maintained fracture system and kinematically maintained fracture systems [5]. The fracture system correspond to the breakdown region because of fault interaction and locked fault intersection is called "dynamically maintained fracture system", while "kinematically maintained fractured systems" refer to slip along fault traces or re-open pre-existing fracture network at slipping fault interaction. This process would affect hydrothermal outflow. We need to analyze data through surface and/or subsurface data to determine the mechanism.

Based on Digital Elevation Model analysis compared with Regional Geology Map [7] (Figure 7), the nearest structure in Sarawet was a pair of NE-SW trending strike slip located in Kualabato (1) and Maliambo (2). Both of them are sinistral. The western-strike slip (1) was trending N60^{o}E and formed as first order and the eastern strike slip (2) was trending N33^{o}E and formed as second order. DEM analysis (Figure 6) showed that those paired of strike slip separated the north compartment of Minahasa to 3 parts; A, B, and C. A was composed of alluvium and Pliocene rocks, B was composed of Pliosen and Quartenary volcanic rocks and characterized by high density of lineament in NW-SE trends and C was composed of Quartenary volcanic rocks. The position of B relatively lower than C.

Dynamic analysis through both sinistral strike slip pattern done by comparing Riedel Shear Model with the main stress in the research area indicated that the western-strike slip is R-Shear and eastern part is P-Shear.

A paired of strike slip would form releasing or restraining step based on its movement. Faults that converge in one direction diverge in the opposite direction [3]. The subsidence of the tip with divergence strike slip faults of the research area could be sketched on Figure 8.Down region was a merged breakdown-region from the interaction of two fault line. The breakdown area has intense fracture and it could be related to the formation of permeability zone. Its form and size were influenced by geometry and kinematics of each fault [5].

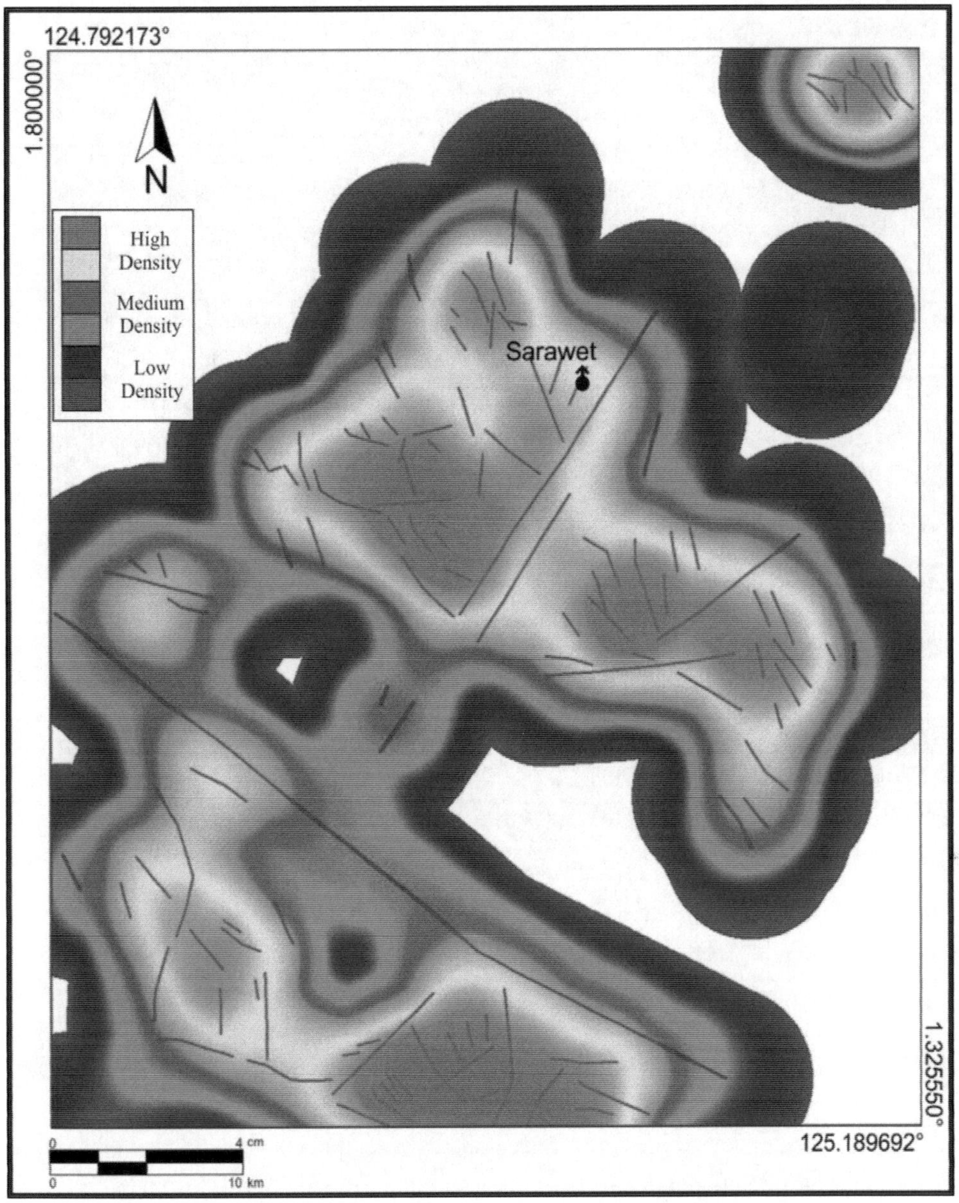

Figure 6. Fault and Fracture Density Map of North Minahasa Area.

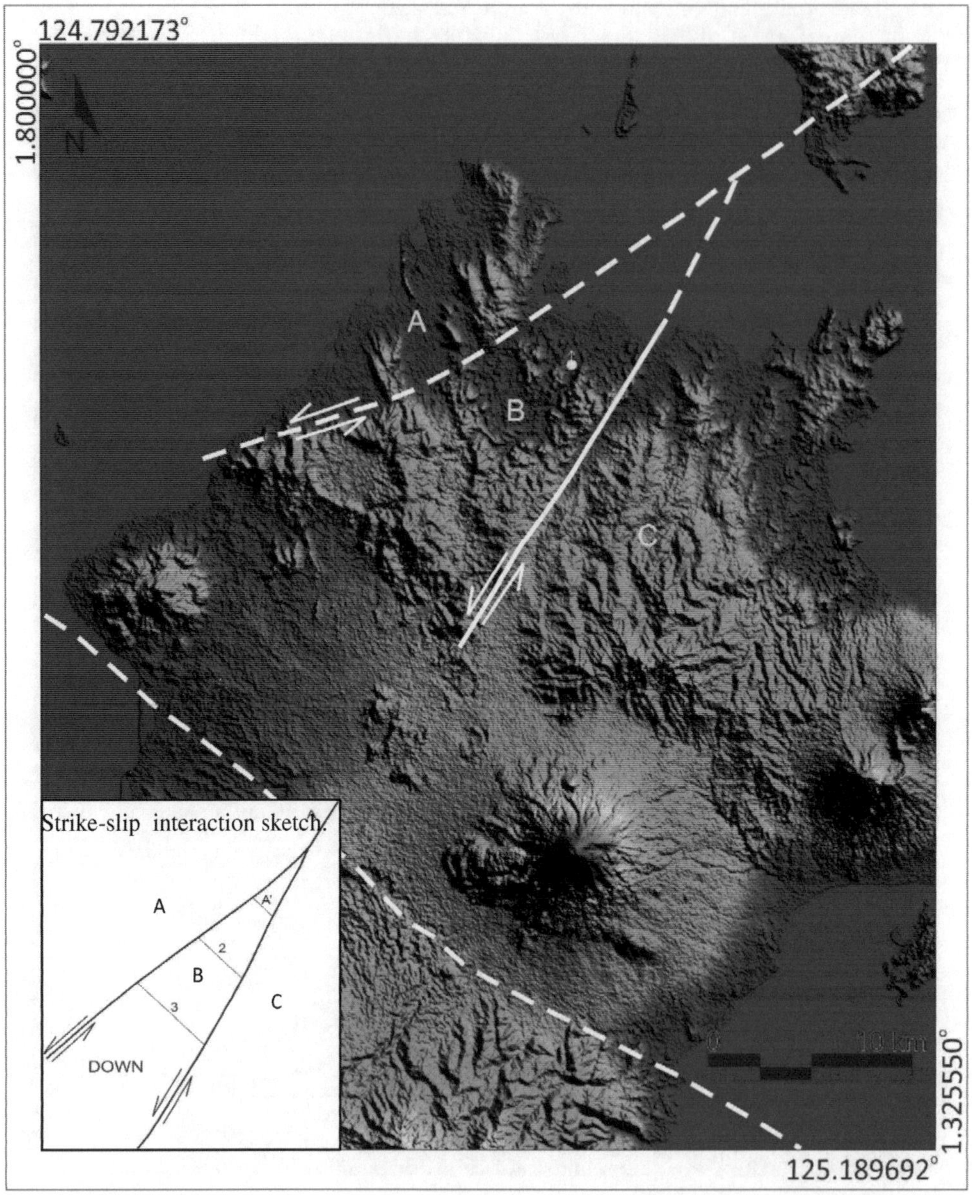

Figure 7. Fault and Fracture Analysis Based on Digital Elevation Model.

6. Heat Characteristic

Heat characteristic figure can affected by structural systems, for example in discontinuity area near fault system. In this paper, heat characteristic analysis were done to get the relationship between heat characteristic pattern and structural control of the research area. Heatflow map, thermal gradient map, and conductivity map was made based on 4 well data named Lapangan-well, Bor 1-well, Bor.1-well, and Bor 2-well. The location of each well were showed at the map.

6.1 Heat Flow

Figure 8 below showed the heat flow within the research area. Heat flows constantly from the mantle to the surface through conduction. Mean heat flowin Minahasa area was 300 mW/m^2 over continental crust . The heat flow values were significantly higher in the western part and there was anomaly in the centre part of Minahasa. Blue-purple area near LAPANGAN-Well were interpreted as damage zone

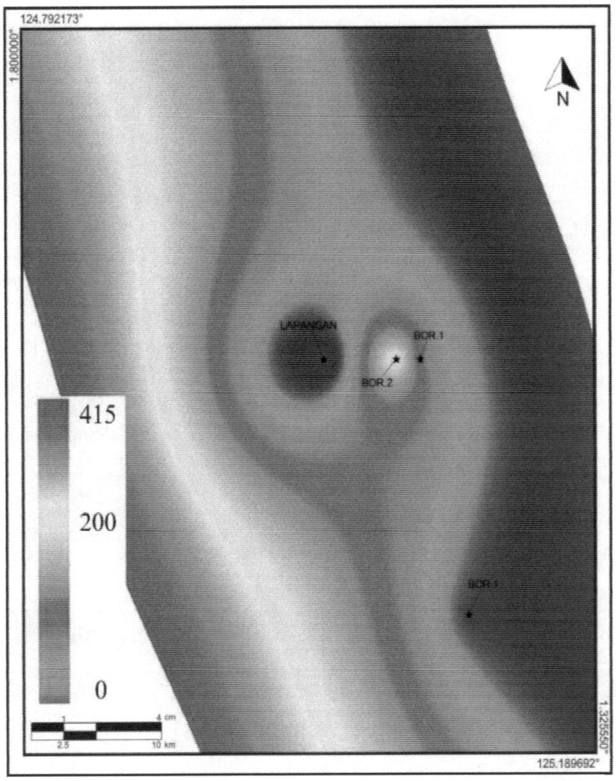

Figure 8. Heatflow Map of North Minahasa area showed the anomaly in the central part near Sarawet hotspring.

were Mode I fracture are concentrated.

6.2 Thermal Gradient

Geothermal gradient is the changes of temperature per depth. Figure 9 showed gradient data from the exploration wells based on "All Heatflow Data" from Royal Holloway South East Asia Research Group [1]. Based on Thermal Gradient Map below, the thermal gradient value was lowered to the South-western part of Minahasa. The distribution of the same thermal values were trending NW-SE but there is small anomaly point which has higher geothermal gradient than its surrounding. The anomaly point was located near the Sarawet hot spring manifestation.

6.3 Thermal Conductivity

IOP Conf. Series: Earth and Environmental Science **42** (2016) 012026 doi:10.1088/1755-1315/42/1/012026

Conductivity is an ability of a material to transfer heat energy. Based on Thermal Conductivity Map (Figure 10), the thermal gradient value was lowered to the South-western part of Minahasa. The distribution of the same conductivity values were trending NW-SE. Different with thermal gradient and heatflow map which showed anomaly point near the manifestation, the distribution of thermal conductivity value in the research area was relatively homogen along NW-SE trends.

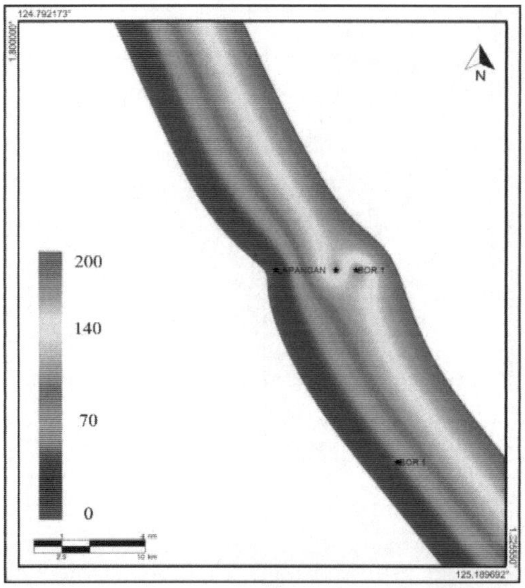

Figure 9. Thermal Gradient Map showed anoaly distribution value near Sarawet hotspring (Between Lapangan dan Bor-well).

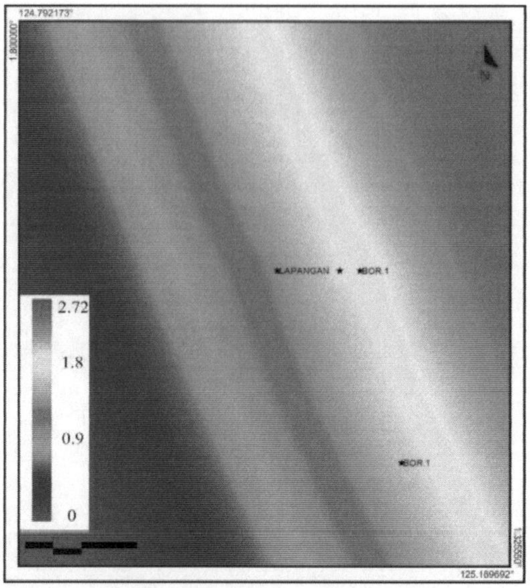

Figure 10. Thermal Conductivity Map **showed the NW-SE trending of conductivity value distribution**.

7. Discussion

According to the host rocks, geochemistry data and isotope data, Sarawet hot spring was classified to bicarbonate and immature water. The hot spring manifestation was located at Sandstone and Breccia (Tps) [7] formation which predominantly composed of sandstone but geochemically its reservoir was volcanic rocks. It was interpreted that there were near surface reaction between hydrothermal fluids and its surrounding and it showed the possibility of fluid mixing with ground water or surface water. It was consist of 2 fluid components, the meteoric water and groundwater. Volcanic gasses from North Minahasa Volcanic Complex were absorbed by meteoric water and then mixing with groundwater along the path to generate thermal water. The thermal water then flows through its outflow system that conducted by fault and fracture.

It's normal because Sarawet hot spring located in the coastal lowland area with sedimentary rock formation which has higher primary permeability than its surrounding. The origin of this water was composed Sarawet hot spring outflow system was mainly controlled by fault and fracture. Generally, the regional distribution of hydrothermal activity may be controlled by the large scale fault geometry while the local hot spring may be controlled by the fracture or lineament around the main fault. Lateral outflow of manifestations is localized by the fracture in damaged zone. Sarawet hot spring can be referred as local hot spring located between two fault zones. Primarily mechanism of forming and sustaining fracture permeability around the research area were related to breakdown region between two faults. So, it's forming and sustaining fracture permeability system was primarily controlled by dynamic maintained fracture systems related to fault interaction. Fractured formed within the breakdown region and provide high permeability fluid flow conduits. The formation of hot spring manifestation was related to extension fracture and releasing step.

8. Conclusion
Sarawet hotspring was found in sedimentary host rock but the reservoir was volcanic rocks. The chemical composition of the geothermal water suggested it was bicarbonate peripheral fluid and influenced by groundwater mixing. The lateral outflow of the thermal water were influenced by fault and fracture that produce permeability zone through dynamic maintained fracture systems related to fault interaction and releasing step.

9. References
[1] All Heat Flow Data (Open Access) Royal Holloway South East Asia Research Group (http://searg.rhul.ac.uk/current-research/heat-flow/, accessed Desember 23rd, 2015)
[2] Beardsmore G, Rybach L, Blackwell D and Baron C 2010 A Protocol for Estimating and Mapping Global EGS Potential *Australian Geothermal Conference 2010* 319
[3] Christie-Blick N and Biddle K T 1985 Strike Slip, Deformation, Basin Formation, and Sedimentation, SEPM Spec. Pub **37**
[4] Clauser C and Huenges E 1995 Thermal Conductivity of Rocks and Minerals in Ahrens, T.J. ed., Rock Physics and Phase Relations: A Handbook of Physical Constants (United States : American Geophysical Union Shelf)
[5] Curewitz D and Karso J A 1997 Structural Setting of Hydrothermal Outflow: Fracture Permeability Maintained by Fault Propagation and Interaction *Journal of Volcanology and Geothermal Research* **79** 149-168
[6] Dolgorjay O 2009 Geochemical Characterization of Thermal Fluids from the Khangay Area, Central Mongolia (United States : Geothermal Training Programme United Nation University)
[7] Effendi A C and Bawono S S 1997 *Peta Geologi Lembar Manado Sulawesi Utara Skala 1:250.000* (Bandung: Pusat Penelitian dan Pengembangan Geologi)
[8] Giggenbach W F 1988 Geothermal Solute Equilibria; Derivation of Na-K-Mg-Ca Geoindicators *Geochim. Cosmochim. Acta* **52**
[9] Mnjokava T T 2007 Interpretation of Exploration Geochemical Data for Geothermal Fluids from the Geothermal Field of the Rungwe Volcanic Area, SW-Tanzania (Tokyo : United Nation University)
[10] Setiawan D I, Mulyadi E, Sundoro H 2012 Penyelidikan Pendahuluan Geologi dan Geokimia Daerah Panas Bumi Kabupaten Minahasa Utara dan Kota Bitung, Provinsi Sulawesi Utara (Bandung : Pusat Sumber Daya Geologi)
[11] Siahaan E E, Soemarina S, Fauzi A, Silitonga T, Azimudin T, Raharjo I B 2005 Tectonism and Volcanism Study in the Minahasa Copartment of the North Arm of Sulawesi Related to Lahendong Geothermal Field, Indonesia *Proceedings World Geothermal Congress, Turkey*
[12] Smith D L and Fishkin L 1988 S New Heat Flow Investigations in Arkansas. Contributions to the Geology of Arkansas, *Arkansas Geological Commission. Misc. Pub 18-C* **III**

IOP Conf. Series: Earth and Environmental Science **42** (2016) 012027 doi:10.1088/1755-1315/42/1/012027

Volcanostratigraphic Study in Constructing Volcano Chronology and Its Implication for Geothermal Resource Estimation; Case Study Mount Sawal, West Java

F A Dermawan, H Hamka, R T A Malik, J Y Sianipar and Q S Ramadhan

Magister Program of Geothermal Engineering, Faculty of Mining and Petroleum Engineering, Bandung Institute of Technology (ITB), Jl. Ganesha no. 10, Indonesia 40132

Email: fikri.adam.derma@gmail.com

Abstract. One of the researches that should be done before carrying out a preliminary survey on the geothermal exploration with a volcanic system or volcanic-hydrothermal is by studying the volcanic stratigraphy. Determining the center of the volcanic eruption and its distribution based on the volcanostratigraphic study will be very helpful in a direct mapping that will be implemented, given that the type and characteristics of volcanic rocks are nearly the same between one source of the eruption and the other. On this case, volcanostratigraphic study had been done on Mount Sawal, where a topographic map with a scale of 1: 100,000 is used to determine the center of eruption of each crowns, while another map with a scale of 1: 50,000 is used to identify the distribution of the monogenetic (Hummock) eruption products and crowns border in detail. It is found approximately three crowns, which are Langlayang, Sawal big crown, Pamokolan, and the Cikucang Hummock that is located on the southern edge of the Langlayang crater. These Hummock and Crowns collection will be grouped into Tasik Bregade. Based on the volcanostratigraphic analysis, DEM, and geology, the chronology of how Tasik Bregade is formed is originally from the Langlayang, Sawal big Crowns, and Pamokolan. Tasik Bregade is classified into sub-mature potential geothermal system, from the analysis results, the potential magnitude of the electrical capacity contained in the system is around 0.74 to 1.24 MWe for 30 years, but further research needs to be done because of the detailed geological and other support data that are still lacking.

1. Introduction

Almost all geothermal system in Indonesia is associated with volcanic heat source which are distributed along subduction zone from Sumatera, Java, Bali, Flores, North Sulawesi, and Maluku archipelago. Generally these systems have high temperature (~250°C) and high enthalpy also large potential (> 50 MW) [3].

Surface geological mapping is a method to investigate and examine the geological condition of this area in particular to determine the volcanic center which often associates with intrusive body inferred as heat source[4]. However, mapping in volcanic area is a tricky task because it is difficult to distinguish the origin of various volcanic product due to their similarity. The problem will become worse when ones encounter volcanic complex in an intensive erosion stage [5]. Therefore volcanostratigraphic study is really helpful to classify and group the eruption product to interpret its eruption center.

This research was conducted in northern Tasikmalaya area, Mount Sawal and its surrounding, using 1:100.000 topographic map for regional analysis and 1:50.000 topographic map for detail analysis. This volcanic complex is classified into type "B" that is a volcano where its eruption activity hasn't been recorded or never shows any activities since 1600, A.D. but it still has forming a volcanic cone [6]. The topographic analysis will be grouped based on volcanostratigraphic classification, namely Hummock (Gumuk), Crown (Khuluk), and Bregade (Bregada) [7,8]. These groups are based on ridge lineament analysis, drainage pattern, eruption source with converge contour, circular feature that indicate an eruption crater.

2. Methodology

The first step of volcanostratigraphic method according [5] and Bronto, S (Personal communication) begins with determining the peaks of volcano marked by geodetic elevation. This determination is followed by digital elevation model (DEM) analysis because not all peaks from the highest topography represent the mountain peak, may be it is a crater rim or the remnants of past volcanic eruption (Bronto, S., Personal communication). The next step is identifying some circular features which mark by more or less concentric depressions. These circular features could be a guidance for distribution of respective eruption products. The next step is tracing every drainage pattern by using blue color as the mark of river flow. In general, the common drainage system encountered in volcanic setting is more or less a radial pattern, this pattern also shows the volcanic center/peak and may last well after the volcano has been eroded [9]. Tracing the ridge patterns can be done at the same time as drainage pattern or after. The ridge lineament is particularly flanked by rivers on each side, indicating volcanic product flow direction. The boundary among units were determined by the previous steps and usually cut by the river. The boundaries of individual units were then marked with black dashed lines.

Eventually, the result of topographic map analysis will be correlated with digital elevation model to assure the peak of volcano. This is because in old volcanic systems which have undergone advance erosional state, conical shapes are remnant of hardest rock or resistant part from erosional process, so it is necessary to correlate with digital elevation model (Bronto, S., Personal communication). Digital elevation model can also assist in analyzing the age of volcano through morphological shape. Volcanoes with smooth morphological texture are classified into young age whereas old age with rough texture.

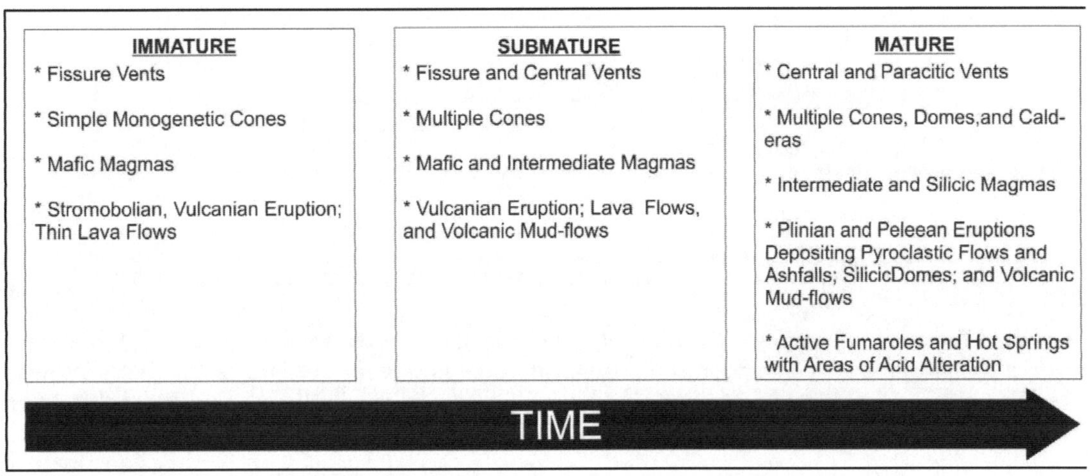

Figure 1. Geothermal Potential based on maturity degree modified from [10].

Geothermal resource potential assessment of Mount Sawal adopts several methods which were published by [10]. The first method is to classify the maturity of stratovolcano based on age (Figure 1). Every replenishment number of eruption will enhance the silicate content of magma, shallow the magma chamber, and enlarge the energy content.

The greater the potential of the volcano, stratovolcano will be more mature. This potential is supported by long-lasting heat source, which is sustained by multiple intrusions and broad volume of magma chamber (the magma's diameter is equal or bigger than caldera ring faults (Smith and Shaw (1975) in [10]).

Another method is by using flow chart guideline for geothermal potential which was proposed by [10,11]. There are five main factors to evaluate the geothermal potential in stratovolcano complex, namely (1) size and elevation of stratovolcano complex, (2) degree of magma evolution, (3) age of volcano, (4) stress regime or vents distribution, and (5) the presence of surface manifestation (Figure 2).

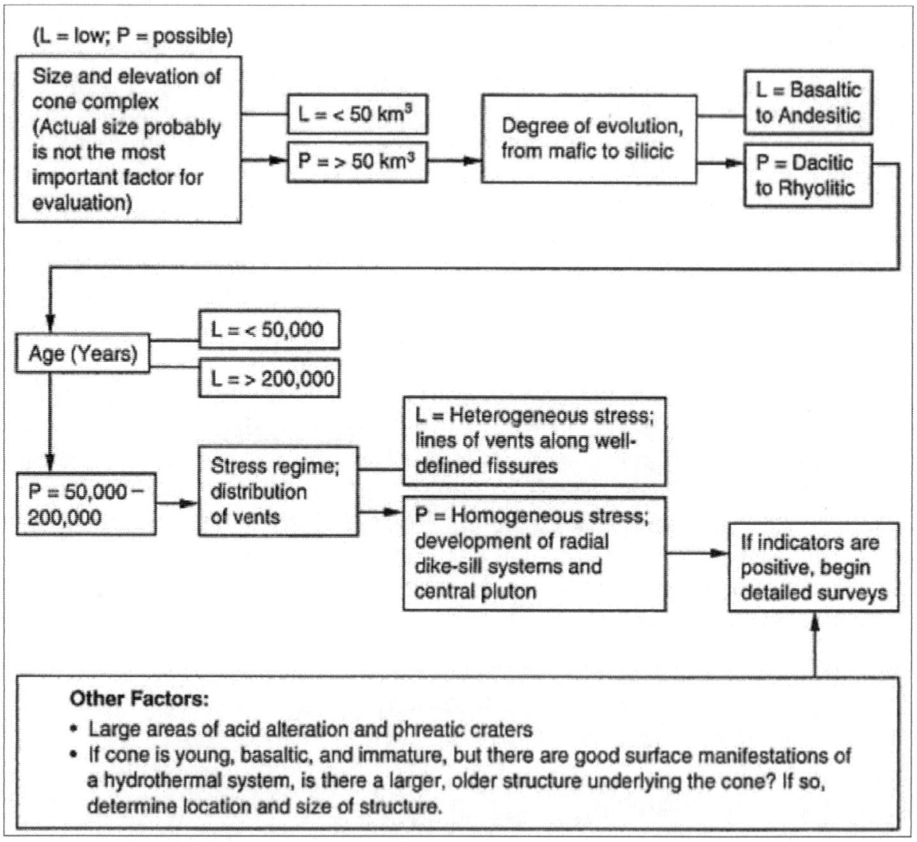

Figure 2. The evaluation guideline for geothermal potential in stratovolcano [10].

The last method of estimating geothermal potential is by evaluating the magmatic thermal energy contained within Mount Sawal's magma chamber. Based on Smith (1979) cited by [10] in their research, the volume of volcanic eruption product is only 10% from the volume of magma chamber. It is assumed that the product of volcanic eruption comprises with equal proportions of lava and pyroclastic (nearly 50% each). Based on all these assumptions, the resource estimation of thermal energy within Mount Sawal could be calculated by the following formulas:

$$V_{DRE}\ Pyroclastics = 0{,}6 \times 0{,}5 \times V\_Volcano \quad (1)$$

$$V_{DRE} Lava = 0{,}5 \times V_{Volcano} \quad (2)$$

$$Q = \rho \times (V_{DRE} Pyroclastic + V_{DRE} Lava) \times 10 \times H \quad (3)$$

Where V_{DRE} is volume of dense rock equivalent, means volume of igneous rock after all the pore space caused by vesiculation, fracturing, and intergranular porosity has been subtracted, ρ is rock density (kg/m3), H is magmatic heat content (kJ/kg), and Q is thermal energy (J) [5,10]. $V_{Volcano}$ in this equation is the volume of per-eruption and in this case the $V_{volcano}$ is from monogenetic eruption product (Cikurai Hummock).

3. Geological Setting

Regionally, the volcanic arc in Java Island was formed by subduction process between Indo-Australia oceanic plates with Eurasia continental plate (convergent plate boundaries) with nearly vertical orientation. This subduction process produce magma as partial melting with asthenosphere in depth and this magma will penetrate through the weak zone such as fracture and fault zone with basaltic to andesitic composition.

The interaction between Indo-Australia plate and Eurasia plate has been active since Late Cretaceous [12]. Eventually, this process leads to the regional structure evolution in Java Island. Based on Pulunggono and Martodjojo (1994) cited by [12], there are three dominant Java structure patterns, namely (1) Meratus direction (SW-NE) with Late Cretaceous epoch, (2) Sumatra direction (NW-SE) with Late Cretaceous to Paleocene epoch, (3) Sunda direction (N-S) with Eocene to Late Oligocene epoch and (4) Java direction (W-E) with Early Miocene epoch. West Java itself has been controlled by Sumatra, Sunda, and Java patterns (Figure 3).

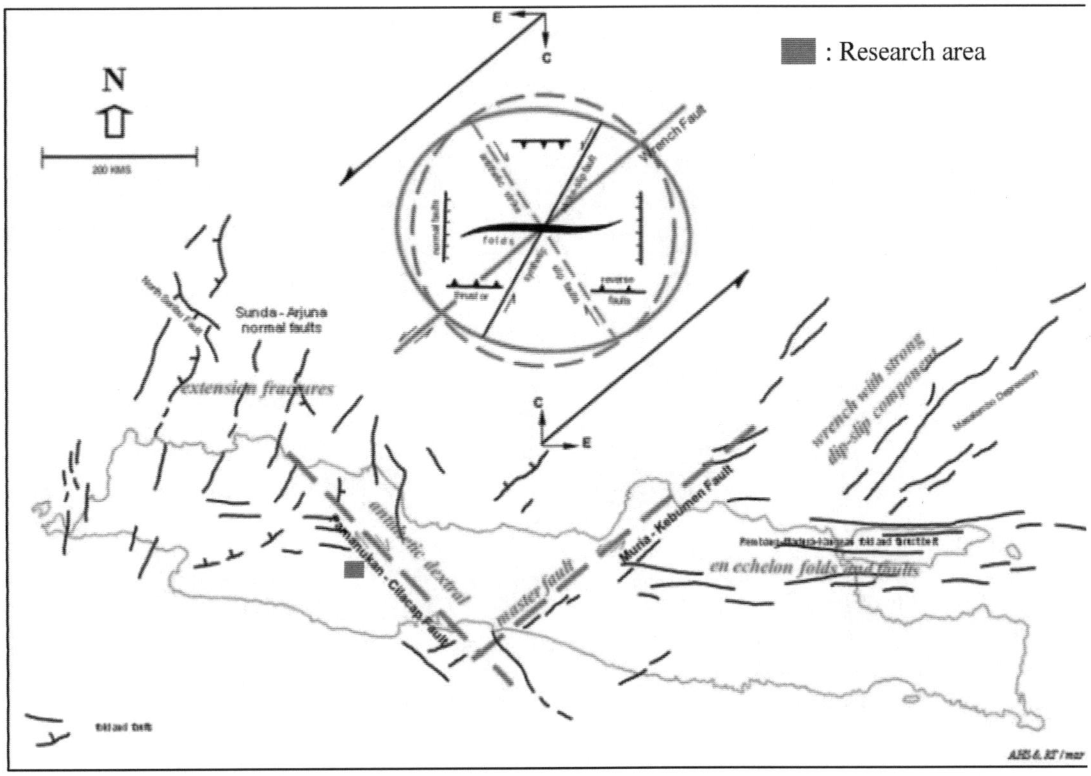

Figure 3. Geologic structure map of Java Island purposed by [12] based on strain ellipsoid kinematic. The main structure is Muria-Kebumen strike-slip in the eastern side of Island on the other side Pamanukan-Cilacap strike-slip as the antithetic.

[13] and [14] mention in their research that the extensional condition, which is produced by the regional structure, will (1) provide conduit for magma to ascend into the upper crust that may act as a

heat source for geothermal and (2) control the secondary permeability of the system. The regional structure that control volcanism and geothermal process in West Java is the West Java Fault system which major trend is N 300°E or NW-SE [13]. This regional structure has two main splay faults that are Salak-Gede-Patuha-Papandayan fault and Tangkuban Parahu-Karaha-Galunggung fault, where in the middle of them pull apart basin occurs as extensional conduit for magma intrusion (Figure 4). Extensional zone is formed by N-S orientation in West Java.

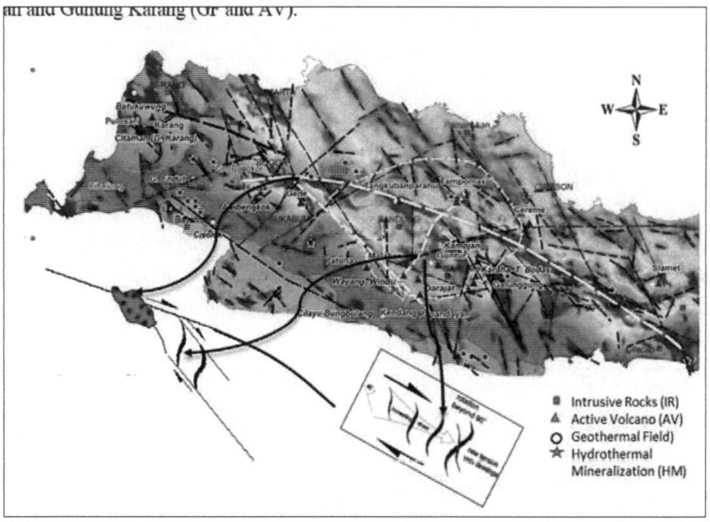

(a)

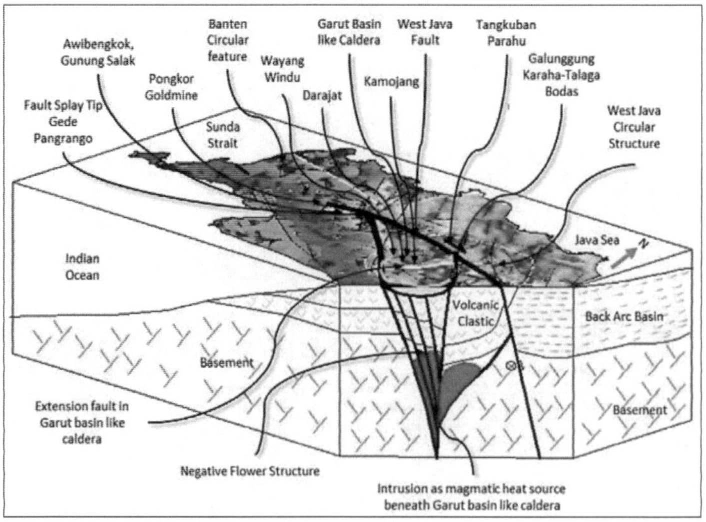

(b)

Figure 4. (a) Overlay map of volcanic product and geologic structure from anomaly bouger analysis. (b) 3D block diagram shows the negative flower structure which control West Java. This structure has two main splays, namely Salak-Gede-Patuha-Papandayan splay fault and Tangkuban Parahu-Karaha-Galunggung splay fault [13].

Jampang Formation, the oldest formation, lies in the southern area, also known as Old Andesitic Formation, was formed at Late Oligocene to Early Miocene and unconformity deposited with Halang Formation (Early Miocene) and Tapak Formation (Early Pliocene). Physiographically, Mount Sawal is categorized as Quarternary Volcanic that is composed of volcanic breccia, flow breccia, tuff, and andesitic to basaltic lava [15]. It is interpreted that Mount Sawal overlies Tapak and Halang Formation. The volcanic product of Mount Sawal is distributed in the E-NE map and inferred as Old and Young Sawal Crown eruption products (fFigure 5).

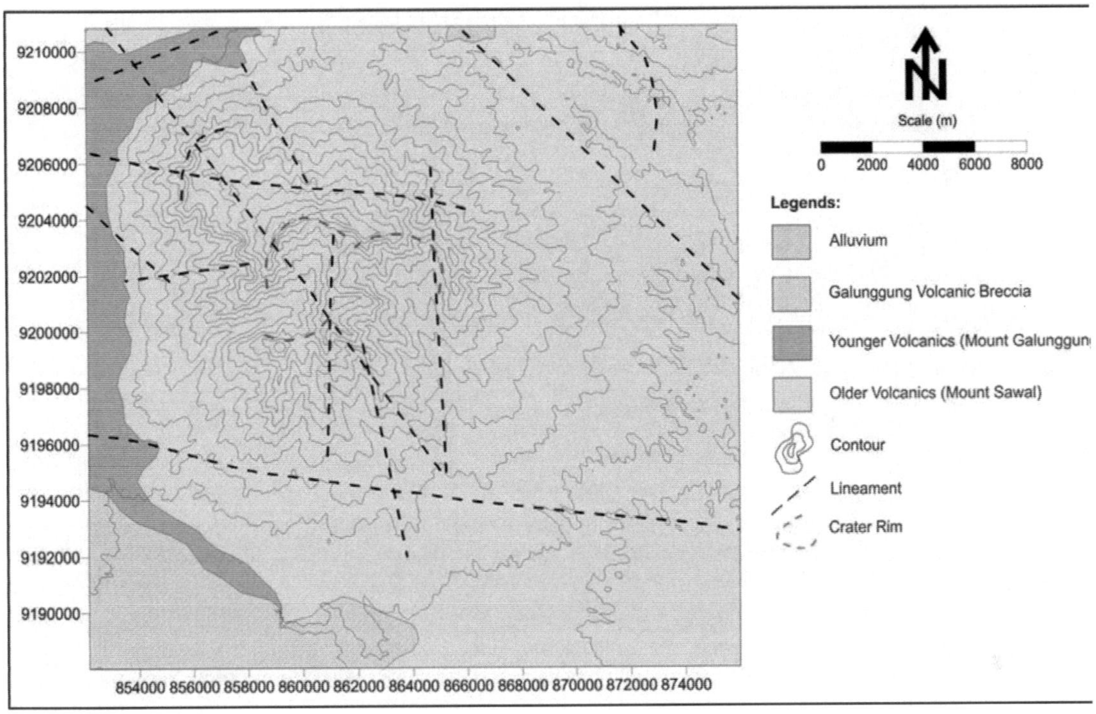

Figure 5. Geological map of Mount Sawal and its vicinity (Modified from [15].

The dominant orientations of geologic structures in Mount Sawal complex are NW-SE, N-S and W-E which resemble Java's regional structure (Sumatra, Sunda and Java directions). This orientation is interpreted as the result of Salak-Gede-Patuha-Papandayan splay fault and triggering the volcanism in this area.

4. Result and Discussion

There are at least three crowns which could be identified using 1:100.000 scale topographic map, namely Langlayang, Sawal and Pamokolan. Later on, Sawal Crown is divided into two sub-crowns based on rims and morphology roughness from DEM data. These sub-crowns are Old Sawal sub-Crown on east side of main Sawal Crown and Young Sawal sub-Crown on the west side. The smallest unit such as hummock (Cikuray Hummock) is also identified on this map where it's located in the southern rim of Langlayang crater (figure 6). The Hummock identification was also supported by DEM for validation. Consequently, this Mount Sawal complex is classified into Tasik Brigade [7,8].

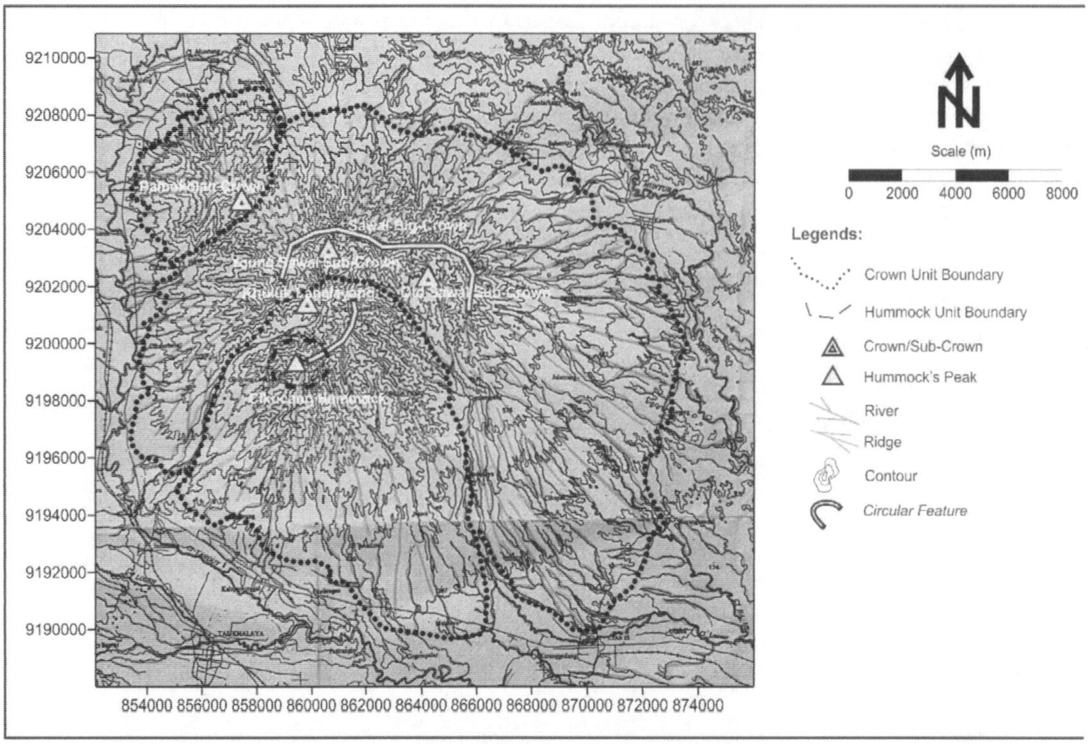

Figure 6. Volcanic stratigraphy distribution map based on topographic map (1:100,000 scale) analysis which consist of three crowns and one hummock. This Mount Sawal complex is classified into Tasik Brigade.

A more detailed topographic map, 1:50.000 scale, was analyzed to discover Cikucang Hummock boundary. This boundary will be used for $V_{volcano}$ calculation where the hummock is interpreted as the latest or a single body of monogenetic eruption product (Figure 7). Once obtained, $V_{volcano}$ value can be applied to estimate thermal energy of magma and geothermal potency for 30 years. Further analysis on DEM data indicates that in the southeastern part of Sawal Crown lies an eruption product or avalanche product (may be pyroclastic flow, or lahar or debris avalanche) from Sawal Crown eruption (Figure 8). It is important to conduct detailed geological mapping of Mount Sawal in the future to comprehend the geological condition and unit boundaries utterly.

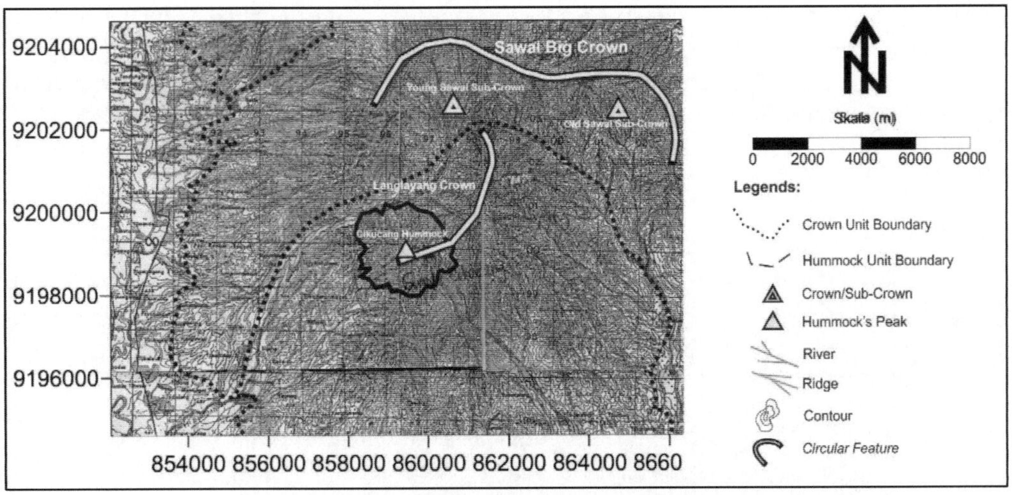

Figure 7. 1:50,000 topographic map was used for tracing unit boundaries of Cikucang Hummock.

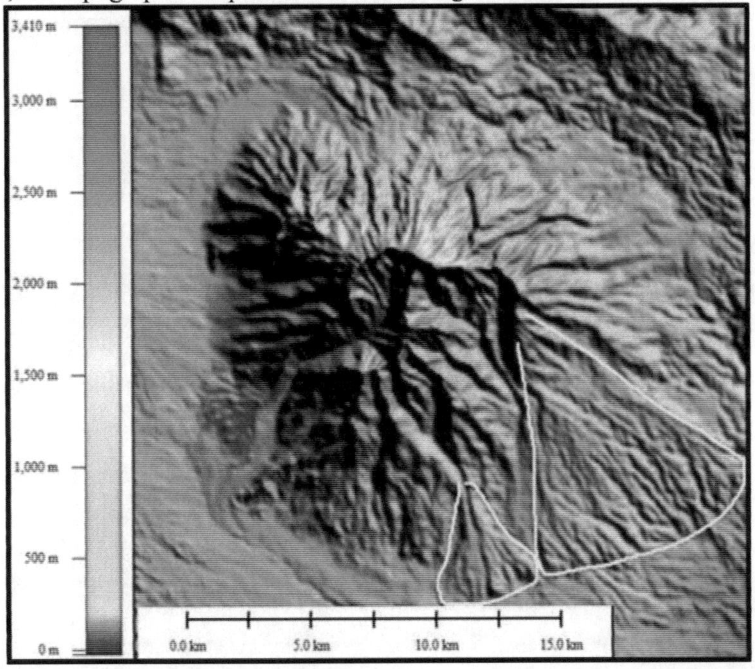

Figure 8. The distribution of Tasik Brigade volcanic product. Based on DEM analysis is interpreted into debris avalance from Sawal Crown eruption.

Hereafter, the volcanostratigrapy distribution result is overlaid with Mount Sawal geological map (Figure 9). The deployment of Mount Sawal product shows similarity with geological map, which are distributed in the center and eastern area map, but with the interpretation result from volcanic stratigraphy and DEM help to divide in detail each eruption product from eruption source.

The chronology of Tasik Brigade formation begins with Langlayang Crown formation and Old Sawal Hummock formation on the northeastern part of crown. N-S and NW-SE structure orientation allegedly played as the weak zone for magma to intrude (Bronto, S., Personal communication). As time goes by, Langlayang Crown erupted and extruded its material on the west flank while the Old Sawal Hummock evolves into sub-Crown. This evolution of Old Sawal was followed by Young Sawal Hummock on the west flank volcano. It is still questionable whether both of them erupted at the same time or the Old

Sawal first. Supposedly, Young Sawal sub-Crown eruption product was covered by Galunggung volcanic product. Subsequently, Pamokolan Hummock grew and arises on the northwestern flank of Sawal Crown, where today the Pamokolan Hummock has evolved into crown. The last eruption product is interpreted to occur at the southern rim of Langlayang Crown, Cikucang Hummcok. This interpretation is based on smooth morphology and clear radial drainage pattern. The eruption trend of Tasik Brigade is northeastern to northwestern direction (Figure 10).

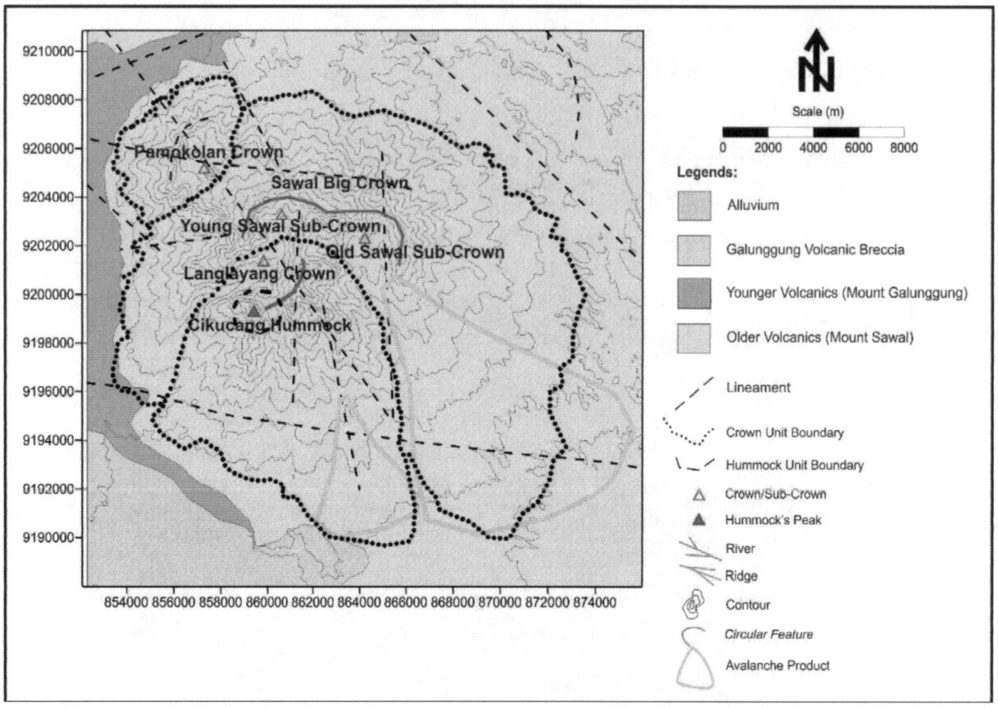

Figure 9. Overlay map between volcanic stratigraphy units, DEM analysis, and geological map of Mount Sawal and vicinity.

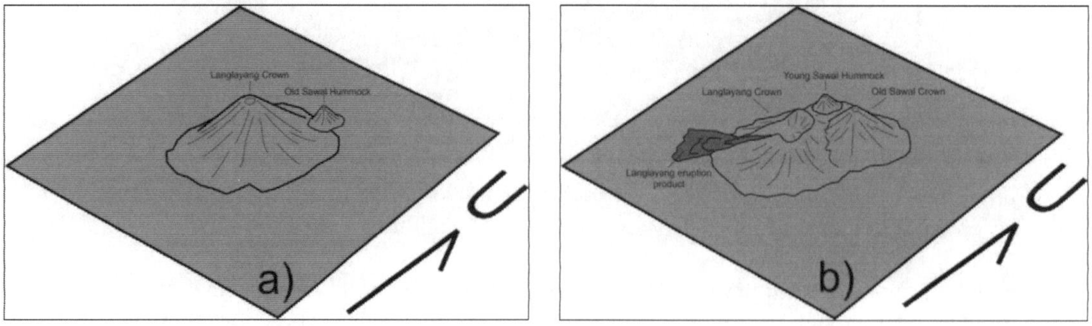

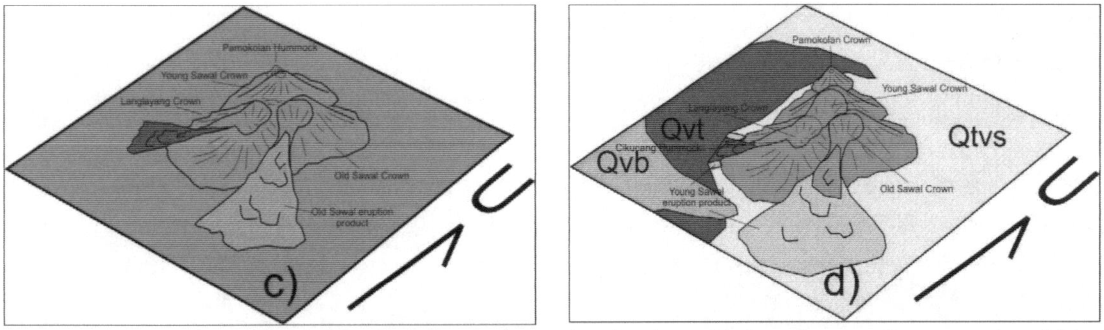

Figure 10. The chronology of Tasik Brigade formation begins with Langlayang Crown as the oldest formation, the Old sub-Crown Sawal, the Young sub-Crown Sawal, Pamokolan Crown and ends with Cikucang Hummock. The eruption trend of Tasik Brigade is NE and NW which is controlled by N-S and NW-SE orientation. Figure (d) is current Tasik Brigade.

According to eruption product composition (basalt and andesitic lava), eruption center, multiple cone (Langlayang, Sawal, Pamokolan Crown) with lava dome (Cikucang Hummock) and vulcanian to pelean eruption type (eruption product was distributed on western side and southeastern side of Brigade), the geothermal potential of Tasik Brigade is categorized into submature type [10]. Detail research is still needed to evaluate the geothermal work flow potential as proposed by [10] in Tasik Brigade. Lack of data in Mount Sawal complex, such as the age of the last eruption product and the occurrence of geothermal manifestation, are affecting the certainty of this Brigade potential.

The last method uses thermal energy calculation which is contained in Brigade's magma chamber. The calculation was using Cikucang Hummock data as the last eruption product of Tasik Brigade. Cikucang Hummock is categorized into homogenetic product, which product consist of one type (Lava or pyroclastic only).

Cikucang Hummock's volume was obtained from the average diameter against its marked peak where the height is from the highest altitude in cross section profile. The hummock volume from calculation is 0,613 km3. Since Hummock is a homogenic eruption product, therefore the V_{DRE} is V_{Volume} of itself (In this case, there are two possibilities of lithology which construct the Hummock, lava or pyroclastic). The thermal energy calculation result is using formula (1) to (3), where the composition is considered as lava, yields the energy contained in Tasik Brigade magma chamber about 11,77 x 1015J or equal to 12.4 MWthermal with 30 years life time production. If the Hummock composition is pyroclastic then the thermal energy contains about 7.06 x 1015J or 7.4 MWthermal/ 30 years life time. Both of these calculation results will be multiplied 0.1 as electrical conversion factor [16] and yield electric capacity ranged from 0.74 MWe up to 1.24 MWe, yet this conversion factor can be range from 0.08 to 0.2 [17]. However, the result was still *an underestimation* since the calculation was only based on morphology without heat loss calculation from geothermal manifestation.

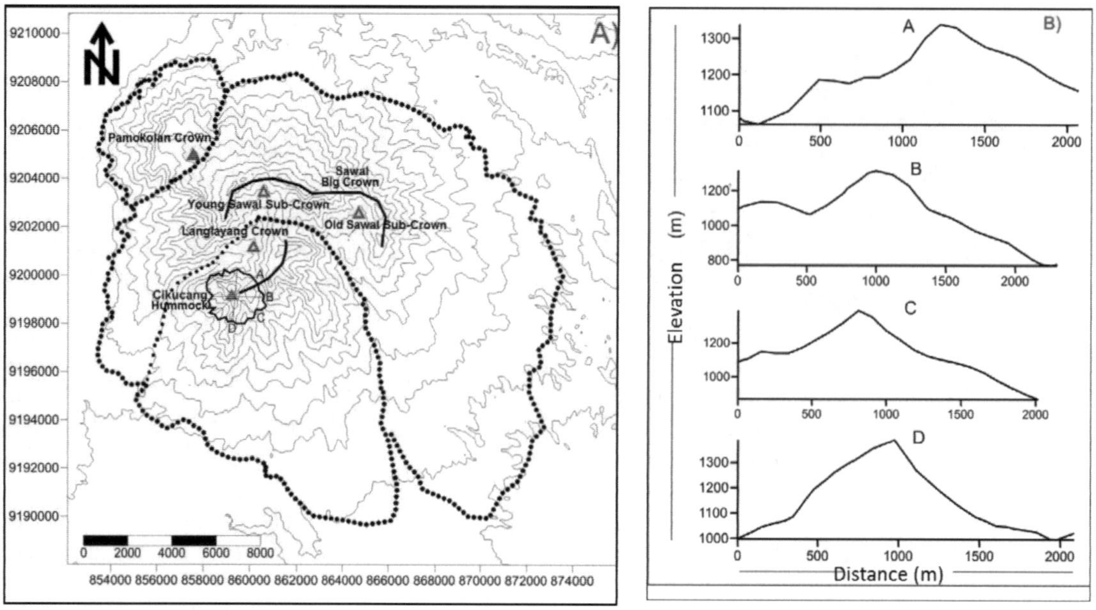

Figure 11. (A) Topographic map with cross section line in Cikucang Hummock. (B) Distance vs elevation profile of Cikucang Hummock which shows a relatively conical shape.

5. Conclusion

Mount Sawal complex has three eruption center (Crown) and one eruption side/monogenetic eruption (Hummock) based on 1:100,000 and 1:50,000 scale topographic map which classified into Tasik Brigade. The distribution of volcanic stratigraphy also correlates with geological map, but the volcanic stratigraphy interpretation has more detailed distribution and boundaries from each eruption product. Digital elevation model analysis shows a debris avalanche/pyroclastic flow on the southeastern flank of Mount Sawal.

Regional structure with N-S and NW-SE orientation presumably control the magma movement in depth or categorized as extensional zone. These structures are also affecting the distribution of volcano eruption pattern (NE and NW pattern). Amalgamating all of the analysis, the chronology of Tasik Brigade formation starts from Langlayang Crown as the oldest, followed by Sawal Crown, Pamokolan Crown and Cikucang Hummock.

Following the evaluation guideline for geothermal potential in stratovolcano, Tasik Brigade is categorized into submature type which possesses electric capacity range from 0.74 MWe up to 1.24 MWe with 30 years life time of production.

Acknowledgement

The writers would like to thank to Dr. Eng. Suryantini for the guidance during volcanology and geothermal system lectures and inspiring us about this research. The writers also would like to thank to Prof. Sutikno Bronto for providing us the opportunity, knowledge, times and discussion which is beneficial to this research. Thanks are given to students of geothermal magister program for the advice and sharing.

References

[1] Poernomo A, Satar S, Effendi P, Kusuma A, Azimudin T and Sudarwo S 2015 *An Overview of Indonesia Geothermal Development – Current Status and Its Challenges*, Proceedings World Geothermal Congress, Melbourne, Australia.

[2] Hochstein M P and Sudarman S 2008 *History of Geothermal Exploration in Indonesia from 1970 to 2000,* Geothermics, **37**: 220-266.

[3] Kasbani 2011 *Tipe Sistem Panasbumi di Indonesia dan Estimasi Potensi Energinya*, Kelompok Penelitian Panasbumi PMG-Badan Geologi, Bandung.

[4] Badan Standardisasi Nasional (BSN) 1998 *Klasifikasi Potensi Energi Panas Bumi di Indonesia,* SNI 13-5012-1998, ICS 73.020.

[5] Putra S D H, Rizki R and Akbar A F 2014 *Volcanostratigraphic Study and Its Implication to The Geothermal Resource Estimation of Mount Wilis, East Java,* Proceedings, 3rd International Geothermal Workshop, ITB, Bandung, Indonesia.

[6] Pratomo I 2006 *Klasifikasi Gunung Api Aktif Indonesia Studi Kasus dari Beberapa Letusan Gunung Api Dalam Sejarah,* Jurnal Geologi Indonesia, **1**: 209-227.

[7] Martodjojo S and Djuhaeni 1996 *Sandi Stratigrafi Indonesia*, Komisi Sandi Stratigrafi Indonesia, IAGI, Jakarta, 25.

[8] Bronto S, Sianipar J Y and Pratopo A K In press *Volcanostratigraphy for supporting geothermal exploration,* 5th International Geothermal Workshop, ITB, Bandung, Indonesia

[9] Huggett R J 2007 *Fundamentals of Geomorphology,* Routledge, Taylor and Francis Group. New York, pp 135-138.

[10] Wohletz K and Heiken G 1992 *Volcanology and Geothermal Energy,* Los Alamos Series in Basic and Applied Sciences. California.Sdf

[11] Suryantini 2015 *Volcanological Approach for Evaluation of Geothermal Potential in Volcanic Associated Hydrothermal System at the Early Stage of Exploration,* Proceedings World Geothermal Congress, Melbourne, Australia.

[12] Satyana A H and Purwaningsih M E M 2002 *Lekukan Struktur Jawa Tengah: Suatu Segmentasi Sesar Mendatar,* Indonesian Association of Geologist (IAGI), Yogyakarta-Central Java Section, Indonesia.

[13] Fauzi A, Permana H, Indarto S and Gaffar E Z 2015 *Regional Structure Control on Geothermal Systems in West Java, Indonesia,* Proceedings World Geothermal Congress, Melbourne, Australia.

[14] Corbett G J and Leach T M 1997 *Southwest Pacific Rim Gold-copper Systems: Structure, Alteration and Mineralization,* Short Course Manual Draft, 5/87 Edn.

[15] Budhistrina T 1986 *Peta Geologi Lembar Tasikmalaya, Jawa Barat Skala 1: 100.000,* Pusat Penelitian dan Pengembangan Geologi.

[16] Badan Standardisasi Nasional (BSN) 2000 *Angka parameter dalam estimasi potensi energi panasbumi,* SNI 13-6482-2000, ICS 07.060.

[17] Williams C F, Reed M J and Marnier R H 2008 *A review of methods applied by the US Geological Survey in the assessment of identified geothermal resources,* US Geological Survey Open-File Report 1296(2008): 27

Volcanostratigraphic Approach for Evaluation of Geothermal Potential in Galunggung Volcano

Q S Ramadhan, J Y Sianipar and A K Pratopo

Geothermal Master Program, Institut Teknologi Bandung, Indonesia

E-mail : qodrisyahrurramadhan@gmail.com

Abstract. The geothermal systems in Indonesia are primarily associated with volcanoes. There are over 100 volcanoes located on Sumatra, Java, and in the eastern part of Indonesia. Volcanostratigraphy is one of the methods that is used in the early stage for the exploration of volcanic geothermal system to identify the characteristics of the volcano. The stratigraphy of Galunggung Volcano is identified based on 1:100.000 scale topographic map of Tasikmalaya sheet, 1:50.000 scale topographic map and also geological map. The schematic flowchart for evaluation of geothermal exploration is used to interpret and evaluate geothermal potential in volcanic regions. Volcanostratigraphy study has been done on Galunggung Volcano and Talaga Bodas Volcano, West Java, Indonesia. Based on the interpretation of topographic map and analysis of the dimension, rock composition, age and stress regime, we conclude that both Galunggung Volcano and Talaga Bodas Volcano have a geothermal resource potential that deserve further investigation.

1. Introduction

Determination of areas which have potential geothermal energy depends on the presence of geothermal manifestations at the surface. Existing geothermal manifestations are indicative of the geothermal potential of an area. However, a geothermal area may lack of surface manifestation. Lack of surface manifestation could cause difficulties during exploration to determine the potential of geothermal in an area.

Volcanostratigraphic approach is performed at the early stage of geothermal exploration to determine the geothermal potency before further exploration. Volcanostratigraphic mapping is a method used in the early exploration stage. This method is useful to distinguish and classify eruption products based on the eruption centers. This classification is important to delineate the eruptive sources as they indicate the presence of once active or still active magmatic intrusions related to heat source thus to make the study of volcanic rock mapping becoming more effective, focused, systematic and targeted.

Volcanostratigraphy analysis is based on topographic maps with scale of 1:100.000 and 1:50.000. Determination of stratigraphic units consist of determining centers of eruption, ridge lineaments, drainage pattern and the boundary of the volcano stratigraphic unit [1]. Eventually, topographic analysis will be grouped based on volcanostratigraphic classification, namely hummock (*gumuk*), crown (*khuluk*), and brigade (*bregada*) [2]. Interpretation of volcanostratigraphy is based on the relation between the topographic map analysis and regional geological map.

In this paper, we conducted evaluation of geothermal potential by adopting the evaluation of geothermal exploration in volcanic regions [3]. The evaluation consists of size and elevation of the cone complex, degree of magma evolution, age of the volcano, stress regime or distribution of the vents and other factors such as the occurrence of surface manifestations. The volume of the cone and the distribution of vents can be determined by volcanostratigraphic mapping. The evolution of magma and the maturity of the geothermal system can be analyzed in reference to the regional geological map.

Galunggung Volcano and Talaga Bodas Volcano are active quaternary volcanoes which are located in the southeastern area of Bandung. Galunggung Volcano has a large crater and is interpreted as the upflow zone of the geothermal system. Based on regional geological map, there are only as follows limited surface manifestation at Galunggung crater, such as cold water with bubble near neutral water at crater and Cipanas hotspring. Other manifestations present such as an alteration zone at the seepage location and warm spring located 25 km south of Galunggung Volcano, in Cigunung, and Cibalong while the geothermal potential of Talaga Bodas Volcano has been proven by exploration drilling. This paper will discuss the application of volcanostratigraphic approach in evaluating the potential of proven area such as at Talaga Bodas Volcano and unproven area like Galunggung Volcano.

2. Methodology

The basic volcanic mapping is using volcanostratigraphy [4]. The stratigraphic unit of a formation is equivalent to a crown, while the smaller units are considered as hummocks. Two or more volcanic crowns can be considered a brigade and a volcanostratigraphic unit which consists of many crowns, brigades and super brigades is an arc.

The first step to determine the volcanostratigraphic unit is delineating non-volcanic rock to confine the research area by overlaying the topographic map with regional geological map of the same scale. The next step is volcanostratigraphy mapping by identifying crown and hummock through the interpretation of topographic map scale 1:100.000 and more detailed identification through interpretation of topographic map scale of 1:50.000 so the hummock can be identified. This identification is only carried out on non-sedimentary units. Identifying the units of volcanostratigraphy such as crowns and hummocks was performed by determination of primary and secondary centers of eruptions. This identification can be used as a guide in estimating the distribution of eruption products. Important morphologic features such as circular structures, craters, or calderas are considered primary eruption markers.

The next step is to trace every single river path with blue to mark the drainage pattern. The path of the main river usually bounds two separate crowns. The next step is to trace lineaments which represent ridges with the color brown. The morphology of lineaments traced in brown usually represents one eruption center and will form a radial pattern.

Lineaments and river flow are identified using SRTM image which are aimed to determine the boundaries between hummock and crown. The boundaries are marked by black dashed line. The assumption that is used for determining volcanostratigraphic limit is when distribution of ridges and rivers form a more or less radial pattern, it is considered the same source [5].

The volcanostratigraphic unit, of a crown, is generally characterized by contour pattern of large volcanic cone or groups which may be linked genetically. While a hummock is generally characterized by contour pattern of small volcanic cone located near to a crown. Each crown and hummock that has been identified can be named according to the geography.

The next step is identification on a more detailed topographic map with scale of 1:50.000. The specific identification on a detailed scale is performed due to the possibility that one crown identified on a topographic map of 1:100.000 scale could turn out to be two crowns in 1:50.000 scale. On detailed topographic map of 1:50.000 scale, radial flow patterns of rivers and lineaments of ridges become detailed and clear.

After stratigraphic identification on topographic map, the next step is evaluation of geothermal exploration in volcanic regions using evaluation scheme. The scheme uses a flowchart with a series of parameters. These parameters including volume of volcanic complex, degree of magmatic evolution, age of volcano, stress regime, and other factors such as the presence of manifestations at the surface (figure 1). In the case of Galunggung Volcano, a simplifying assumption is used in that the edifice is cone-shaped such that its volume can be roughly approximated in equation (1).

$$V_{volcano} = \frac{1}{3} \times \pi \times r^2 \times t \quad (1)$$

Where, $V_{volcano}$ is volume of volcano (m^3)
r is radius of volcano (m)
t is elevation difference between the highest marked peak and the surrounding average lowest elevation respectively (m)

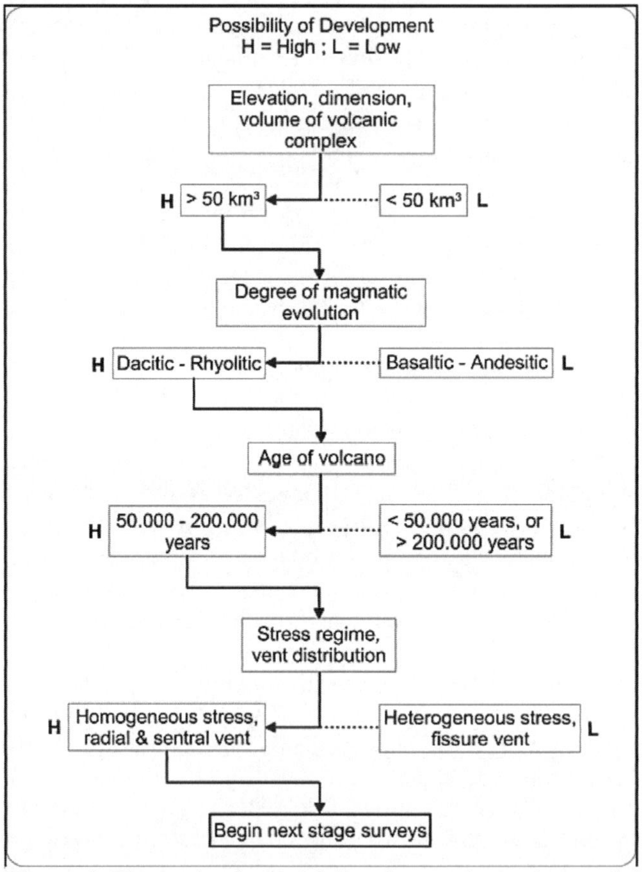

Figure 1. The evaluation guideline for geothermal potential in stratovolcano - modified from [3]

Evaluation of volcano's size is used to estimate the depth of a magma intrusion beneath the surface. The shallower magma intrusion, the greater potential into heat source of the geothermal system. The size of the volcano will help delineate the area of magma chamber that lies beneath the surface. Such area may be associated with an area of reservoir.

Degree of volcano's maturity was estimated by comparing the volcanic parameters which is listed in figure 2. After understanding composition of magma, we can estimate the heat which is contained in volcano. Composition of magma changes from mafic to silicic along with rising magma to the surface. Shallow magma confined close to the subsurface causes assimilation with the surrounding rocks which increase silica content. Besides, a number of intrusions cause mixing between shallow magma and deep magma which can move upward to the surface.

The composition of magma can also be expected from the type of volcanic eruption. Strombolian and vulcanian are types of eruption that can produce monogenetic cones of mafic magma. Central vent and fissure vent can produce complex cone of mafic-intermediate magma from vulcanian eruption type.

While the central vent can produce complex parasitic cone, dome and caldera of intermediate-silicic magma which type of eruption is plinian or pelean. When basaltic magma from crust flows up to the surface caused by fracture or fissure its composition will get changed through differentiation, mixing or assimilation. If fractures or fissures that act as magma discharge path are small, it will prevent magma to emanate to the surface so that its composition will get changed into intermediate or maybe silicic while prevented in subsurface. The more the process of mixing or differentiation that occurs the composition of magma will increasingly intermediates or silicic when emanates to the surface. Figure 2 shows that maturity degree of volcanoes can be characterized by its products that are pyroclastic flows and ash falls.

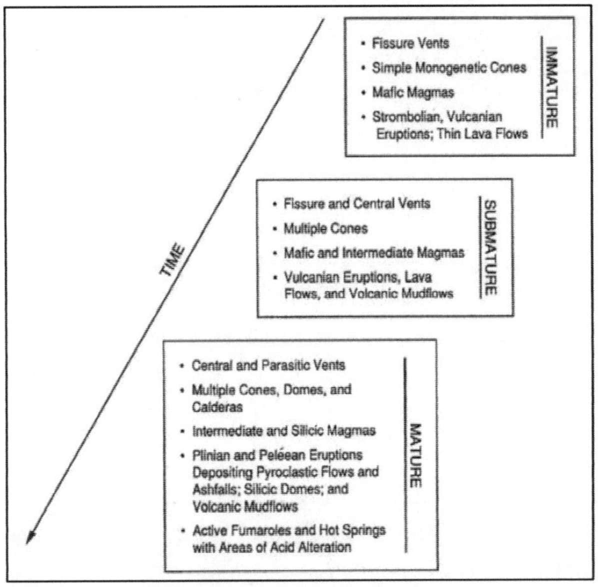

Figure 2. Geothermal potential based on maturity degree in volcanoes [3]

3. Result and discussion

Based on the volcanostratigraphic interpretation, there are five crown units that we have identified from the Tasikmalaya topographic map on 1:100,000 scale. These units are Galunggung, Talaga Bodas, Sawal, Cakrabuana and Sadakeling (figure 3).

IOP Conf. Series: Earth and Environmental Science **42** (2016) 012028 doi:10.1088/1755-1315/42/1/012028

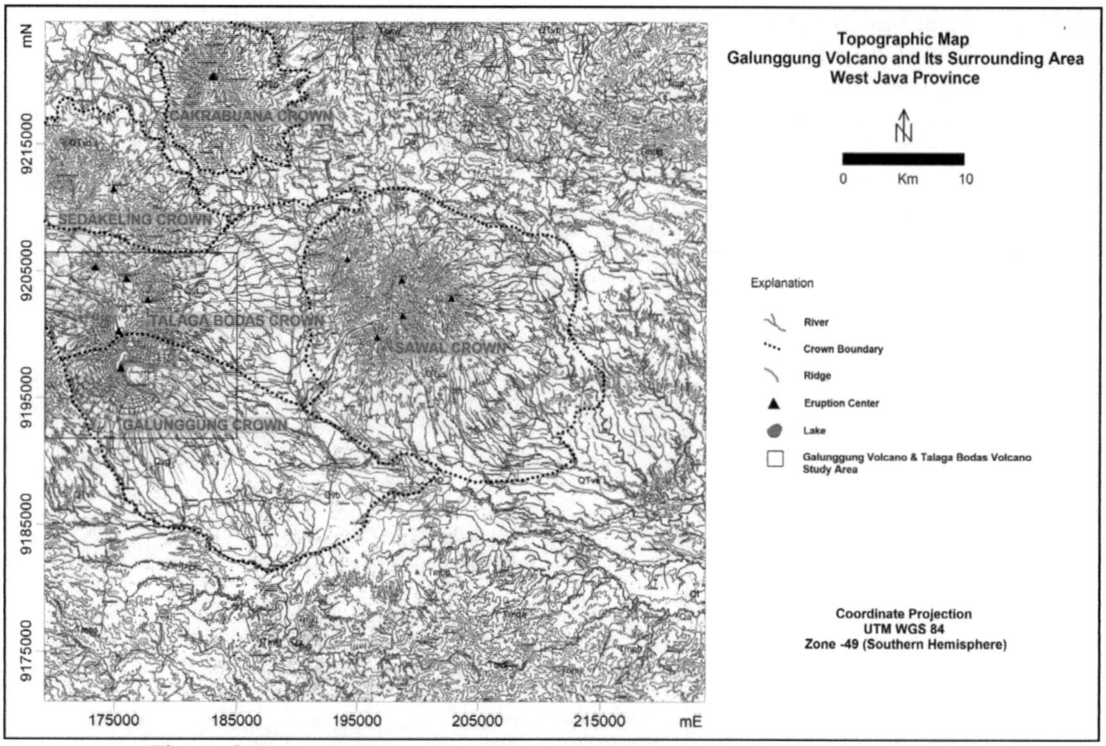

Figure 3. Topographic map of Galunggung Volcano 1:100.000 on scale [6]

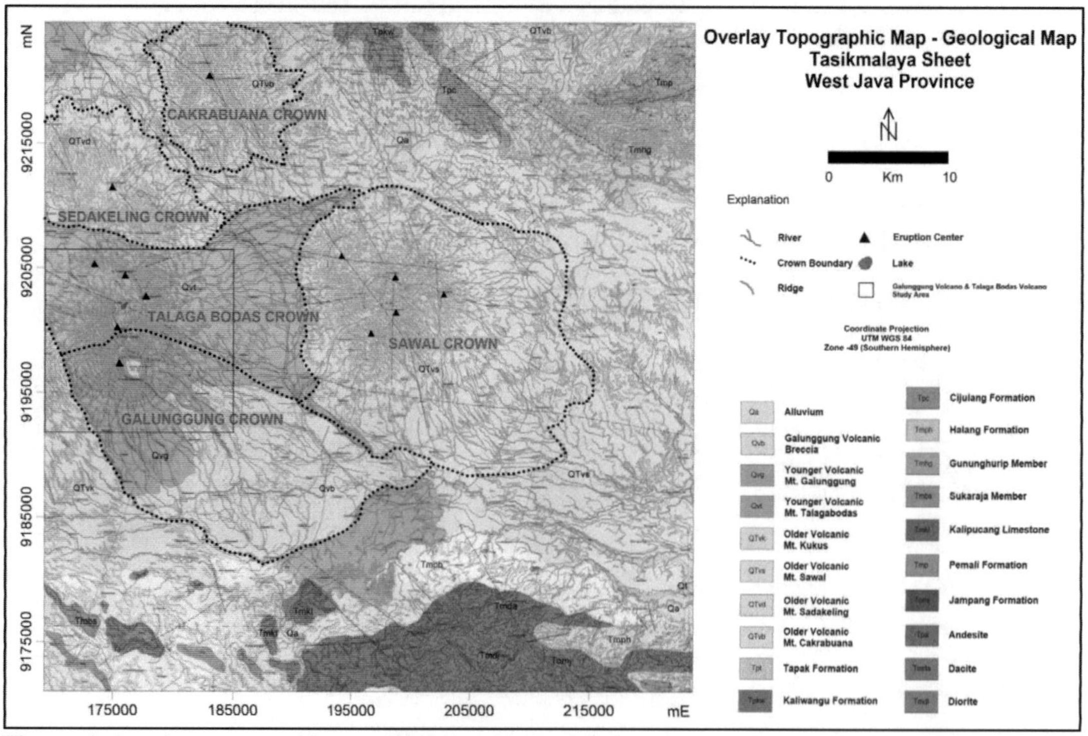

Figure 4. Overlay topographic - geological map [7] of Galunggung Volcano and its surrounding area

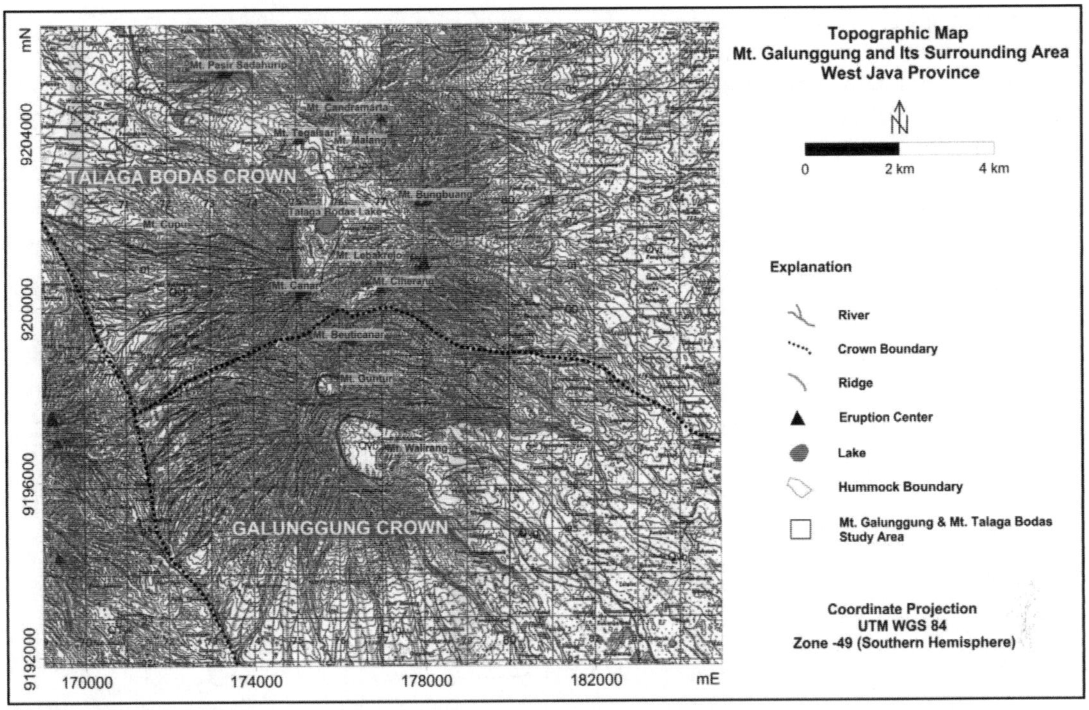

Figure 5. Topographic map of Galunggung Volcano 1:50.000 on scale and its surrounding area [8]

The interpretation results were then compared with published geological map of Tasikmalaya (figure 4) and there is relatively good correlation between the interpreted volcanostratigraphy unit boundaries and the distribution of volcanic lithological units. It also implies that there is a correlation between the distributions of eruption products and the morphology of volcano.

A more detailed topographic map on scale of 1:50,000 and satellite images was also analyzed for Galunggung Crown and Talaga Bodas Crown region on the northwest flank of Tasikmalaya. There are additional hummock units on Galunggung Crown and Talaga Bodas Crown which were previously unidentifiable. The additional hummock units on Galunggung Crown; are Beuticanar, Walirang and Guntur, while Talaga Bodas Crown has nine additional hummock units; there are Ciherang, Cupu, Tegalsari, Canar, Lebakrejo, Bungbuang, Malang, Candramata and Pasir Sadahurip. Guntur crater is an old summit crater of Galunggung composite volcano before collapsed producing horse shoe-shaped crater and Tasikmalaya debris avalanches deposits [9]. These hummock units and crown units are grouped as the Gandewa Brigade (figure 5). The analysis of geothermal resource and potential for development of Galunggung Crown and Talaga Bodas Crown was done by adopting the evaluation scheme in geothermal exploration [3]. The scheme uses a flowchart with a series of geological and physical parameters.

3.1. Galunggung Volcano

For the analysis of Galunggung Volcano dimension, the diameter is taken as the average distances between pairs of slopes with lowest inclination, while the elevation difference is taken from the Digital Elevation Model (DEM). The diameter of Galunggung Volcano is 32 km with elevation difference is 1,591 m; therefore the volume of Galunggung Volcano is 430 km^3. The large volume of the volcano indicates that Galunggung Volcano has a large potential for geothermal energy. Volcanic rocks of Galunggung Volcano consist of Old Galunggung Formation, Tasikmalaya Formation and Cibanjaran Formation. The first rock formation is volcanic rock that formed Galunggung Stratovolcano (age 10.000 - 50.000 years?) while the second formation overlies the rock which formed caldera (150 ± 4.200 years?) and the third formation were the erupted products from 1822, 1894, 1918 and 1982 - 1983 [9].

Old Galunggung Formation consists of intercalation of lava flows, pyroclastic and dyke. By using ^{14}C method, the age of this formation is 20.000 - 25.000 years, implying the entire activity of Galunggung Volcano took place 50.000 - 100.000 years ago. Tasikmalaya Formation consists of debris avalanche and pyroclastic flow. Debris avalanche is rocks in volcano that show contact between lava flows and pyroclastic deposit. Pyroclastic flows of Tasikmalaya Formation are grey-brown in color and unconsolidated. Cibanjaran Formation or Post Caldera consists of eruption products from Galunggung Volcano in 1822 as pyroclastic flow, in 1894 as pyroclastic fall, in 1982-1983 as pyroclastic flow that are rich in ash and bomb fragments with 5,6 x 10^6 m^3 in volume [9].

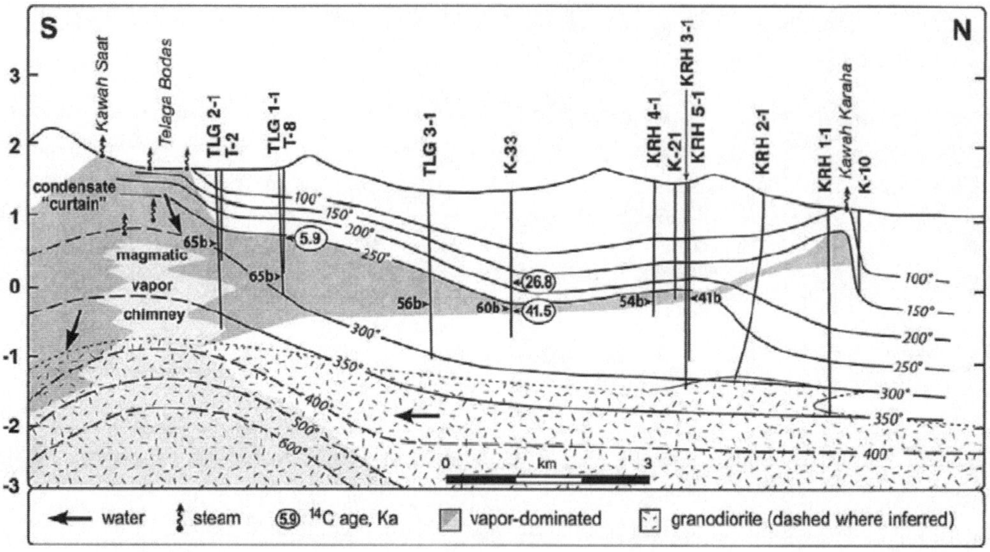

Figure 6. Conceptual model geothermal system of Karaha – Talaga Bodas [12]

Galunggung Volcano has homogeny stress regime, it can be seen since the spread of vent centralizes at peak of Galunggung Volcano. This means the plutonic rocks as heat source are located inside it. There are only as follows limited surface manifestation at Galunggung crater, such as cold water with bubble near neutral water at crater and Cipanas hotspring. Other manifestations present such as an alteration zone at the seepage location and warm spring located 25 km south of Galunggung Volcano, in Cigunung, and Cibalong. The temperature is 60°C. These manifestations are expected as outflow from Galunggung Volcano geothermal system [10]. Based on the evaluation for geothermal potential of stratovolcano [3], Galunggung Volcano has large potential for geothermal energy that can be developed and needs further detailed exploration.

3.2. Talaga Bodas Volcano

The second crown is Talaga Bodas. Talaga Bodas is a volcano located at northern of Galunggung Volcano. The difference between highest elevation and lowest elevation is 1.623 m with total area 126 km^2, thus its volume is 65 km^3. This large volume indicates there is a large heat source. The dominant volcanic rocks in Talaga Bodas are pyroclastic deposit, epiclastic, and the distribution of andesitic-basaltic lava flow. Based on age dating using ^{14}C methods, the age of the rocks of that area is from 41.500 ± 1.200 to 5.910 ± 96 years ago [11]. There are also some manifestations such as hot springs, fumarole and acid crater (Saat Crater).

Talaga Bodas crown also has homogeny stress regime where the dominant trend of geological structure is northwest-southeast and centralized vent. This phenomenon indicates that the maturity level of Talaga Bodas is mature and has good potential for geothermal energy. The potential of Talaga Bodas geothermal area has been proven by drilling results. The drilling was conducted by Karaha Bodas Co.

LLC by the end of 1990. There are 29 exploration wells with 3 km in depth. Figure 6 shows the geothermal system in Talaga Bodas. It is a high temperature system with the temperature at the bottom of well TLG 2-1 is 353°C.

4. Conclusion

Based on the criteria evaluation geothermal exploration [3], the degree of magma evolution, age, stress regime and the distribution of the crater, it can be concluded that Galunggung Volcano and Talaga Bodas Volcano are possible candidates for geothermal exploration and are to be investigated further. Based on the results of the study, show that volcanological approach can be applied in early stage of geothermal exploration activity in volcanic regions.

References

[1] Yuwono Y S 2004 Pemetaan daerah vulkanik (Bandung: Indonesia)
[2] Bronto S, Sianipar J Y, and Pratopo A K 2016 Volcanostratigraphy for supporting geothermal exploration. *On publishing process in J. Enviromental and Earth Science*
[3] Wohletz K and Heiken G 1992 Volcanology and Geothermal Energy (Berkeley University of California) *Press, ISBN : 0-520-07914-0*
[4] IAGI 1996 Sandi Stratigrafi Indonesia *Edition 1996 rev. From 1973* (Jakarta: Indonesia)
[5] Van Zuidam R A 1985 Aeral photo interpretation in terrain analysis and geomorphological mapping *Smits Publisher*
[6] Geological department cartographer 1976 Topographic map of the Tasikmalaya quadrangle 1:100.000 on scale
[7] Budhitrisna T 1990 Geologic map of the Tasikmalaya quadrangle, West Java *Second edition rev. from 1986 Geological Research and Development Centre* (Bandung: Indonesia)
[8] Geological department cartographer. Topographic map of the Tasikmalaya quadrangle 1:50.000 on scale
[9] Bronto S 1989 Volcanic geology of Galunggung, West Java, Indonesia *Dissertation* (Geology University of Canterbury)
[10] Suryantini 2015 Volcanological approach for evaluation of geothermal potential in volcanic associated hydrothermal system at the early stage of exploration. *Proc. Int. Conf. on World Geothermal Congress* (Melbourne: Australia)
[11] Moore J N, Allis R, Renner J L, Mildenhall D and McCulloch J 2002 Petrologic evidence for boiling to dryness in the Karaha – Telaga Bodas geothermal system, Indonesia *Proc. Int. Conf. on 27th Workshop Geothermal Reservoir Engineering* (Stanford University: USA) pp 223-232
[12] Moore J N 2004 The mineralogical consequences and behavior of descending acid – sulfate waters: An example from The Karaha – Telaga Bodas geothermal system, Indonesia. *J. The Canadian Mineralogist* **42** pp 1483 – 1499

Acknowledgements

The author is grateful to Dr. Eng. Suryantini for her suggestion during the research. The author also would like to thank Professor Sutikno Bronto and Anggie Rengganis, MT for providing us the opportunity, knowledge, time and discussion.

Quantitative comparison of two 3-D resistivity models of the Montelago geothermal prospect

W A van Leeuwen[1], Suryantini[2] and G P Hersir[3]

[1] IF Technology, Velperweg 37, 6813DP Arnhem, the Netherlands
[2] Institut Teknologi Bandung, Jl. Ganesha 10, Bandung 40132, Indonesia
[3] Iceland GeoSurvey (ÍSOR), Grensásvegur 9, 108 Reykjavík, Iceland

E-mail: w.vanleeuwen@iftechnology.nl

Abstract. A combined TEM-MT survey was carried out in the Montelago geothermal prospect, situated on Mindoro Island, the Philippines, with the aim to obtain the dimensions and depth of the geothermal reservoir as well as to formulate the prospects' conceptual model. The acquired MT data are static shift corrected using the TEM measurements. Two different 3-D inversion codes are used to create subsurface resistivity models of the corrected MT data set. The similarities and differences between the two resistivity models are quantitatively assessed using a set of structural metrics. Both resistivity models can be generalized by a three-layered model. The model consists of a thin heterogeneous, conductive layer overlying a thick resistive layer, while the basement has a decreased resistivity. Although this is a common characteristic resistivity response for the alteration mineralogy of a volcanic geothermal system, the temperatures at depth are lower than would be expected when interpreting the modelled resistivity model accordingly. Since the last volcanic activity in the area was about one million years ago, it is anticipated that the resolved resistivity structure is a remnant of a hydrothermal system associated with a volcanic heat source. This model interpretation is validated by the alteration minerals present in the exploration wells, where high temperature minerals such as epidote are present at depths with a lower temperature than epidote's initial formation temperature. This generalized description of the resistivity model is confirmed by both resistivity models. In this paper the two inversion models are not only compared by assessing the inversion models, but also by reviewing a set of gradient based structural metrics. An attempt is made to improve the interpretation of the conceptual model by analyzing these structural metrics. Based on these analyses it is concluded that both inversions resolve similar resistivity structures and that the location of the two slim holes drilled are well chosen.

1. Introduction

The Montelago geothermal prospect on Mindoro Island, the Philippines, has been the target of development studies over a period spanning several decades. The prospect is situated in Oriental Mindoro Province along the east coast of Mindoro Island as illustrated in Figure 1. Despite the mountain range being situated along the north-western shoreline of Mindoro, the topography of the prospect area is locally very rugged. The prospect area is about 2 km wide and 3 km long, striking in a north-west south-east direction. In the north-east the prospect area is bounded by Tablas Strait and in the south-west it is adjacent to Lake Naujan.

One of the most prominent components of the geothermal exploration work carried out in the Montelago geothermal prospect is the resistivity survey, carried out in 2014 [1]. The TEM-MT survey was carried out to verify the existence of the inferred geothermal reservoir as well as to

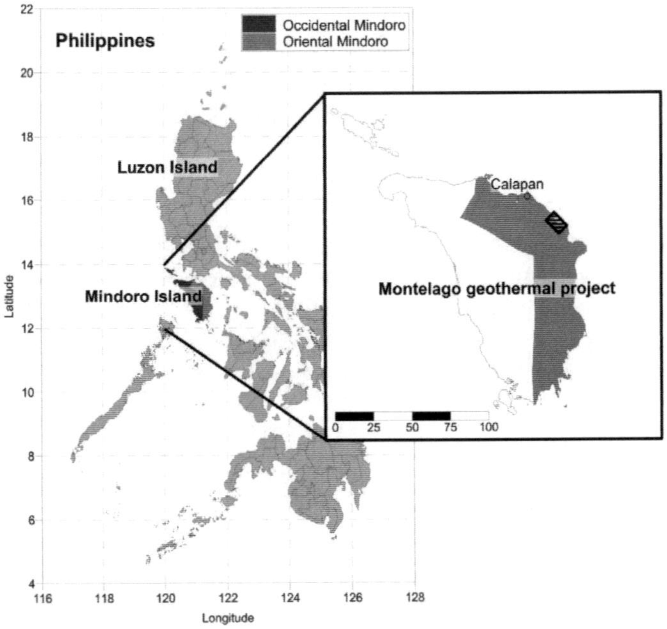

Figure 1. Location of the Montelago geothermal prospect on Mindoro Island, the Philippines.

determine its dimensions for a volumetric assessment and, in case of sufficient potential, the location of the exploration wells.

The acquired resistivity data were 1-D inverted [1] and later 3-D inverted using two different inversion codes ([2] and this paper). The inversion code used by Árnason and Hersir [2] is WSINV3DMT [3], while in this paper the MT data are inverted using ModEM [4]. The structural resistivity differences between the two 3-D inversion models are discussed and later analyzed using a set of gradient based structural metrics. The consequences of these differences for the geological interpretation and the definition of the conceptual model are discussed subsequently. Another aspect discussed is the placement of the exploration wells.

2. Geology

The geology, geochemistry and conceptual model of the Montelago geothermal prospect is extensively discussed by PT LAPI ITB [5] and summarized here.

The volcanism present in the prospect area are part of a narrow north-west trending volcanic chain of Pleistocene-Quaternary age. Locally, this volcanic range overlies Tertiary sediments as well as the Paleozoic-Mesozoic basement consisting of meta-sediments and amphibolites. There are currently no active volcanoes present on Mindoro Island, where the latest volcanic activity is dated between 1.6 and 0.8 Ma. Within the prospect area, the volcanic centres are eroded.

The dominant tectonic setting of the prospect area has a northwest-southeast striking direction. This orientation controls the structural geology on the island, which is dominated by strike-slip faulting. In the prospect area, geological structures are dominated by northeast and northwest trending faults. Since the locations of the geothermal manifestations at the surface coincide with northeast trending faults, it is believed that these faults act as fluid path-ways for geothermal flow to the surface.

Hydrothermal clay alteration mineralogy present at the surface in the prospect area comprises interlayered illite-smectite as well as quartz-epidote-calcite minerals. The former is formed at temperatures of at least 150 °C and is typically found at depths greater than 50 m. The latter is formed at much higher temperature (> 240 °C). The occurrence of these two alteration minerals at surface suggests that either erosion has occurred or that the materials are brought to the surface.

The geochemical analysis of sampled waters from the hot springs indicates that the system is water dominated. Furthermore, the magmatic component in the thermal water is very small, which implies that the geothermal system is not related to shallow magma or recent volcanic activity.

On basis of the findings summarized above, a conceptual model of the geothermal reservoir was defined [5]. The elements of the conceptual model are shown as a north-south orientated cross section of the geothermal system in Figure 2. The Buloc-Buloc bay hot springs with a temperature of 40 °C, possibly originating from another geothermal reservoir. Recharge of the geothermal system is believed to be from Lake Naujan, located southwest of the prospect area. Thermal gradient well NGH-4 measured a temperature of 94 °C at a depth of 305 m. As the thermal gradient in this well is linear the drilled formation is impermeable, possibility a clay cap.

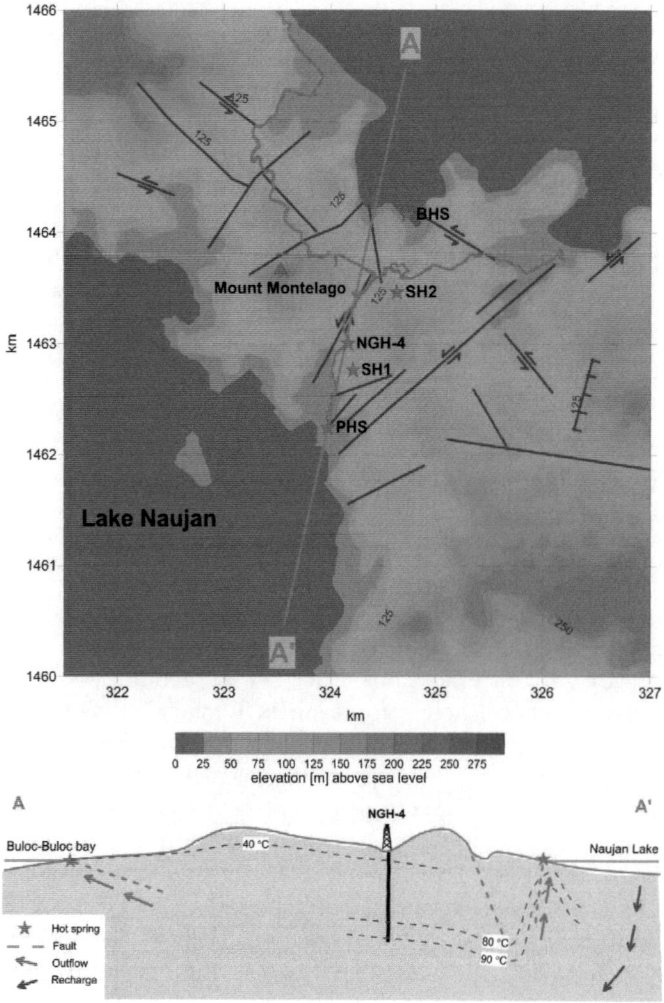

Figure 2. Map of the prospect area of the Montelago geothermal prospect (upper) and a not to scale cross section of the geothermal prospect (lower). Indicated on the map are a gradient temperature well (NGH-4) and which revealed a temperature of 94 °C at a depth of 305 m, Mount Montelago and the location of the cross section. Also the location of the Pungao (PHS) and Buloc-Buloc (BHS) hot springs are given, as well as the faults in the prospect area. BHS is located in the Buloc-Buloc bay. The main road and the two main villages in the area are shown in purple. In the cross section the inferred isotherms are also given. Coordinates of this and all following maps, besides when specifically mentioned otherwise, in this paper are projected in UTM zone 51N using the WGS-84 datum. The cross section indicates inferred outflow areas, faults and recharge.

It is inferred that the heat source of the system is a very deep cooling pluton. Reservoir temperatures are estimated to be at least 200 °C.

Two slim holes down to 1,200 m depth below the surface were drilled to test the conceptual model of the Montelago geothermal prospect. Both wells were fully cored and the drill cuttings were described using hand lens and microscope. Methylene Blue (MeB) analysis was carried out on the cores and the cuttings of both wells to identify the clay type and identify the top of geothermal reservoir when drilled into.

Determination of the clay type can be carried out following a simplified relation between MeB-index and clay type. This relation states that a small MeB index value (< 10) corresponds to illite or kaolinite clay minerals, while a high MeB index value (> 10) indicates smectite, interlayered illite-smectite or interlayered chlorite-smectite clay minerals.

Based on core and cutting description, high temperature alteration clay minerals such as epidote occur in well SH-1 at 900-1,200 m below surface and between 600-1,000 m below surface in well SH-2. The temperature at the depth interval in which epidote is found is below 200 °C [6]. In contrast to the well temperature as shown in Figure 3, epidote is formed at temperatures above 230 °C. This suggests that the present day temperature is lower than the paleo temperature. In other words, the alteration mineralogy found in the present day geothermal system is the remnant of a former geothermal system.

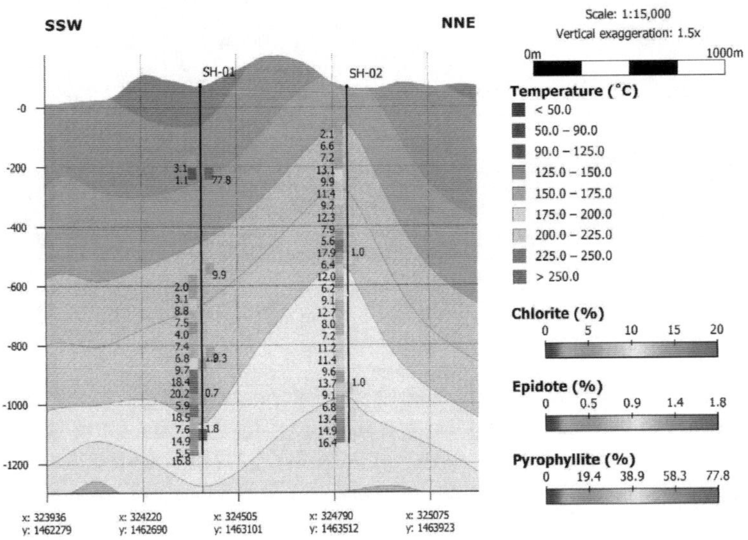

Figure 3. Illustration of chlorite, pyrophyllite, and epidote in slim well SH-01 and SH-02 and rock temperature. (chlorite = left of well line; epidote = center of well line; pyrophyllite = right of well line). The location of the wells is for example given in Figures 7 and 8.

The results of the MeB analysis suggests that at a depth of 500 m below surface the clay alteration type in the geothermal system is changing from interlayered illite-smectite (or smectite) to illite (or kaolinite). Interlayered illite-smectite usually is an indication of clay alteration above the geothermal reservoir, while illite is commonly related to clay minerals formed at higher temperatures near the top of the geothermal reservoir.

3. Data acquisition and processing

Resistivity data were acquired using both MT and time-domain electromagnetics (TEM). The TEM measurements are used to correct for the static shift effect in the MT data [1, 7].

A total of 54 stations were recorded with a remote reference station simultaneously recording about 40 km away from the survey area. Each MT station collected data for approximately 20 hours. During data acquisition every third station was a telluric station. The survey layout is a semi-grid with an interstation spacing of approximately 500 m as shown in Figure 4.

Recorded MT data are processed using Phoenix Geophysics software. Telluric stations are processed with the simultaneously measured horizontal magnetic fields from the nearest station. The MT data are static shift corrected by an automated iterative process, simultaneously inverting the invariant of the MT response and shifting the invariant towards the coincident 1-D TEM response [8].

Despite the presence of elevated power lines and several villages in the survey area, as well as the rugged topography, the data quality is assessed as fairly good. Nevertheless all stations showed some problems around the MT dead-band.

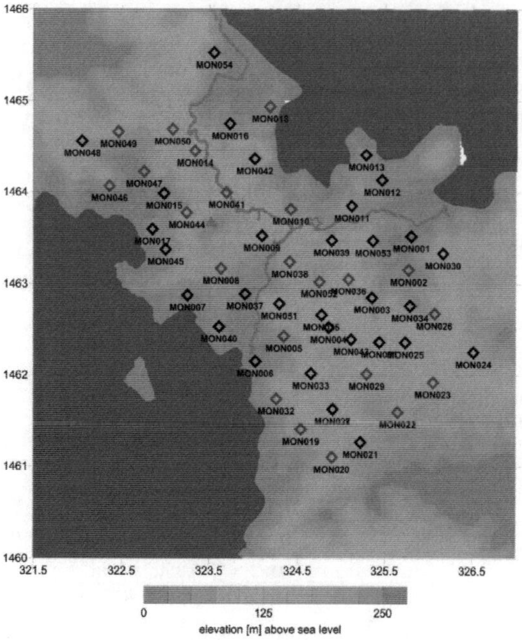

Figure 4. Map of the Montelago geothermal prospect area showing the measuring grid of the MT-TEM survey. Black stations are full MT stations and blue stations are telluric stations. Mount Montelago and the main road are also shown.

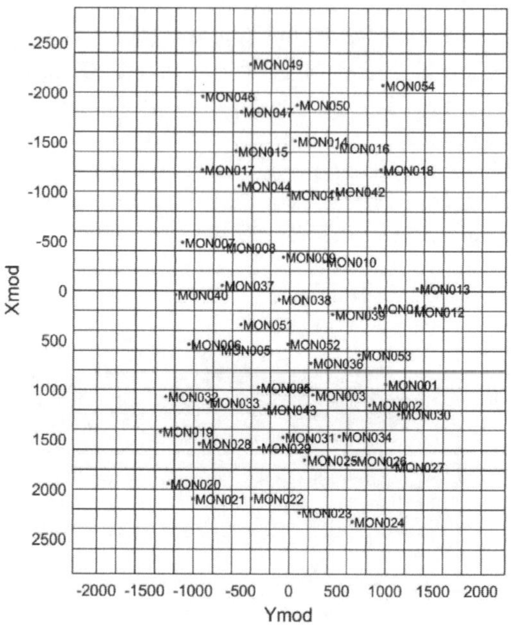

Figure 5. XY-view of the central area of the model mesh used for the 3-D inversion of the MT data using WSINV3DMT and ModEM.

4. Data inversion

The resulting MT data set is inverted using two different 3-D inversion codes. The first code, WSINV3DMT, is developed by Siripunvaraporn et al. [3]. The results of that inversion, WSINV_10_mod, are presented in Árnason and Hersir [2]. The second code is ModEM, developed by Egbert and Kelbert [4]. The results of this inversion, ModEM_10_mod, are presented in this paper. Subsequently a comparison is made between the inversion results of the two models, WSINV_10_mod [2] and ModEM_10_mod (this paper).

To make a meaningful comparison, inversion mesh and inverted frequencies are chosen to be as similar as possible. As the MT data are already corrected for static shift, it is assumed by Árnason et al. [7] that inverting the model without topography, but including bathymetry, and applying a post-inversion correction for topography will deliver the most reliable resistivity model. To produce comparable models, the same inversion strategy was chosen for the ModEM inversion of the MT data.

The model mesh used for both models consists of 52x48x32 cells (x,y,z). The model mesh and the MT coordinate system are rotated to -46 °. In Figure 5 the central area of the model mesh is shown, as well as the location of the recorded MT stations within the mesh. The central area of the model is 5.7 by 4.5 km, while the cell size is at 200 x 200 m. The horizontal cell size increases semi-logarithmically towards the model boundaries. The smallest vertical cell size is 2 meters increasing semi-logarithmically with depth, to a maximum dimension of 110 km at the base of the mesh.

The 2014 MT data are resampled to 31 periods with six values per decade covering the period range 0.01 s to 1,000 s for the inversion using WSINV3DMT. The inversion using the ModEM code uses MT data resampled to 28 periods with five values per decade covering the same range of periods.

In their 3-D inversion Árnason and Hersir [2] run three models using different initial models: *a.)* a homogeneous half-space of 10 m, *b.)* a homogeneous half-space of 100 m, and *c.)* the resistivity model compiled from joint 1-D inversion of individual TEM and MT sounding pairs. It is concluded by Árnason and Hersir [2] that the inversion model using the 10 Ωm initial model shows the best fit to the data and gives the most realistic image of the resistivity structure of the subsurface. Therefore, a 10 Ωm initial model is also used for the 3-D inversion using ModEM. Where the final rms misfit of WSINV_10_mod is 1.74, the final rms misfit of ModEM_10_mod is 1.92.

5. Inversion results

The results based on the two different inversions are shown in Figure 6 and 7, in which the resistivity maps at 500 and 1,000 m b.s.l. are given. As visible in Figure 6, both models show strongly heterogeneous conductivities at a depth of 500 m b.s.l. High conductivities are observed at the boundaries of the survey area. While the resistive anomaly in the south part of the survey area seems to be coincident in the two models, WSINV_10_mod also shows a resistive anomaly below Mount Montelago that is not present in ModEM_10_mod. This resistive anomaly is still visible at a depth of 1,000 m b.s.l., as observed in Figure 7. At this depth, the resistivity maps based on the two inversion models are in pretty good agreement. However, at 1,000 m b.s.l. (Figure 7) the resistive anomaly in the south part in ModEM_10_mod is apparent as a single coherent anomaly, while in WSINV_10_mod the resistive anomaly consists of two discrete parts. Again, beneath Mount Montelago a resistive anomaly is present in WSINV_10_mod.

To consider these observations further, the cross section in the model x-direction (see the model mesh in Figure 5) for both 3-D inversion models is given in Figure 8. See Figures 6 or 7 for the location of the cross section. In Figure 8 it is observed that there is a difference in character of the resistive anomaly at depth between the two models. Although both models require a conductive layer at shallow depth, the deep resistive anomaly is significantly smaller in ModEM_10_mod compared to WSINV_10_mod.

The location of the exploration slim holes SH-1 and SH-2 is based on the resistivity model and tectonics.

6. Quantitative comparison

The bulk of the resistivity structures in the two 3-D inversion models presented in the previous section are related to geological structures, particularly those anomalies consistent between both models. However, some of the features may be artefacts introduced by the inversions. To make a quantitative comparison between the two models, a set of structural metrics, considered to identify those resistivity structures which are required by the model, are computed. This means that in order for the modelled data to fit the observed data, the resolved resistivity structures need to be there. As the solution of a 3-D inversion is non-unique, an anomaly appearing in two different inversions is very reassuring. Those resistivity structures not appearing in both

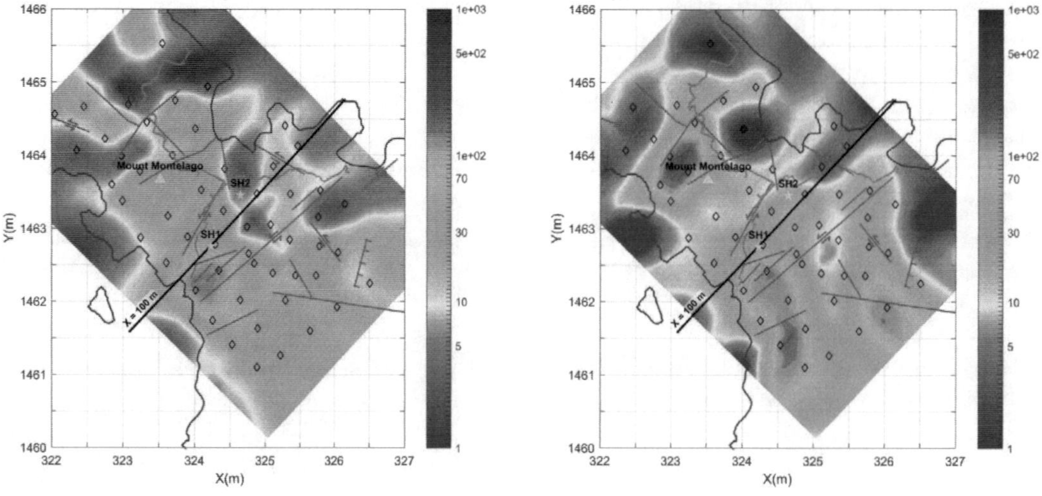

(a) (b)

Figure 6. 3-D inversion results presented as resistivity maps at an elevation of 500 m b.s.l. of (a) ModEM_10_mod and (b) WSINV_10_mod. Resistivity values are in Ωm. Coastlines, faults, Mount Montelago (grey triangle) and main roads are shown as well. The location of slim holes SH-1 and SH-2 are indicated (grey stars). Diamond symbols are MT stations and the black line is the location of the resistivity cross section shown in Figure 8.

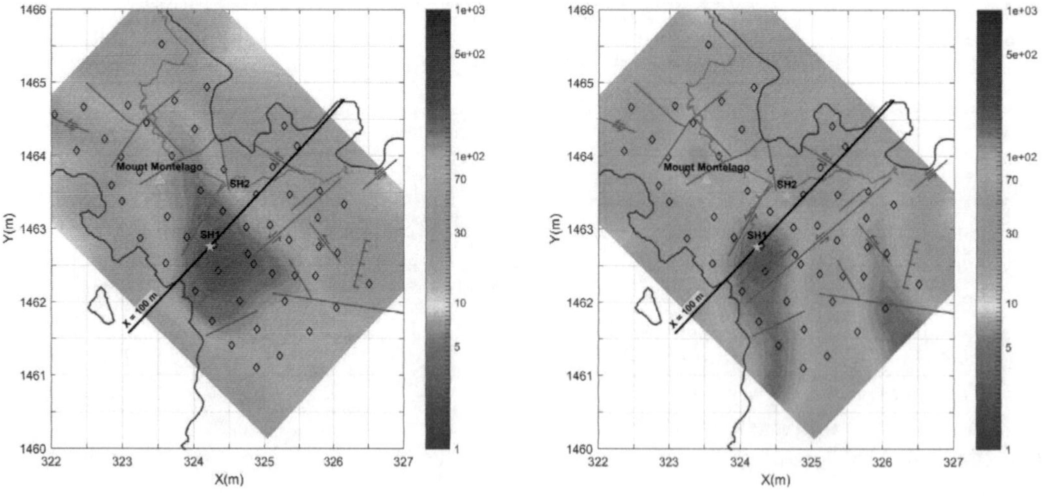

(a) (b)

Figure 7. 3-D inversion results presented as resistivity maps at an elevation of 1,000 m b.s.l. of (a) ModEM_10_mod and (b) WSINV_10_mod. Resistivity values are in Ωm. Coastlines, faults, Mount Montelago (grey triangle) and main roads are shown as well. The location of slim holes SH-1 and SH-2 are indicated (grey stars). Diamond symbols are MT stations and the black line is the location of the resistivity cross section shown in Figure 8.

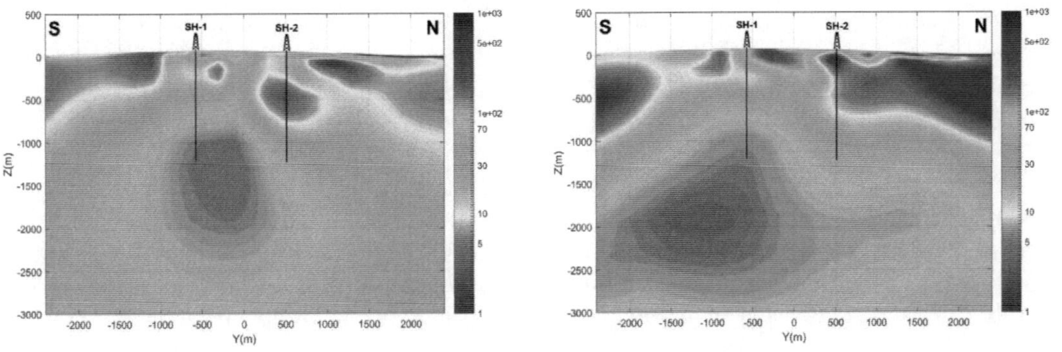

(a) (b)

Figure 8. 3-D inversion results presented as resistivity cross sections in the model X-direction at X = 100 m of (a) ModEM_10_mod and (b) WSINV_10_mod. Resistivity values are in Ωm. The location of slim holes SH-1 and SH-2 are indicated. For the location of the cross section, see Figures 6 and 7.

inversion models might be introduced by the code and not related to geological structures. The robust and required resistivity structures appearing in both models might then be regarded as related to geological structures and consequently suitable for the geothermal interpretation of the resistivity models.

6.1. Method

The structural metrics used in this section are derived from a joint inversion approach where they are used to compare the structural resemblance between models of different geophysical methods. An overview of the available metrics is given in a review paper by Gallardo and Meju [9]. Rosenkjaer et al. [10] experimented with several of these metrics. Here two structural metrics are used to compare the two inversion models.

(i) The difference between the normalized model gradients of the two resistivity models $\delta\overline{\varphi}$, linearly assesses the structural similarities between the models:

$$\delta\overline{\varphi} = \frac{\nabla\overline{m}_1}{\|\nabla m_1\|} - \frac{\nabla\overline{m}_2}{\|\nabla m_2\|}. \tag{1}$$

(ii) The norm of the cross product of the two model gradients, the cross gradient $\overline{\tau}$, delivers a direct comparison between the two resistivity models:

$$\overline{\tau} = \nabla m_1 \times \nabla m_2. \tag{2}$$

Here $\overline{m}$ is the three-dimensional model matrix of the inversion model.

These structural metrics are computed for the WSINV_10_mod and ModEM_10_mod models. The results are presented for the resistivity maps and cross sections as shown in Figures 6, 7 and 8.

6.2. Discussion

The difference between the normalized model gradients of the two inversion models at an elevation of 500 and 1,000 m b.s.l., respectively, are given in Figure 9. At an elevation of 500 m b.s.l. the differences between the two model gradients are large, except for an area at the center of the models. The differences between the gradients are large at the locations of the resistive anomalies and at the boundaries. Experiments with synthetic data showed that small differences in the exact locations of a resistive structure can lead to large differences between the model gradients. This is for example visible at an elevation of 1,000 m b.s.l. The difference between the two models in shape and location of the resistive anomaly in the south part of the area leads to large $\delta\overline{\varphi}$ values. In this case, the large $\delta\overline{\varphi}$ values might confirm the presence of the resistive anomaly in the south part of the prospect area and provide an indication of the actual boundaries of this anomaly. Here, it might be inferred that the minimum edge of the boundary is located at the east side of the corresponding large $\delta\overline{\varphi}$ area. Beneath Mount Montelago the gradient of WSINV_10_mod is variable while the gradient of ModEM_10_mod at this location is relatively constant. This leads the large gradient difference below Mount Montelago.

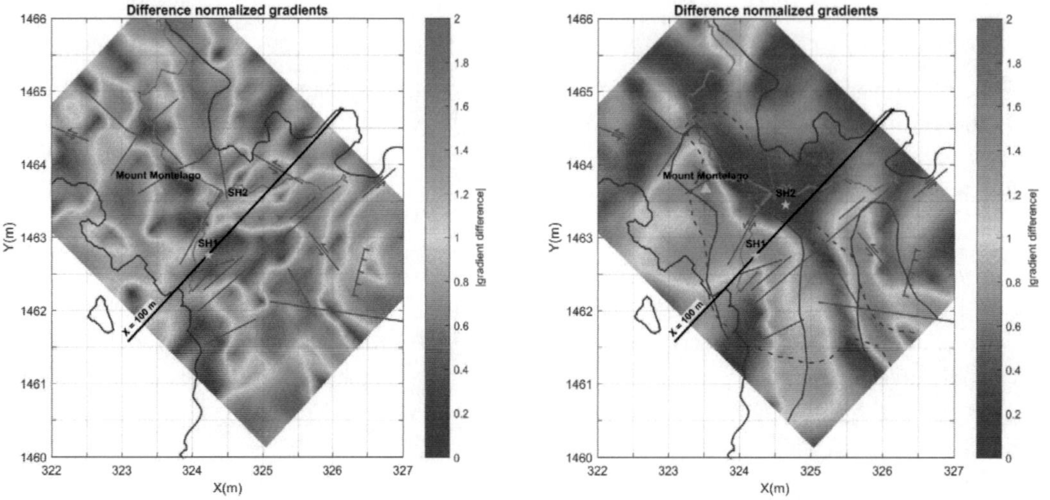

(a) (b)

Figure 9. Magnitude of the difference between the normalized model gradients of the inversion results of WSINV_10_mod and ModEM_10_mod at an elevation of (a) 500 m and (b) 1,000 m b.s.l., respectively. Red colors indicate those locations where the difference between the model gradients is relatively large and blue colors indicate those locations where the model gradients are relatively similar. The coastlines, faults, main roads, as well as the locations of Mount Montelago and the two slim holes drilled in 2015 are shown as well. In the map to the right the 70 Ωm resistivity contour of the two models (dotted line: ModEM_10_mod; solid line: WSINV_10_mod) is given. The black line is the location of cross section shown in Figure 11.

In Figure 10 the norm of the cross product of the model gradients at 500 and 1,000 m b.s.l., respectively, is plotted. Where the resistivity values of two models are significantly different, even when the structures are comparable, experiments with synthetic data showed that the cross gradient will show large differences between models. This is the case at an elevation of 500 m b.s.l. Similar shapes at different locations can be identified by a high cross gradient, while a low cross gradient indicates areas where the resistivity shows little variation in both models. At an elevation of 1,000 m b.s.l., the minimum area covered by the resistive structure in the south part of the survey area is indicated by the low cross gradient area. Although, resistivity values are significantly different, the resistive anomaly below Mount Montelago is present in both models, indicated by an area of "stable" resistivity.

In Figure 11 the difference between the model gradients and the norm of the cross product of the model gradients of the resistivity cross section as shown in Figure 8 is plotted. In the

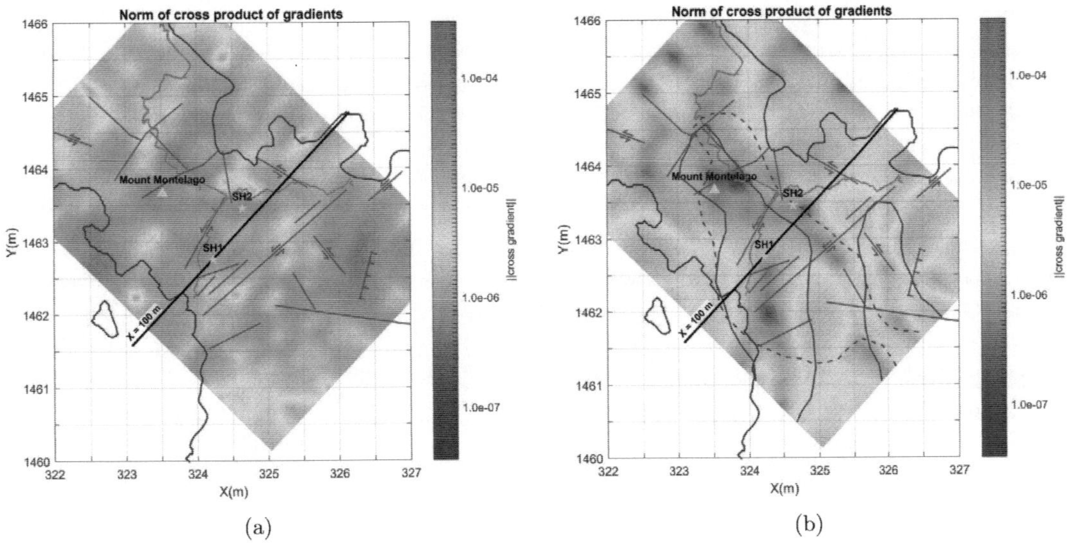

Figure 10. Norm of the cross product of the model gradients of the inversion results of WSINV_10_mod and ModEM_10_mod at an elevation of (a) 500 m and (b) 1,000 m b.s.l., respectively. Red colors indicate those locations where the cross gradient of the models is relatively large and blue colors indicate those locations where the cross product of the model gradients is relatively small. The coastlines, faults, main roads, as well as the locations of Mount Montelago and the two slim holes drilled in 2015 are shown as well. In the map to the right the 70 Ωm resistivity contour of the two models (dotted line: ModEM_10_mod; solid line: WSINV_10_mod) is given. The black line is the location of cross section shown in Figure 11.

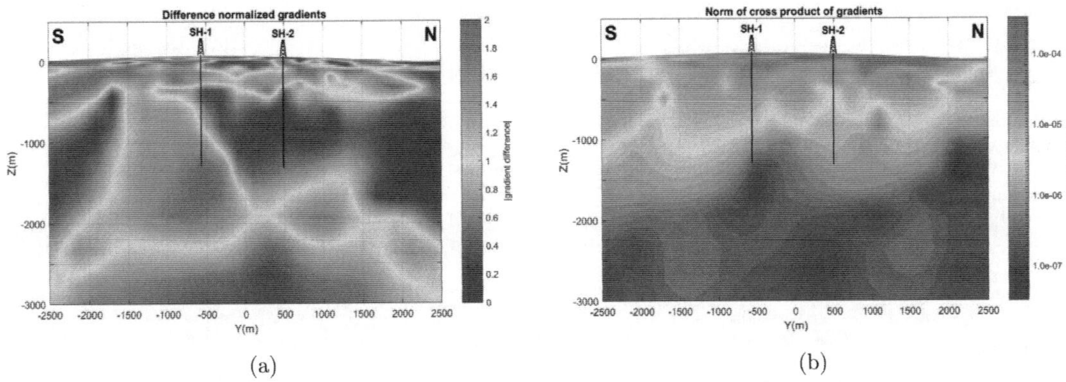

Figure 11. Cross section of the magnitude of (a) the difference between the normalized model gradients and of (b) the norm of the cross product of the model gradients of the inversion results of WSINV_10_mod and ModEM_10_mod at X = 100 m. The location of wells SH-1 and SH-2 are indicated on the cross sections.

difference between the model gradients, the differences in the resistivity structures at a depth of 2,000 m are easily observed. On the other hand, the sharp boundary at Y = -300 m indicates the probable position of the resistive anomaly. The low cross gradient values below -1,000 m indicate that both models agree on the resistive structures at this depth. The resistive heterogeneity at shallower depth is reflected in the high cross gradient values.

The anomalies in the resistivity models and the cross gradient structural metric at greater depths coincide with the location of the faults in the prospect area, indicating a certain degree of fault control of the geothermal prospect. This finding is confirmed by the structural geology data in the prospect area.

The analysis of these structural metrics becomes especially interesting when an attempt is made to interpret them in relation to the positioning of wells. Slim holes SH-1 and SH-2 were drilled in 2015 and proved the existence of the geothermal reservoir. Based on the structural metrics, the location of SH-1 is well chosen, drilling right the by the structural metrics confirmed presence of the resistive anomaly. SH-1 reached the reservoir at a depth of 1,200 m b.s.l.

7. Conclusions

The two 3-D inversion models of the same MT data set from the Montelago geothermal prospect show roughly similar resistivity structures, although on a detailed scale differences are noticed. This gives a confidence in the present resistivity model.

Although having significantly different characteristics in both models, a resistive anomaly is present below Mount Montelago. The presence of the anomaly is confirmed by the analyses of the structural metrics. At this point it is unclear why this anomaly is relatively conductive in ModEM_10_mod and relatively resistive in WSINV_10_mod.

By evaluating the inversion models using two structural metrics, the existence of most of the main resistivity anomalies could be validated. The metrics presented here highlight other aspects of the similarities and differences between the two models.

Although a classic clay alteration resistivity profile is present, the actual temperatures are not matching with the clay alteration mineralogy found in the exploration wells. Consequently it can be concluded that the clay alteration of the Montelago geothermal system is the product of a paleo hydrothermal system.

Both resistivity models can be generalized by a three-layered model. This model consists of a thin heterogeneous, conductive layer overlying a thick resistive layer, while the basement has a decreased resistivity.

Based on the structural metrics, the location selected for slim hole SH-1 is well chosen.

Acknowledgments

Acknowledgements The authors would like to thank Emerging Power Inc. for making the MT data and other information available for publication. Knútur Árnason, geophysicist at Iceland GeoSurvey read the manuscript and made some valuable comments. The authors acknowledge his contribution.

References

[1] Hersir G P, Árnason K and Vilhjálmsson A M 2014 Reistivity survey in montelago on mindoro island, philippines; data acquisition, processing and 1d inversion Tech. rep. Iceland Geosurvey
[2] Árnason K and PHersir G 2014 3d inversion of mt data from montelago on mindoro island, philippines Tech. rep. Iceland Geosurvey
[3] Siripunvaraporn W, Egbert G, Lenbury Y and Uyeshima M 2005 *Phys. of the Earth and Planetary Interiors* **150** 11
[4] Egbert G D and Kelbert A 2012 *Geophys. J. Int.* **89** 16
[5] PT LAPI ITB 2014 Monte carlo volumetric assessment and p90 study for montelago geothermal concession mindoro philippines Tech. rep. PT LAPI ITB
[6] Sigurgeirsson M A, Helgadóttir H M, Jónsson S J and Egilson T 2015 Drilling of slimholes sh-01 and sh-02 from surface down to 1250 and 1200 m montelago in mindoro - philippines Tech. rep. Icelandic GeoSurvey
[7] Árnason K, Eysteinsson H and Hersir G P 2010 *Geothermics* **39** 21
[8] Árnason K 1989 Central loop transient electromagnetic soundings over a horizontally layered earth Tech. Rep. OS-89032/JHD-06 Orkustofnun; national energy authority geothermal division Reykjavik

[9] Gallardo L A and Meju M A 2011 *Rev. of Geophys.* **49** rG1003
[10] Rosenkjaer G K, Gasperikova E, Newman G A and Árnason K 2015 *Geothermics* **57** 16

Performance Prediction of Two-Phase Geothermal Reservoir using Lumped Parameter Model

F Nurlaela and Sutopo

Master Program of Geothermal Technology, Faculty of Mining and Petroleum Engineering, Institut Teknologi Bandung, Jalan Ganeca 10, Bandung, 40132, Indonesia

Email : famelia.nurlaela@gmail.com

Abstract. Many studies have been conducted to simulate performance of low-temperature geothermal reservoirs using lumped parameter method. Limited work had been done on applying non-isothermal lumped parameter models to higher temperature geothermal reservoirs. In this study, the lumped parameter method was applied to high-temperature two phase geothermal reservoirs. The model couples both energy and mass balance equations thus can predict temperature, pressure and fluid saturation changes in the reservoir as a result of production, re-injection of water, and/or natural recharge. This method was validated using reservoir simulation results of TOUGH2. As the results, the two phase lumped parameter model simulation without recharge shows good matching, however reservoir model with recharge condition show quite good conformity.

1. Introduction

Geothermal reservoir simulation is conducted to predict a geothermal field reservoir behavior for a certain time. Traditionally, reservoir simulation is needed, especially in the early stage geothermal field development and performance monitoring stage. But in early stage development, geothermal reservoir simulation process constrained by incomprehensive and limited data. While distributed parameter numerical model that commonly used requires a comprehensive data to perform simulations. This reservoir numerical modeling requires long run-times especially when the model becomes more complex. The history-matching process or the model calibration process take time because it needs to be run many times for adjustment parameters of reservoir. Lumped parameter model is an alternative method to reservoir simulation due to less parameters involved and shorter run-times.

Researches on lumped parameters to simulate geothermal reservoirs have been widely applied. Several examples include use lumped parameter models to simulate the pressure response of the reservoir [1]; conducted a reservoir simulation for a low temperature geothermal systems [6]; used a lumped parameter model for reservoir applications of low temperature liquid-dominated geothermal system [4]; conducted a study on high temperature geothermal reservoirs [8]; and studied a lumped parameter model for a liquid dominated low temperature reservoir containing carbon dioxide [3]. Many researches has been done for single-phase low-temperature geothermal system but only a few for high-temperature.

In this paper, the lumped parameter method is described for two-phase high-temperature geothermal reservoirs model. The model couples both energy and mass balance equations thus can predict temperature, pressure and fluid saturation changes. The basic flow equations are formulated with the lumped-parameter method. Example cases of simulation are presented to illustrate the capabilities of the model developed based on this approach.

2. Non-Isothermal Lumped Parameter Model

The lumped parameter model of non-isothermal system can predict changes of pressure and temperature. Pressure and temperature change occur because of production, mass will be reduced, pressure will drop and therefore internal energy will decline as well and causing a slight drop in temperature value. Temperature reservoir is a function of reservoir volume, production rate, injection rate and injection temperature, natural recharge rate and recharge temperature [4].

This study was used a single tank lumped parameter model with a non-isothermal system. The single tank model represents a geothermal system where there are reservoir, production, injection, and recharge. Reservoir in the model is a two-phase high-temperature reservoir which there are steam, hot water, and rock. Production mass are steam and water while for injection is only water. Lumped parameter model as illustrated in Figure 1.

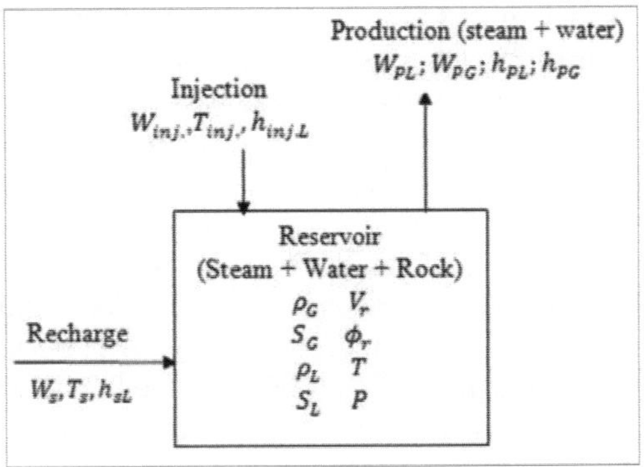

Figure 1. Single tank two-phase high-temperature lumped parameter model.

In single tank reservoir model there are density and saturation of steam, (ρ_V, S_V); density and saturation of water, (ρ_L, S_L); and rock. This two-phase reservoir has pressure, P and temperature T. Reservoir is produced with steam production rate, $W_{pV}(kg/s)$ and water production rate, $W_{pL}(kg/s)$. At the reservoir also injected water with injection rate, $W_{inj}(kg/s)$. Net production rate is the difference of total production rate with total injection rate. Mathematically, it can be written as follows:

$$W = W_{pL}(t) + W_{pV}(t) - W_{inj}(t) \tag{1}$$

Recharge mass flow rate can be approximated by [1][4]:

$$W_s = \alpha_s(P_i - P(t)) \tag{2}$$

Where $\alpha_s(kg/bar{\cdot}s)$ is an index recharge, $P_i(bar)$ is the initial pressure and $P(t)(bar)$ is the pressure of reservoir as a function of time. While recharge index value based on the equation of the relationship with permeability, fluid viscosity, fluid density, length of reservoir, and width and height of reservoir [2]. Equation (3) shows water recharge index $(\alpha_{s,L})$ and steam recharge index $(\alpha_{s,V})$.

$$\alpha_{s,L} = 8\frac{wz}{L_x}\frac{k}{\mu_L}\rho_L, \quad \alpha_{s,V} = 8\frac{wz}{L_x}\frac{k}{\mu_V}\rho_V \tag{3}$$

Based on these assumptions the mass conservation equation can be written as follows:

$$V_r \frac{d}{dt}[\rho_L S_L \phi_r + \rho_V S_V \phi_r] - \alpha_{s,L}(P_i - P(t)) - \alpha_{s,V}(P_i - P(t)) + W_{pL}(t) + W_{pV}(t) - W_{inj}(t) = 0 \quad (4)$$

Where V_r is volume reservoir and ϕ_r is reservoir porosity. The mass conservation equation encompasses mass accumulation in the reservoir, mass flow rate from recharge source, and net production rate.

There are two kinds of heat transfer in geothermal system; convection and conduction. Temperature changes occur by the fluid movement such as production, injection, or natural recharge. Therefore, convection is more dominant in system, in this model conduction and heat loss is negligible. The energy conservation encompasses energy accumulation in the reservoir, heat flow from injection fluid, heat flow from recharge, and heat flow as a result of production.

Energy conservation equation can be written as follows:

$$\frac{d}{dt}[(1-\phi_r)V_r\rho_m C_m T + V_r\phi_r(\rho_L S_L u_L + \rho_V S_V u_V)] - W_{inj}(t)h_{inj,L}(t) - \alpha_{s,L}(P_i - P(t))h_{s,L}(t) - \alpha_{s,V}(P_i - P(t))h_{s,V}(t) + W_{pL}(t)h_{pL}(t) + W_{pV}(t)h_{pV}(t) = 0 \quad (5)$$

Where $\rho_m, C_m, T, u_L, u_V, h_{inj,L}, h_{s,L}, h_{pL}, h_{pV}$ respectively are density of rock matrix, specific heat capacity of rock matrix, temperature of reservoir, specific internal energy of water, specific internal energy of steam, specific enthalpy of injected water, specific enthalpy of recharge water, specific enthalpy of production water, and specific enthalpy of production steam.

In this model, porosity changes depend on pressure and temperature as in the following equation [4]:

$$\phi_r(P,T) = \phi_i[1 + c_r(P - P_i) - \beta(T - T_i)] \quad (6)$$

Where c_r, β, ϕ_i, T_i respectively are compressibility of porosity at constant temperature, thermal expansion coefficient of porosity at constant pressure, initial porosity at initial pressure and temperature, and initial temperature of reservoir.

Mass and energy conservation equation are non-linear differential equation. The solution of couple equation is solved by using Newton-Raphson method. A finite difference discretization scheme is used for the terms involving the derivative of the time.

3. Example Synthetic Application

In this section, we provide a few synthetic applications of the non-isothermal two-phase model. The lumped parameter models is validated to the reservoir simulation results of TOUGH2.

Two-phase condition occurs when pressure and temperature value are on the saturation line. Temperature value is temperature saturation of the pressure. In this study, the initial temperature is set as temperature saturation of the initial pressure ($T_i = T_{sat}@P_i$).

The simulation two-phase model is performed in two case conditions: first, there are no fluid flow from the recharge source; and second, the recharge source contribute to system. Several reservoir properties that used in this model are shown in Table 1.

Table 1. Properties and initial condition of lumped parameter two-phase model.

Parameters	Symbol	Value	Units
Compressibility	β	0	1/°C
Initial porosity	ϕ_i	0.2	$fraksi$
Thermal expansion	c_r	1.33×10^{-4}	$1/bar$
Initial pressure	p_i	50	bar
Initial temperature	T_i	263.91	°C
Injection temperature	T_{inj}	60	°C
Density of rock matrix	ρ_m	2650	kg/m^3

Specific heat capacity of rock matrix	C_m	1000	$J/kg°C$
Volume reservoir	V_r	1.0×10^7	m^3
Permeability	k	2.719×10^{-14}	m^2

Parameter values such as; density, internal energy, enthalpy of recharge, enthalpy of injection, enthalpy of production are obtained from the calculation of steam table [7]. Density and internal energy are calculated based on pressure in the solution of the equation, while enthalpy production is calculated based on pressure and temperature of production. The enthalpy of recharge is calculated based on initial pressure and temperature, while enthalpy of injection is based on initial pressure and temperature of injection.

Simulations are conducted to observe historical pressure, temperature, and saturation for 30 days. There are three periods in production and injection scenario that used in this simulation. First and third 10 days conducted production and injection with production rate 20 kg/s and injection rate 15 kg/s. While in the second 10 days or second period, there are no mass produced and injected.

The results are compared with simulation results using numerical simulator TOUGH2 [5]. A model without recharge condition is described as a single grid block with volume $V_r = 1.0 \times 10^7 m^3$ and there are no fluid flow and heat flow from the surrounding (Figure 2a). Production and injection well are in the middle of reservoir.

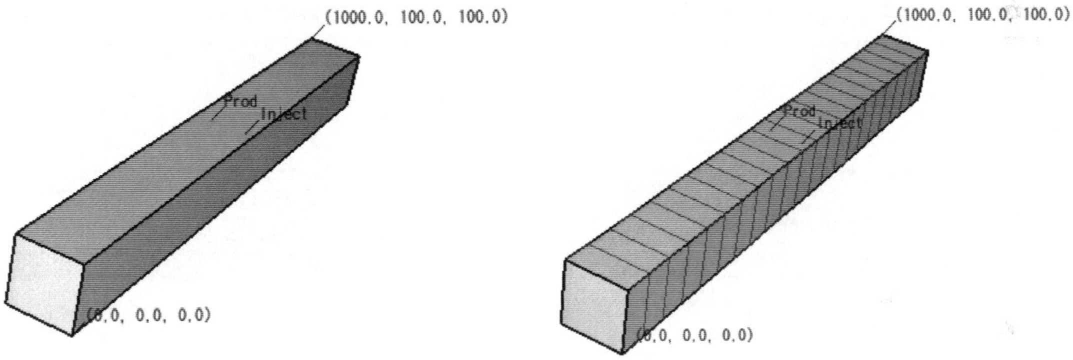

(a) **(b)**

Figure 2. TOUGH2 model without **(a)** recharge and **(b)** with recharge.

For recharge condition model is described as a reservoir with 21 grid blocks with fixed pressure and temperature on the sides (see Figure 2b). To minimize discretization errors, the system consists of 21 grid blocks. This boundary condition allow the fluid flow from the boundary towards the well. Pressure and temperature value are compared to the average value of the whole blocks in the TOUGH2 model.

4. Simulation Results

4.1. Without recharge condition

In this two-phase simulation, the results that observed are pressure, temperature, water saturation, and steam saturation change. There is a decline for pressure and temperature in first 10 days or first period of production. In the second period pressure and temperature tend to be constant and there is decline again when in the third period. Figure 3a. shows that lumped parameter simulation result and TOUGH2 pressure curve are quite match and has a difference 0.15 bar. In temperature curve, simulation result of lumped parameter and TOUGH2 temperature curve also quite match as well with difference 0.2°C.

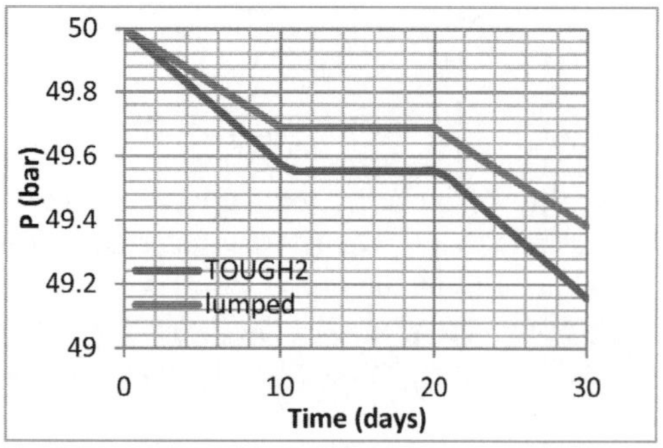

(a)

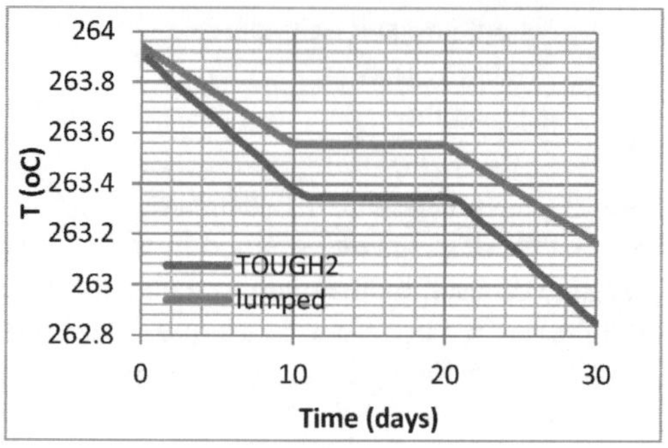

(b)

Figure 3. (a) Pressure change and **(b)** temperature change without recharge.

Water saturation and steam saturation change can be seen in Figure 4. Water saturation curve indicate a decline in first and third period and a constant value during second period. While in steam saturation curve, there is an increase value when produced and tends to a constant when the well is shut-in then there is an increase when well is produced again. The increase of steam saturation represents the increasing steam phase in the reservoir. Lumped parameter simulation result and TOUGH2 water saturation curve are quite match and only has difference 0.0002. Lumped parameter simulation result and TOUGH2 steam saturation also has difference 0.0002 and quite match as shown in Figure 4b.

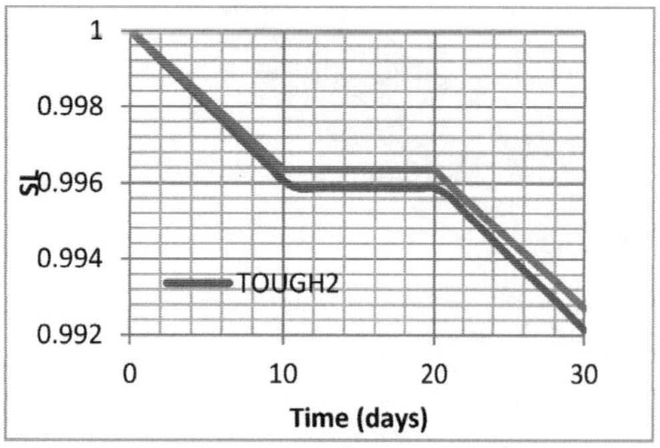

(a)

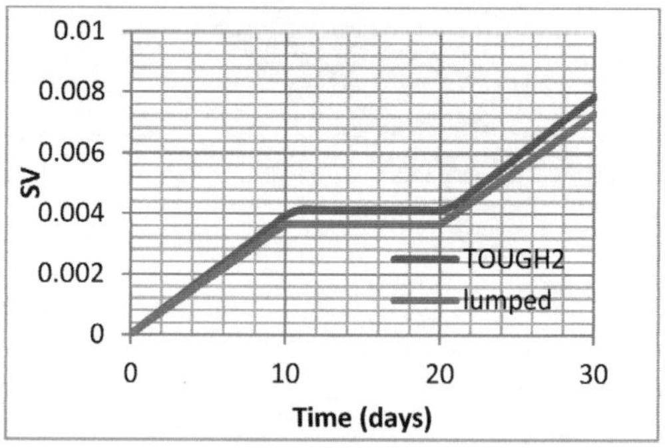

(b)

Figure 4. (a) Water saturation change and **(b)** steam saturation change without recharge.

Based on lumped parameter simulation result, the curve quite match, and the difference are very small. The lumped parameter models for two-phase condition without recharge has shown conformity and validated to be applied.

4.2. With recharge condition

In the two-phase lumped parameter model is also conducted a simulation with there is a fluid flow from recharge source into the reservoir. Figure 5a shows pressure change in system with recharge condition. Pressure curve in lumped parameter simulation result is decreased in first period, tends to be constant in second period and decreased again in third period. TOUGH2 simulation result pressure curve also decreased in first period but not as much lumped parameter. But in second period, the pressure curve is increase to the initial pressure and tends to be constant on day 13 to day 20. In third period, pressure is decreased again with the same gradient as in first period. Lumped parameter and TOUGH2 pressure curve has a difference 0.4 bar. In temperature curve, lumped parameter and TOUGH2 simulation result are quite match with the difference 0.15°C.

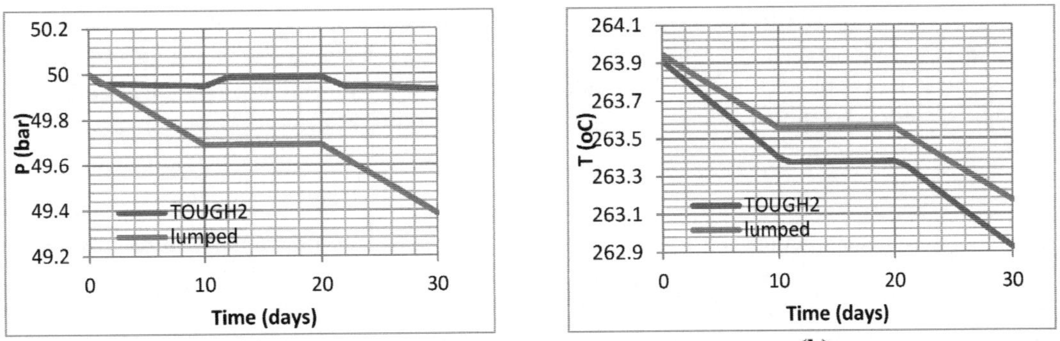

(a) **(b)**

Figure 5. (a) Pressure change and **(b)** temperature change with recharge.

There is an increase value of water saturation in lumped parameter simulation result during second period. This occurs because in the second period there is a fluid flow from recharge that into the reservoir. With no fluid produced and only water from recharge thus will increase the water phase and decrease the steam phase. TOUGH2 simulation results in water saturation curve relatively constant during second period and decreased in third period. Lumped parameter and TOUGH2 water saturation curve are quite match with difference 0.001. Figure 6b shows that there is a difference 0.001 between lumped parameter and TOUGH2 steam saturation curve.

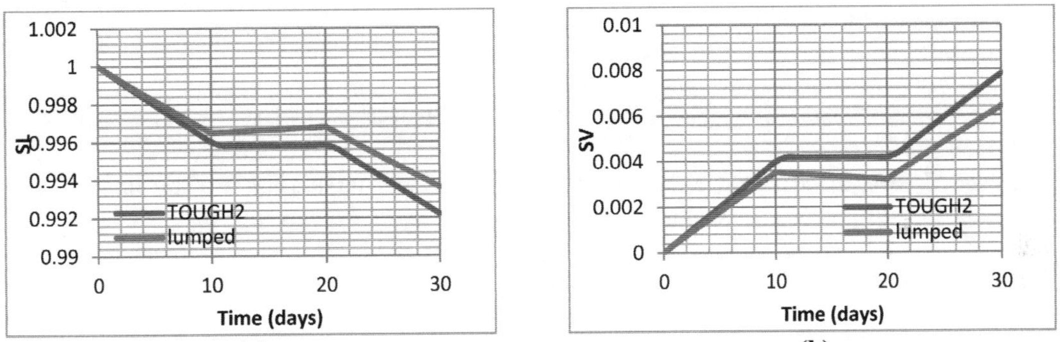

(a) **(b)**

Figure 5. (a) Pressure change and **(b)** temperature change with recharge.

Lumped parameter model is model that based on mass and energy conservation equation so as TOUGH2 software. The difference between these two methods is the equation solving method. In lumped parameter, the equation is ordinary differential equations (ODE) and simpler to be solved because the dependent variable is only function of time. While in TOUGH2, it is a bit more complicated because the dependent variables are function of time and one or more spatial variable. Thus, the equation need to be solved for TOUGH2 is the partial differential equation (PDE). The differences equation solver method allows difference value on the simulation results but still rapprochement.

Reservoir simulation using the lumped parameter method has been developed for high-temperature reservoir. Reservoir simulation program with this method has been made and the results have been validated. This simulation program can be used in early stage of development where there are limited data available. The results of reservoir simulation by this method can predict the pressure and temperature reservoir condition where there are production and injection in the future. Lumped parameter reservoir simulation method is simpler and can be run in short time so it can quickly provide an estimate of reservoir pressure and temperature. The estimates of pressure and temperature is important to know at the early stage of development due to the construction of a power plant is influenced by reservoir conditions. The estimates of pressure and temperature also can provide a reference criterion for separator, turbine, or other power plant equipment as well as pipes installation from well pad to power plant. If reservoir condition can be known with certainty, power plant capacity can be immediately determined.

5. Conclusions

Lumped parameter method has been developed and applied to two-phase high-temperature geothermal reservoir model. Mass and energy conservation equation are solved using Newton-Raphson method with pressure, temperature and saturation as the solution. Two-phase high-temperature reservoir simulation program with the lumped parameter method with and without recharge has been made and can predict the condition reservoir due to production and injection. The simulation results could be a reference to make a strategic plan in the early stage development of geothermal field. The result of two phase lumped parameter model simulation without recharge shows good matching, however the result is quite good conformity for reservoir model with recharge condition. The model has not been validated using actual data history. Therefore the next study will be validated with actual data so that lumped parameter model of the reservoir can get closer to the actual situation.

References

[1] Axelsson G 1989 Simulation of Pressure Response Data from Geothermal Reservoirs by Lumped Parameter Models *Proc. 14th Workshop on Geothermal Reservoir Engineering* (California: Stanford University) pp 257-263

[2] Ay E 2005 *Comparison of Lumped-Parameter and 1D Linear Distributed Models Used in Reservoir Simulation* Undergraduate Thesis (ITU Petroleum and Natural Gas Engineering Dept)

[3] Hosgor F B et al 2013 A New Lumped parameter Model for Reservoirs Containing Carbon Dioxide *Proc. 38th Workshop on Geothermal Reservoir Engineering* (California: Stanford University)

[4] Onur M et al 2008 A New Non-Isothermal Lumped-Parameter Model for Low Temperature, Liquid Dominated Geothermal Reservoirs and Its Application *Proc. 33rd Workshop on Geothermal Reservoir Engineering* (California: Stanford University)

[5] Pruess K 1999 *TOUGH2 User's Guide* ver 2.0 Earth Sciences Div Lawrence Berkeley National Laboratory (California: University of California)

[6] Sarak H et al 2003 New Lumped Parameter Models for Simulation of Low-Temperature Geothermal Reservoirs *Proc. 28th Workshop on Geothermal Reservoir Engineering* (California: Stanford University)

[7] Steam Tables 1967 *Thermodynamic Properties of Water and Steam Thermal Conductivity of Water and Steam* (New York: St. Martin's press)

[8] Sveinbjörnsson B 2011 *A Lumped parameter Modelling Method For High-Temperature Geothermal Reservoirs* Master's thesis (Iceland: The School for Renewable Energy Science)